COMPLEX ANALYSIS

Complex Analysis

(THE HITCHHIKER'S GUIDE TO THE PLANE)

IAN STEWART

Lecturer in Mathematics, University of Warwick

DAVID TALL

Senior Lecturer in Mathematics Education, University of Warwick

CAMBRIDGE
UNIVERSITY PRESS

PUBLISHED BY THE PRESS SYNDICATE OF THE UNIVERSITY OF CAMBRIDGE
The Pitt Building, Trumpington Street, Cambridge, United Kingdom

CAMBRIDGE UNIVERSITY PRESS
The Edinburgh Building, Cambridge CB2 2RU, UK   http://www.cup.cam.ac.uk
40 West 20th Street, New York, NY 10011–4211, USA   http://www.cup.org
10 Stamford Road, Oakleigh, Melbourne 3166, Australia

First published 1983
Reprinted 1984, 1985 (with corrections), 1987, 1988, 1990, 1992,
1993, 1996, 1997, 1999

Printed in the United Kingdom at the University Press, Cambridge

A catalogue record for this book is available from the British Library

ISBN 0 521 24513 3 hardback
ISBN 0 521 28763 4 paperback

Contents

	Preface	ix
	Acknowledgement	ix
0	**The origins of complex analysis, and a modern viewpoint**	1
	1. The origins of complex numbers	1
	2. The origins of complex analysis	5
	3. The puzzle	7
	4. A modern view	8
1	**Algebra of the complex plane**	10
	1. Construction of the complex numbers	10
	2. The $x + iy$ notation	12
	3. A geometric interpretation	13
	4. Real and imaginary parts	14
	5. The modulus	14
	6. The complex conjugate	16
	7. Polar coordinates	17
	8. The complex numbers cannot be ordered	18
	Exercises 1	18
2	**Topology of the complex plane**	22
	1. Open and closed sets	24
	2. Limits of functions	25
	3. Continuity	28
	4. Paths	33
	5. The Paving Lemma	37
	6. Connectedness	40
	Exercises 2	45
3	**Power series**	48
	1. Sequences	48
	2. Series	52
	3. Power series	56
	4. Manipulating power series	58
	5. Appendix	60
	Exercises 3	61

4   Differentiation   64
   1. Basic results   64
   2. The Cauchy–Riemann equations   67
   3. Connected sets and differentiability   71
   4. Hybrid functions   72
   5. Power series   73
   6. A glimpse into the future   76
   Exercises 4   79

5   The exponential function   82
   1. The exponential function   82
   2. Real exponentials and logarithms   84
   3. Trigonometric functions   84
   4. The analytic definition of π 86
   5. The behaviour of real trigonometric functions   86
   6. Complex exponential and trigonometric functions are periodic   88
   7. Other trigonometric functions   90
   8. Hyperbolic functions   91
   Exercises 5   92

6   Integration   95
   1. The real case   96
   2. Complex integration along smooth paths   97
   3. The length of a smooth path   100
   4. Contour integration   105
   5. The Fundamental Theorem of Contour Integration   108
   6. The Estimation Lemma   111
   7. Consequences of the Fundamental Theorem   114
   Exercises 6   117

7   Angles, logarithms, and the winding number   120
   1. Radian measure of angles   120
   2. The argument of a complex number   122
   3. The complex logarithm   125
   4. The winding number   126
   5. The winding number as an integral   130
   6. The winding number round an arbitrary point   131
   7. Components of the complement of a path   132
   8. Computing the winding number by eye   133
   Exercises 7   137

8   Cauchy's Theorem   141
   1. The Cauchy Theorem for a triangle   143
   2. Existence of an antiderivative in a star-domain   146
   3. An example – the logarithm   148

4. Local existence of an antiderivative 149
5. Cauchy's Theorem 149
6. Applications of Cauchy's Theorem 153
7. Simply connected domains 157
Exercises 8 157

9 Homotopy versions of Cauchy's Theorem 159
1. Integration along arbitrary paths 159
2. The Cauchy Theorem for a boundary 162
3. Homotopy 165
4. Fixed end point homotopy 167
5. Closed path homotopy 169
6. The Cauchy Theorems compared 172
Exercises 9 175

10 Taylor series 177
1. Cauchy's integral formula 178
2. Taylor series 179
3. Morera's Theorem 183
4. Cauchy's Estimate 184
5. Zeros 185
6. Extension functions 188
7. Local maxima and minima 189
8. The Maximum Modulus Theorem 191
Exercises 10 192

11 Laurent series 195
1. Series involving negative powers 195
2. Isolated singularities 201
3. Behaviour near an isolated singularity 202
4. The extended complex plane, or Riemann sphere 205
5. Behaviour of a differentiable function at ∞ 207
6. Meromorphic functions 208
Exercises 11 209

12 Residues 212
1. Cauchy's residue theorem 212
2. Calculating residues 215
3. Evaluation of definite integrals 217
4. Summation of series 228
5. Counting zeros 231
Exercises 12 234

13 Conformal transformations   238
  1. Real numbers modulo 2π 238
  2. Conformal transformations   241
  3. Möbius mappings   246
  4. Potential theory   249
  Exercises 13   252

14 Analytic continuation   257
  1. The limitations of power series   257
  2. Comparing power series   260
  3. Analytic continuation   261
  4. Multiform functions   265
  5. Riemann surfaces   268
  6. Complex powers   271
  7. Conformal mapping using multiform functions   273
  8. Contour integration of multiform functions   274
  9. The road goes ever on . . .   281
  Exercises 14   282

  Index   287

Preface

Students faced with a course on 'Complex Analysis' often find it to be just that – complex. In the sense of 'complicated'.

It is true, of course, that the proofs of some of the major theorems in the subject can demand a certain technical versatility. But in many ways, on a conceptual level, complex analysis is actually *easier* than real analysis; it just isn't always taught that way.

This book is intended for use at the level of second or third year undergraduates, and it is based on experience accumulated from such courses over the past decade. To exhibit this inherent simplicity of complex analysis we have organized the material around two basic principles: (1) generalize from the real case; (2) when that reveals new phenomena, use the rich geometry of the plane to understand them. Our aim throughout is to encourage geometric thinking, with the proviso that it must be adequately backed up by analytic rigour.

The opening chapter sets the work in its historical context; and the history is often alluded to later as partial motivation. However, we feel that cultural changes often affect the status of conceptual problems: what was once an important difficulty can become a triviality when viewed with hindsight. It is not always necessary to drag today's students through yesterday's hang-ups. We argue the point at greater length below: it is fundamental to our entire approach.

INS & DOT
Barby and Kenilworth
October 1981

Acknowledgement
Parts of Chapters 11, 12, 13 have appeared in the *Handbook of Applicable Mathematics*, vol. 4, chapter 9, published by Wiley and Sons Ltd., 1982. They are here reprinted with permission.

0

The origins of complex analysis, and a modern viewpoint

If, as Kronecker asserted, the integers are made by God and all the rest is the work of Man, then complex numbers are certainly one of Man's most intriguing mathematical artefacts. For centuries they have been a wonder to mathematicians and philosophers alike. It took nearly 300 years from their first appearance in Cardano's *Ars Magna* to the publication of a formal definition which satisfies our modern standards of rigour. Building on such foundations the uninitiated reader might be forgiven for thinking that complex analysis must be an incredibly complicated theory. Yet here we come to a historical puzzle. Although it took nearly three centuries to obtain a satisfactory treatment of complex *numbers*, it then took less than a tenth of that time to complete a major part of complex *analysis*.

Obviously the numbers must come first, or there is nothing to do analysis with, but the timescale is surprising. A possible explanation is that setting up the foundations adequately involved deep problems of a philosophical nature: it took a long time to come to grips with them, but once the 'breakthrough' had occurred, the further development was easy by comparison.

History suggests otherwise.

1. The origins of complex numbers

The celebrated *Ars Magna* of Girolamo Cardano in 1545 treated the simultaneous equations

$$x + y = 10$$
$$xy = 40,$$

obtaining a solution (in modern notation) of the form

$$x = 5 + \sqrt{-15} \qquad y = 5 - \sqrt{-15}.$$

Cardano gave no interpretation for the square root of a negative number,

1

but he did observe that, on the assumption that the quantities obey the usual algebraic rules, one may check that they do satisfy the equations. His attitude to this discovery was dismissive: 'So progresses arithmetic subtlety, the end of which . . . is as refined as it is useless.'

In the same book he noted that the application of the Tartaglia formula for solving cubic equations to

$$x^3 = 15x + 4$$

leads to the expression

$$x = \sqrt[3]{2 + \sqrt{-121}} + \sqrt[3]{2 - \sqrt{-121}}$$

in contrast to the obvious answer $x = 4$.

Raphael Bombelli (1526–73) suggested a way to reconcile the two solutions by manipulating the 'impossible' roots *as if they were ordinary numbers*: since

$$(2 \pm \sqrt{-1})^3 = 2 \pm \sqrt{-121},$$

Cardano's expression becomes

$$x = (2 + \sqrt{-1}) + (2 - \sqrt{-1}) = 4.$$

The 'impossible' root is just the familiar root in a complex disguise. Bombelli's work was the first hint that complex numbers could prove useful in solving real mathematical problems.

In *La Géométrie* (1637), René Descartes made the distinction between 'real' and 'imaginary' numbers, interpreting the occurrence of imaginaries as a sign that the problem concerned was insoluble, an opinion shared by Newton at a later date.

John Wallis represented a complex number geometrically in his *Algebra* of 1673. On a fixed line the real part of the number was measured off (in the direction given by its sign); then the imaginary part was measured off at right angles (Fig. 0.1).

For some reason this proposal seems to have been subsequently ignored.

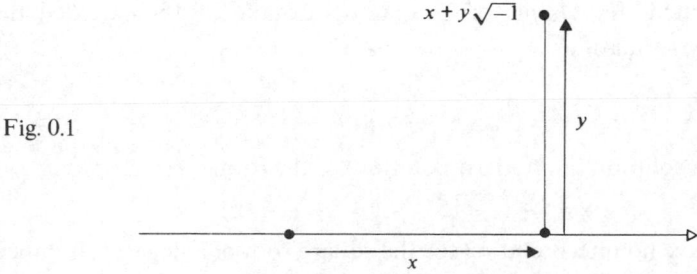

Fig. 0.1

In 1702 John Bernoulli was evaluating integrals of the form

$$\int \frac{dx}{ax^2 + bx + c}$$

by partial fractions. Using the philosophy that complex numbers could be manipulated like real ones, he wrote the integrand in the form

$$\frac{1}{ax^2 + bx + x} = \frac{A}{x - \alpha} + \frac{B}{x - \beta}$$

(using modern notation) where α, β are the roots of the quadratic denominator and found the integral in the form

$$A \log (x - \alpha) + B \log (x - \beta).$$

His bold decision to use the same method when the quadratic had no real solutions led to logarithms of complex numbers. But what were they? Both Bernoulli and Leibniz used the method and by 1712 were engaged in controversy; Leibniz asserted that the logarithm of a negative number was complex whilst Bernoulli insisted it was real.

Bernoulli argued that, since

$$\frac{d(-x)}{-x} = \frac{dx}{x},$$

it follows by integration that $\log(-x) = \log(x)$. Leibniz, on the other hand, insisted that the integration was only correct for positive x.

Leonhard Euler resolved the controversy in favour of Leibniz in 1749, pointing out that the integration required an arbitrary constant,

$$\log(-x) = \log(x) + c,$$

a point which Bernoulli had ignored.

By formally manipulating expressions involving complex numbers, Euler derived a host of theoretical relations, including the famous formula of 1748:

$$e^{i\theta} = \cos \theta + i \sin \theta.$$

(As if to emphasize that there's nothing totally new in this world, it should be noted that an equivalent formula was known to Roger Cotes in 1714!) Putting $\theta = \pi$ we find

$$e^{i\pi} = -1,$$

a fantastic relation that blends the three mathematical symbols e, i and π in one surprising equation.

Extending the theory of logarithms to the complex case by defining

$$\log z = w \quad \text{if and only if } e^w = z,$$

we find more intriguing results. Formal manipulation gives

$$e^{\log z + m\pi i} = e^{\log z}(e^{\pi i})^m = z \cdot (-1)^m.$$

For an even integer $m = 2n$ this gives

$$e^{\log z + 2n\pi i} = z,$$

and so $\log z + 2n\pi i$ is also a logarithm of z; the complex logarithm is *many-valued*. For an odd integer $m = 2n + 1$, we have

$$e^{\log z + (2n+1)\pi i} = -z,$$

whence

$$\log(-z) = \log z + (2n+1)\pi i.$$

This gives the resolution of the Leibniz-Bernoulli controversy: for real positive x, $\log(-x)$ must be complex.

The theory of complex numbers was beginning to grow more and more fascinating. What was lacking was a concrete interpretation that explained precisely what these entities were.

In 1797 Caspar Wessel published a paper in Danish describing the representation of a complex number as a point in the plane. It went almost totally unnoticed until a French translation was published a hundred years later. Meanwhile the idea was attributed to Jean-Robert Argand who wrote it up independently in 1806. Since that time the geometric interpretation of complex numbers has become known as the Argand diagram.

Another pioneer in the theory of complex numbers was Carl Friedrich Gauss. In his doctoral dissertation of 1799, he addressed himself to a problem which had concerned mathematicians since the early eighteenth century. Initially it had been widely believed that, just as the solutions of real quadratic equations could lead to new 'complex' numbers, so would solutions of equations with complex coefficients lead to even more kinds of new numbers. Jean D'Alembert (1717–83) conjectured that complex numbers alone would suffice. Gauss confirmed this in the 'fundamental theorem of algebra' – every polynomial equation has a complex root. At first he proved it in the purely 'real' form that any real polynomial factorizes into linear and quadratic factors, avoiding explicit use of imaginaries; later he treated the general case. By 1811 he viewed the complex numbers as points in the plane, saying so in a letter to Bessel. In 1831 he published full details of the geometric representation of complex numbers, which had begun to acquire an air of respectability.

In 1837, nearly three centuries after Cardan's use of 'imaginary numbers', William Rowan Hamilton published the definition of complex numbers as ordered pairs of real numbers subject to certain explicit rules of manipulation. (In the same year Gauss wrote to Wolfgang Bolyai that he had developed the same idea in 1831.) At last this placed the complex numbers on a firm *algebraic* basis.

2. The origins of complex analysis

Unlike the gradual emergence of the complex *number* concept, the development of complex *analysis* seems to have been the direct result of the mathematician's urge to generalize. It was sought deliberately, by analogy with real analysis.

As noted above, there are early traces of analytic operations on complex functions in the work of Bernoulli, Leibniz, Euler, and their contemporaries.

In his 1811 letter to Bessel, Gauss shows that he knew the basic theorem on complex integration around which complex analysis was subsequently built. In real analysis, when one integrates a function f between limits a and b,

$$\int_a^b f(x)\,dx,$$

the limits fully specify the integral. But in the complex case, with a and b represented as points in the plane, it is also necessary to specify a definite curve from a to b, and to 'integrate along the curve'. The question is, to what extent does the value of the integral depend on the curve chosen?

Gauss says the following. 'I affirm now that the integral $\int f(x)\,dx$ has only one value even if taken over different paths provided $f(x) \ldots$ does not become infinite in the space enclosed by the two paths. This is a very beautiful theorem whose proof . . . I shall give on a convenient occasion.'

It seems that the occasion never arose. The crucial step of publishing a proof of this result was taken in 1825 by the man who was to occupy centre stage during the first flowering of complex analysis: Augustin-Louis Cauchy. After him it is called 'Cauchy's Theorem'. In Cauchy's hands the basic ideas of complex analysis rapidly emerged. For a complex function to be differentiable it had to be very specialized in nature; its real and imaginary parts had to satisfy certain properties called the Cauchy–Riemann equations. Contour integrals of differentiable functions were seen to have the property noted privately by Gauss. Furthermore, if an integral was computed along a path which wound round points where the function became infinite, then Cauchy showed how this integral could be computed using the 'theory of residues'. The latter simply required the calculation of a constant, called the 'residue' of the function, at each exceptional point; and the knowledge of how many times the path wound round it. The precise route of the path did not actually matter at all!

Power series proved to be important in the theory and other workers extended the ideas. Pierre-Alphonse Laurent introduced 'Laurent series' involving negative powers in 1843. In this formulation, near an exceptional point z_0, a differentiable function was expressed in the form of a sum of two series

$$f(z) = [a_0 + a_1(z - z_o) + \cdots + a_n(z - z_o)^n + \cdots]$$
$$+ [b_1(z - z_o)^{-1} + \cdots + b_n(z - z_o)^{-n} \ldots].$$

The residue of the function at z_o was then just the coefficient b_1. Using the theory of residues the computation of complex integrals often proved to be far simpler than one could ever have dreamed.

Cauchy's definitions of analytic ideas such as continuity, limits, derivatives and so on, were not the same as we use today. He based them on infinitesimal notions which subsequently fell into disrepute. A rigorous treatment was then developed by Karl Weierstrass (1815–97) using definitions which are still regarded as fundamental (though recent developments in 'non-standard analysis', which legitimize infinitesimals, show us that we may have been over-hasty in judging Cauchy's ideas). Weierstrass founded his whole approach on power series. However, the geometric viewpoint was sorely lacking in his work (at least as published); this deficiency was remedied by far-reaching ideas introduced by Bernhard Riemann (1826–66). In particular the concept known as a 'Riemann surface', which dates from 1851, treats many-valued functions by making the complex plane have multiple layers, on each of which the function is single-valued. The topological way in which the layers join up is the crucial thing.

From the mid-nineteenth century onwards the progress of complex analysis has been strong and steady with many far-reaching developments. The fundamental ideas of Cauchy remain, now refined and clothed in more recent topological language. The abstruse invention of complex numbers, once described by our mathematical forbears as 'impossible' and 'useless', has become part of an aesthetically satisfying theory with eminently practical applications in aerodynamics, fluid mechanics and many other areas.

3. The puzzle

We return to our 'historical puzzle'. *Why was the development of complex numbers so laboured and hesitant, whereas that of complex analysis was explosive?*

We shall suggest a possible answer (it is of course only personal opinion and thus open to argument), somewhat different from the 'foundations + breakthrough' explanation suggested earlier.

Looking at the early history of complex numbers, the overall impression is of countless generations of mathematicians beating out their brains against a brick wall in search of – what? A triviality. The definition of complex numbers as ordered pairs (x, y), or as points of the plane, was obtained over and over and over again. It is even implicit in Bombelli's work; it is there for all to see in Wallis's; it crops up again by way of Wessel, Argand, Gauss. Morris Kline once remarked: 'That many men – Cotes, de Moivre, Euler, and Vandermonde – really thought of complex numbers as points in the plane follows from the fact that all, in attempting to solve $x^n - 1 = 0$, thought of the solutions . . . as the vertices of a regular polygon.'

If the problem has such a simple solution, why was this not recognized sooner?

The early mathematicians were not so much seeking a *construction* for complex numbers as a *meaning*, in the philosophical sense: 'What *are* complex numbers?' However, the development of complex *analysis* showed that the complex number concept was so useful that no mathematician in his right mind could possibly ignore it. The unspoken question became 'what can we *do* with complex numbers?', and once that had a satisfactory answer, the original philosophical question evaporated.

There was no jubilation at Hamilton's incisive answer to the three-hundred year old foundation problem – it was 'old hat'. Once mathematicians had woven the notion of complex numbers into a powerful coherent theory, the fears that they had concerning the existence of complex numbers became unimportant and mathematicians lost interest in them.

There are other cases of this nature in the history of mathematics but perhaps none more clear-cut. As time passes, the cultural world-view changes. What one generation sees as a problem or a solution is not interpreted in the same way by another generation. We would do well to bear this in mind when we view the historical development of mathematics. To view history solely from the viewpoint of the current generation may easily lead to distortion and misinterpretation.

4. A modern view

The conclusion which we have just enunciated has implications for those studying complex analysis for the first time. Though sometimes it is useful to see the development of the theory in its historical context, it is not always necessary to fight the historical battles again. In this text we give honour where we can to those pioneers who carved their way through virgin mathematical territory. But more recent developments allow us to see the theory itself in a new light. To the modern ear the very *name* 'complex analysis' carries misleading overtones: it suggests complexity in the sense of complication. The older meaning, 'composite', was perhaps appropriate when the 'real part' of a complex number had a quite different status from the 'imaginary part'. But nowadays a complex number is a perfectly integrated whole: to think of complex analysis as if it were, so to speak, two copies of real analysis, is to place undue emphasis on the algebra at the expense of the geometry, which in the long run is far more influential. And in fact complex numbers are *not* more complicated than reals: in some ways they are simpler. For instance, polynomials always have roots. Likewise complex analysis is often simpler than real analysis: for example, every differentiable function is differentiable as often as we please and has a power series expansion.

In preparing our approach to the subject we have adopted two basic organizing principles. The first is the direct generalization to the complex case of real analysis. Definitions of limits, continuity, differentiation and integration are the natural extension of the real notions, provided that we view them in suitable terms. Since nowadays any student taking a course in complex analysis may be assumed to have made a lengthy study of the real counterpart, *many battles have already been won*: we can refer students to their accumulated knowledge, pausing only to phrase it in a form appropriate for our use. This saves time as well as energy and allows us to proceed to the heart of the subject where the interesting differences occur. Invariably this happens because the plane has a richer geometry than the line, and this brings us to our second major organizing principle: geometric insight is valuable and should be cultivated. Of course, this insight must be translated into sound formal arguments; this can be done using modern topological notions.

Using these two principles, a straightforward approach to complex analysis emerges. First, complex numbers are defined formally as ordered pairs of real numbers, giving them an interpretation as points in the plane. The topology of complex numbers is then given in terms of plane topology. In quick succession it is possible to derive complex generalizations of the notions of continuity, limits, and differentiation, with particular emphasis on power series which will occupy a central role later on. A study of the complex exponential function, given in terms of the usual power series, reveals the intimate connection between this function and the trigonometric functions (also considered as power series). After generalizing the notion of integration, the logarithm can be viewed either as the inverse function of the exponential, or as the integral

$$\log z = \int 1/z \, dz,$$

suitably interpreted. This interpretation yields intimate links between geometric intuition and formal analysis.

At this stage Cauchy's Theorem is presented in various guises and the use of integration arguments leads to a proof that every differentiable function can be expressed as a power series. More generally, Laurent series (using positive and negative powers) take care of isolated points where functions become infinite and lead to the powerful 'theory of residues' for the calculation of complex integrals.

Returning to geometric ideas, complex analysis proves invaluable in two-dimensional potential theory. Finally the geometric ideas of Riemann can be viewed in terms of modern topology to give a global insight into 'many-valued' functions (such as the logarithm) and open up new areas of progress.

1

Algebra of the complex plane

'The Divine Spirit found a sublime outlet in that wonder of analysis, that portent of the ideal world, that amphibian between being and not-being, which we call the imaginary root of negative unity.' So said Gottfried Leibniz in 1702 – though he may have let his eloquence run away with him. The current view of $\sqrt{-1}$ is a little more prosaic, though the uses made of it are at least as inspiring. The logical status of complex numbers, which caused so much distress during the eighteenth century, is now seen to be very much on a par with that of the 'real' numbers. What puzzled the ancients was the obvious artificiality and abstraction of the complex number system, in contrast to the apparently natural and concrete *real* number system; but the mathematician of today sees even the real numbers as possessing a similar artificiality and abstraction.

In this chapter we shall discuss the construction of a system of numbers, containing the familiar real numbers, and also permitting the solution of the equation $x^2 = -1$. This system is known as the *complex numbers*. Many readers will already know the contents of this chapter: they should read it through rapidly to check such items as notation, and pass on at once to the next.

There is a natural geometric representation of complex numbers as a plane, analogous to that of the reals as a line. The extra freedom inherent in the plane gives the whole subject a very geometric flavour, which it is our intention to keep to the fore in the development of the theory.

1. Construction of the complex numbers

We define a *complex number* to be an ordered pair (x, y) of real numbers. Addition and multiplication of complex numbers are defined by:

$$(x_1, y_1) + (x_2, y_2) = (x_1 + x_2, y_1 + y_2), \tag{1}$$

$$(x_1, y_1)(x_2, y_2) = (x_1 x_2 - y_1 y_2, x_1 y_2 + y_1 x_2). \tag{2}$$

For example,

$$(3, 5)(2, 7) = (3.2 - 5.7, 3.7 + 5.2) = (-29, 31).$$

This definition is the culmination of several centuries of struggle to understand complex numbers, and it shows how elusive a simple idea can be. Before we see what these pairs have to do with $\sqrt{-1}$, however, let us establish some simple properties.

THEOREM 1.1. The set of complex numbers, with the operations defined above, is a field. That is, the following axioms hold: if $z_1 = (x_1, y_1)$, $z_2 = (x_2, y_2)$, and $z_3 = (x_3, y_3)$ are complex numbers, then

Addition and multiplication are commutative:

$$\left. \begin{array}{l} z_1 + z_2 = z_2 + z_1 \\ z_1 z_2 = z_2 z_1. \end{array} \right\} \quad (3)$$

Addition and multiplication are associative:

$$\left. \begin{array}{l} (z_1 + z_2) + z_3 = z_1 + (z_2 + z_3) \\ (z_1 z_2) z_3 = z_1 (z_2 z_3). \end{array} \right\} \quad (4)$$

There is an additive identity $(0, 0)$:

$$z_1 + (0, 0) = z_1. \quad (5)$$

There is a multiplicative identity $(1, 0)$:

$$z_1 (1, 0) = z_1. \quad (6)$$

Each element has an additive inverse:

$$(x, y) + (-x, -y) = (0, 0). \quad (7)$$

Each element other than $(0, 0)$ has a multiplicative inverse:

$$(x, y) \left(\frac{x}{x^2 + y^2}, \frac{-y}{x^2 + y^2} \right) = (1, 0). \quad (8)$$

Multiplication distributes over addition:

$$z_1 (z_2 + z_3) = z_1 z_2 + z_1 z_3. \quad (9)$$

Proof. All assertions (3)–(8) are direct consequences of (1) and (2), using only the field properties of the set $\mathbb{R}$ of real numbers. For example in (9) we have

$$\begin{aligned} z_1 (z_2 + z_3) &= (x_1, y_1)(x_2 + x_3, y_2 + y_3) \\ &= (x_1 (x_2 + x_3) - y_1 (y_2 + y_3), x_1 (y_2 + y_3) + y_1 (x_2 + x_3)) \\ &= (x_1 x_2 + x_1 x_3 - y_1 y_2 - y_1 y_3, x_1 y_2 + x_1 y_3 + y_1 x_2 + y_1 x_3), \end{aligned}$$

and

$$z_1z_2 + z_1z_3 = (x_1, y_1)(x_2, y_2) + (x_1, y_1)(x_3, y_3)$$
$$= (x_1x_2 - y_1y_2, x_1y_2 + y_1x_2) + (x_1x_3 - y_1y_3, x_1y_3 + y_1x_3)$$
$$= (x_1x_2 - y_1y_2 + x_1x_3 - y_1y_3, x_1y_2 + y_1x_2 + x_1y_3 + y_1x_3)$$

which, by real algebra, is the same ordered pair.

The reader should provide similar proofs for the remaining assertions. $\square$

The symbol $\mathbb{C}$ is used for the field of complex numbers.

2. The $x + iy$ notation

The symbol commonly used for a complex number is not (x, y) but $x + iy$. This goes back to Euler, who used i to denote $\sqrt{-1}$ in 1777, though the notation was first used consistently by Gauss.

To recover this notation, we proceed as follows. First note that since

$$(x_1, 0) + (x_2, 0) = (x_1 + x_2, 0),$$
$$(x_1, 0)(x_2, 0) = (x_1x_2, 0),$$

we may identify a complex number $(x_1, 0)$ with the real number x_1. More pedantically, the map $(x_1, 0) \to x_1$ defines an isomorphism between the set of complex numbers of the form $(x_1, 0)$ and the field $\mathbb{R}$ of real numbers. Now define

$$i = (0, 1).$$

Then

$$x + iy = (x, 0) + (0, 1)(y, 0)$$
$$= (x, y) \text{ (by (1) and (2) above).}$$

Finally, observe that

$$i^2 = (0, 1)(0, 1)$$
$$= (0 \cdot 0 - 1 \cdot 1, 0 \cdot 1 + 1 \cdot 0)$$
$$= (-1, 0)$$
$$= -1.$$

In this sense we may say that $i = \sqrt{-1}$.

The $x + iy$ notation is more convenient, and will be used henceforth. Algebraic computations with it are easy. They use all the normal algebraic rules, *plus* the rule $i^2 = -1$. So to multiply, we work out

$$(x_1 + iy_1)(x_2 + iy_2) = x_1x_2 + x_1iy_2 + iy_1x_2 + iy_1iy_2$$
$$= x_1x_2 + i(x_1y_2 + y_1x_2) + i^2y_1y_2.$$

But $i^2 = -1$, so this becomes

$$x_1 x_2 - y_1 y_2 + i(x_1 y_2 + y_1 x_2).$$

This computation, of course, explains the choice of the multiplication formula (2); the addition formula (1) comes the same way but is easier. The definition by (1) and (2) is thus a very sneaky piece of hindsight.

The formula (8) for inverses may also be derived as follows:

$$\frac{1}{x+iy} = \frac{1}{x+iy} \cdot \frac{x-iy}{x-iy} = \frac{x-iy}{x^2+y^2}.$$

EXAMPLE

Express $\dfrac{2+3i}{1+2i}$ in the form $x+iy$.

We have

$$\frac{2+3i}{1+2i} = \frac{(2+3i)(1-2i)}{(1+2i)(1-2i)} = \frac{2-6+i(-4+3)}{5} = -\frac{4}{5} - \frac{i}{5}.$$

3. A geometric interpretation

Since ordered pairs (x, y) provide coordinates on the plane $\mathbb{R}^2$ we can visualize $\mathbb{C}$ as a plane, with the number $x+iy$ corresponding to the point (x, y) as in Figure 1.1.

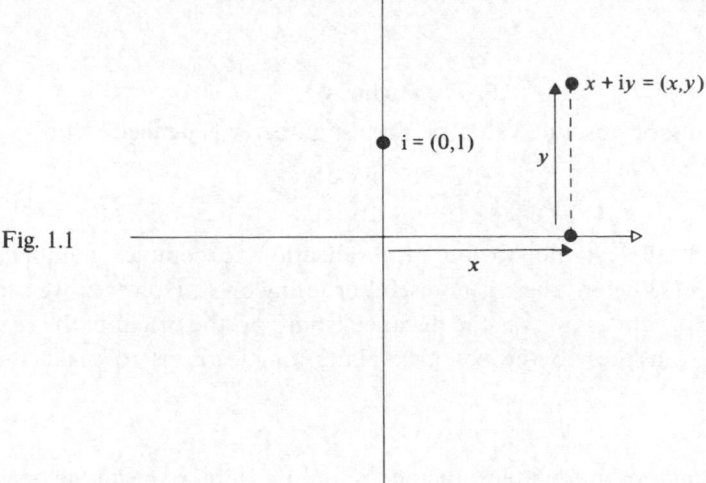

Fig. 1.1

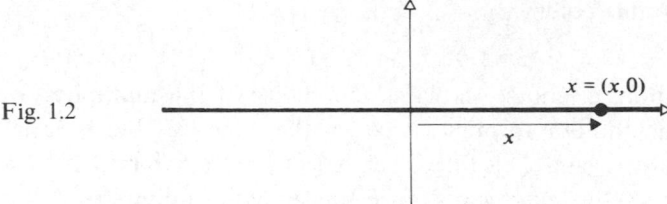

Fig. 1.2

The identification of $(x, 0)$ with x then amounts to considering the real numbers as forming the *real axis* on the plane, as in Figure 1.2.

The y-axis, at right angles to this, is the *imaginary axis*.

This geometric representation is often called the Argand Diagram or the Gauss Plane. Since so many other mathematicians have justifiable claims on it, we shall avoid the danger of giving undue credit to any of them and simply refer to it as the *complex plane*. In purely geometric terms, of course, it is just the *real* plane $\mathbb{R}^2$; but interpreted as $\mathbb{C}$ it has the additional algebraic structure of a field, and it is this that gives the complex plane its special qualities.

4. Real and imaginary parts

Given a complex number $z = x + iy$, we call x the *real part* of z and y the *imaginary part*, using the notation

$$x = \mathrm{re}(z)$$
$$y = \mathrm{im}(z).$$

Both of these are *real* numbers: the coordinates of z in the complex plane.

5. The modulus

The modulus, or absolute value, of a real number x is defined to be

$$|x| = \begin{cases} x & \text{if } x \geqslant 0 \\ -x & \text{if } x < 0. \end{cases}$$

As it stands there is no obvious generalization to complex numbers, because (see §8 below) there is no useful ordering on $\mathbb{C}$. However, we can interpret $|x|$ geometrically as the distance from x to the origin of the real line. This translates to the complex plane and leads us to make the definition

$$|z| = \sqrt{(x^2 + y^2)}$$

for the *modulus* of the complex number $z = x + iy$. Here we mean the positive square root: note that $x^2 + y^2$ is a positive real number so the formula makes sense.

The modulus has the following properties:

$$|z_1 + z_2| \leqslant |z_1| + |z_2|, \tag{10}$$

$$|z_1 z_2| = |z_1||z_2|. \tag{11}$$

Property (11) follows at once from the definitions. The *triangle inequality* (10) is a little harder to prove directly, although its geometric interpretation (Fig. 1.3) is the obvious fact that one side of a triangle is no longer than the sum of the lengths of the other two sides.

To prove (10) we note that since both sides are positive, it is equivalent to

$$|z_1 + z_2|^2 \leqslant (|z_1| + |z_2|)^2$$

which takes the form

$$(x_1 + x_2)^2 + (y_1 + y_2)^2 \leqslant |z_1|^2 + 2|z_1||z_2| + |z_2|^2$$

where $z_1 = x_1 + iy_1$, $z_2 = x_2 + iy_2$. Simplifying, this holds if and only if

$$x_1 x_2 + y_1 y_2 \leqslant |z_1||z_2|.$$

Since the right-hand side of this is positive, the desired inequality follows from

$$(x_1 x_2 + y_1 y_2)^2 \leqslant |z_1|^2 |z_2|^2.$$

But

$$|z_1|^2 |z_2|^2 - (x_1 x_2 + y_1 y_2)^2 = (x_1^2 + y_1^2)(x_2^2 + y_2^2) - (x_1 x_2 + y_1 y_2)^2$$
$$= (x_1 y_2 - x_2 y_1)^2$$

which is positive. This completes the proof.

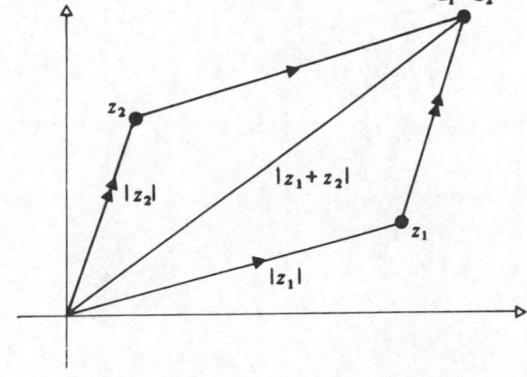

Fig. 1.3

6. The complex conjugate

If $z = x + iy$, then its *complex conjugate* is

$$\bar{z} = x - iy.$$

Geometrically, this is obtained by reflecting z in the x-axis (Fig. 1.4). The following properties are easy to verify directly:

$$\overline{z_1 + z_2} = \bar{z}_1 + \bar{z}_2 \tag{12}$$

$$\overline{z_1 z_2} = \bar{z}_1 \bar{z}_2 \tag{13}$$

$$\mathrm{re}(z) = \tfrac{1}{2}(z + \bar{z}) \tag{14}$$

$$\mathrm{im}(z) = \tfrac{1}{2}(z - \bar{z}) \tag{15}$$

$$|z|^2 = |z\bar{z}| \tag{16}$$

$$z \text{ is real if and only if } z = \bar{z}. \tag{17}$$

Properties (12) and (13) have the important implication that the complex conjugate of any polynomial expression in complex numbers $z_1, \ldots, z_n$ may be obtained by writing a bar over each individual coefficient or variable in the expression. For example,

$$\overline{5 z_1 z_2 - z_3^7 + 2 z_1} = \bar{5} \bar{z}_1 \bar{z}_2 - \bar{z}_3^7 + \bar{2} \bar{z}_1$$

$$= 5 \bar{z}_1 \bar{z}_2 - \bar{z}_3^7 + 2 \bar{z}_1$$

since 5, 2 are real, hence $\bar{5} = 5$, $\bar{2} = 2$.

Fig. 1.4

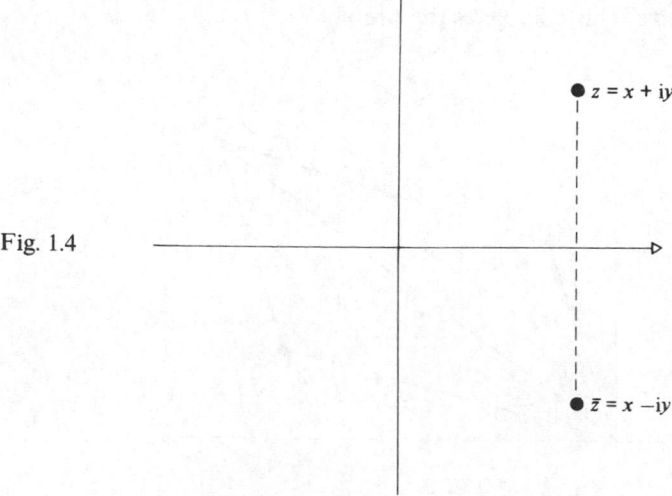

7. Polar coordinates

The expression $x+iy$ for a complex number is intimately related to *Cartesian* coordinates (x, y) in the plane. It turns out often to be useful to work with *polar* coordinates (r, θ), which we recall correspond to a point distance r from the origin along a ray making an angle θ measured from the positive x-axis in an anticlockwise direction (Fig. 1.5). Of course we measure θ in radians. The coordinate systems are related as follows:

$$x=r \cos \theta$$
$$y=r \sin \theta. \tag{18}$$

It follows that

$$r=\sqrt{x^2+y^2}=|z|$$

where $z=x+iy$.

Any value of θ for which (18) holds is called an *argument* of z. The article 'an' is used because θ is not unique: if θ is an argument so is $\theta+2k\pi$ for any integer k. With the understanding that θ is unique only up to multiples of 2π, we may use the notation

$$\theta = \arg(z).$$

Often the choice of θ is rendered unique by some convention: for example, we may insist that θ is chosen in the interval $[0, 2\pi)$, or in $(-\pi, \pi]$. The unique value of θ in the interval $(-\pi, \pi]$ is known as the *principal value* of the argument. (We follow standard practice in taking this particular

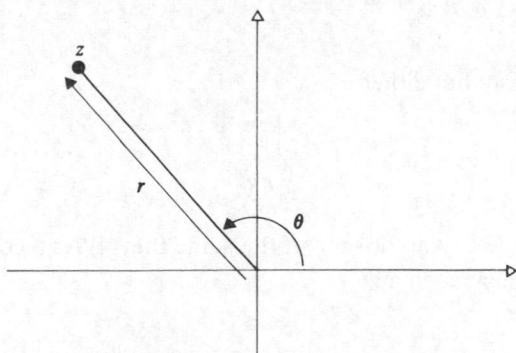

Fig. 1.5

interval. Its main advantage is that θ then behaves nicely near the positive real axis, where $\theta = 0$. But this is a technical point that only acquires importance much later. The non-uniqueness of θ is a phenomenon with tremendous ramifications in the theory, as we shall see.)

With r, θ defined as above, we have

$$z = x + iy = r(\cos \theta + i \sin \theta).$$

The expression $\cos \theta + i \sin \theta$ is of considerable importance in complex analysis. In Chapter 5 we relate it to the complex exponential function.

8. The complex numbers cannot be ordered

The real numbers may be given an ordering (the usual one, $<$) which has among its properties the following:

$$\text{If } x \neq 0 \text{ then either } x > 0 \text{ or } -x > 0, \text{ but not both.} \tag{19}$$

$$\text{If } x, y > 0 \text{ then } xy > 0, x + y > 0. \tag{20}$$

No such ordering can be defined on the complex numbers. Suppose for a contradiction that one can. Then since $i \neq 0$ it follows from (19) that either $i > 0$ or $-i > 0$. Then (20) implies that $-1 = i \cdot i > 0$ or $-1 = (-i)(-i) > 0$. At the same time, $1 = (-1)^2 > 0$. But then both 1 and -1 are greater than 0, contrary to (19).

It is therefore not possible to use inequalities, analogous to those for reals, when discussing complex numbers. Any inequality that occurs must involve only *real* numbers, possibly related to the given complex numbers. For example, if $z \in \mathbb{C}$, then

$$z > 1$$

makes no sense; but either of

$$|z| > 1$$

or

$$\text{re}(z) > 1$$

is acceptable. (They do not mean the same thing!) As a convention, if we write a statement such as

$$\varepsilon > 0$$

this will automatically imply that ε is a *real* number.

Exercises 1

1. Check in full detail that the complex numbers $\mathbb{C}$ form a field under the operations of addition and multiplication defined above.

2. If $z_1, z_2 \in \mathbb{C}$ prove that
   (i) $|z_1 + z_2| \leqslant |z_1| + |z_2|$,
   (ii) $||z_1| - |z_2|| \leqslant |z_1 - z_2|$.

3. In Figure 1.6 the black dots represent three complex numbers u, v, w (as marked). The circle is the unit circle $|z| = 1$. The open dots a, b, c, d, e, f, g, h represent (in some order) the numbers $u + v$, $u + w$, $v + w$, $u + v + w$, uv, uw, vw, uvw. Which is which?

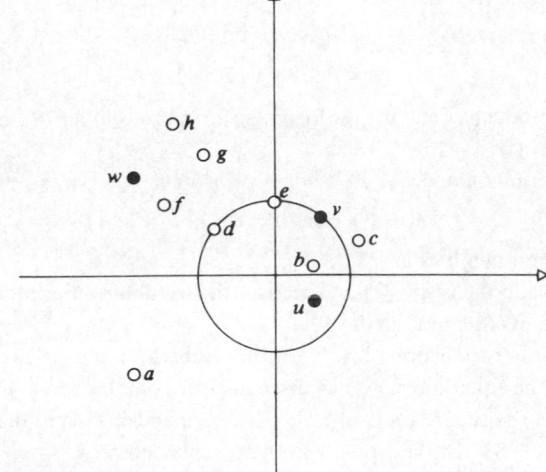

Fig. 1.6

4. By writing z in the form $z = a + b\mathrm{i}$, find all solutions z of the following equations:
   (i) $z^2 = -5 + 12\mathrm{i}$
   (ii) $z^2 = 2 + \mathrm{i}$
   (iii) $(7 + 24\mathrm{i})z = 375$
   (iv) $z^2 - (3 + \mathrm{i})z + (2 + 2\mathrm{i}) = 0$
   (v) $z^2 - 3z + 1 + \mathrm{i} = 0$.

5. If λ is a positive real number, show that
$$\{z \in \mathbb{C} \mid |z| = \lambda|z - 1|\}$$
is a circle, *unless* λ takes one particular value (which?).

6. Draw the set of points
$$\{z \in \mathbb{C} \mid \operatorname{re}(z + 1) = |z - 1|\}$$
by substituting $z = x + \mathrm{i}y$ and computing the real equation relating x and y.

   Now note that $\operatorname{re}(z + 1)$ is the distance from z to the line $y = -1$; and $|z - 1|$ is the distance between z and 1. Compare with the classical 'focus–directrix' definition of a parabola: the locus of a point equidistant from a fixed line (here $y = -1$) and a fixed point $((x, y) = (1, 0))$.

7. Draw the set of all $z \in \mathbb{C}$ satisfying the following conditions:
   (i) $\operatorname{re} z > 2$ (ii) $1 < \operatorname{im} z < 2$
   (iii) $1 < \operatorname{im}(z - \mathrm{i}) < 2$ (iv) $|z| < 2$

(v) $|z|>1$ (vi) $1<|z|<2$

(vii) $|z-1|<1$ (viii) $|z-1|<|z+1|$.

Interpret the expressions $|z-1|$, $|z+1|$ as the distances between certain points, and explain (g) and (h) in terms of the distances from z to 1 and -1.

8. Draw the set of all $z \in \mathbb{C}$ satisfying the following conditions:

   (i) $z\bar{z}=1$ (ii) $z+i\bar{z}+1+i=0$

   (iii) $z+\bar{z}+2=0$ (iv) $z+\bar{z}+2i=0$.

9. Let r, s, θ, ϕ be real. Let

$$z=r(\cos\theta+i\sin\theta)$$

$$w=s(\cos\phi+i\sin\phi).$$

Form the product zw and use the formulae for $\cos(\theta+\phi)$, $\sin(\theta+\phi)$ to show that $\arg(zw)=\arg(z)+\arg(w)$.

By induction on n, derive *De Moivre's Theorem*

$$(\cos\theta+i\sin\theta)^n=\cos n\theta+i\sin n\theta$$

for all natural numbers n.

Specialize to the case $n=3$ and recover the usual formulae for $\cos 3\theta$ and $\sin 3\theta$ in terms of $\cos\theta$ and $\sin\theta$.

10. Use De Moivre's Theorem (ex. 9) and the substitution $z=r(\cos\theta+i\sin\theta)$ to show that the equation $z^3=1$ has three distinct complex roots. Find them.

Compute the square roots of $1+\sqrt{3}i$, $\sqrt{3}-i$, and $1+i$; and the cube roots of $\sqrt{3}+i$, $1-i$, i. Sketch these points in the complex plane.

11. In earlier textbooks, multiplication of complex numbers is often defined as follows. Given two complex numbers z_1, z_2, represent them by points A and B in the complex plane; and let 0, U be the points $z=0$, $z=1$ respectively (see

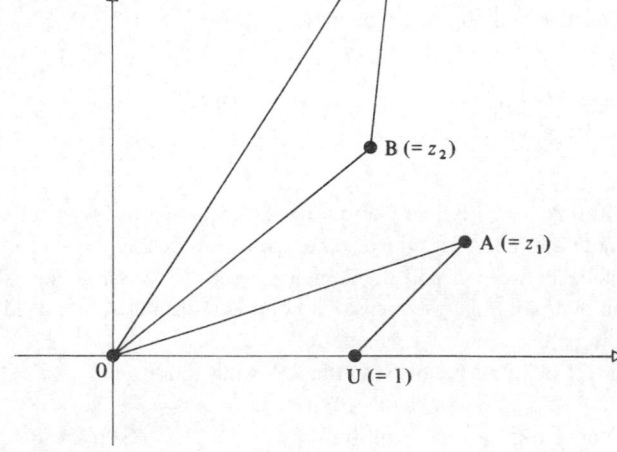

Fig. 1.7

Fig. 1.7). Draw triangle OBC similar to triangle OUA (where $\angle BOC = \angle UOA$, $\angle OBC = \angle OUA$). Then $z_1 z_2$ is represented by the point C so constructed.

Using the fact that $|z_1 z_2| = |z_1||z_2|$, and the result of exercise 6, show that this agrees with the formula definition given in this text.

12. Define a square root $\sqrt{z}$ of a complex number z to be any complex number w such that $w^2 = z$. Prove that every non-zero complex number has exactly two complex square roots, and give formulae for them in terms of re z and im z.

If $a, b, c \in \mathbb{C}$, with $a \neq 0$, show that the solutions of the quadratic equation

$$az^2 + bz + c = 0$$

are precisely

$$z = \frac{-b \pm \sqrt{(b^2 - 4ac)}}{2a}.$$

13. Use De Moivre's Theorem (exercise 9) to compute $\cos 5\theta$ and $\sin 5\theta$ in terms of $\cos \theta$ and $\sin \theta$.

14. Prove that De Moivre's Theorem remains true if n is a negative integer.

15. Define a kth root $\sqrt[k]{z}$ to be any w such that $w^k = z$. Use De Moivre's Theorem to find an expression for $\sqrt[k]{(r(\cos \theta + i \sin \theta))}$.

16. Let $\alpha, \beta, \gamma, \delta \in \mathbb{C}$ with $\gamma \neq 0$, $\alpha\delta \neq \beta\gamma$. Define ϕ by

$$\phi(z) = \frac{\alpha z + \beta}{\gamma z + \delta}.$$

Prove that ϕ defines a bijection between $\mathbb{C} \setminus \{-\delta/\gamma\}$ and $\mathbb{C} \setminus \{\alpha/\gamma\}$. Show that it takes circles into circles. What about circles through $-\delta/\gamma$?

2

Topology of the complex plane

In this chapter we collect together all the basic topological ideas we require for our study of complex analysis. The list is not very demanding. Some items are needed to handle differentiation neatly, and some are needed for integration. Differentiation is naturally set against a background of *limits* and *continuity* and these are best dealt with on *open* sets. On the other hand, an integral from one complex number to another is computed along a specified *path* between them. A set within which any two points can be joined by a path is said to be *connected*. To be able to cope with both integration and differentiation in the simplest possible manner later on, we shall restrict our complex functions to those defined on open connected sets. Such a set is called a *domain*.

Domains can have exotic shapes and paths can wiggle around a great deal. To be able to appeal to geometric intuition without our imagination having to work overtime thinking about complications like this, we use a carefully conceived technical device called the *Paving Lemma*. We show in this lemma that a path in an open set (in particular in a domain) can be subdivided into a finite number of smaller pieces in such a way that each piece is contained in a disc within the open set (thus 'paving' the path with discs). (Fig. 2.1) Now a disc is geometrically very simple; for instance any two points in it can be joined by a straight line. Joining the endpoints of the pieces of the original path in each disc paving it, we obtain a new path made up of straight line segments, still lying in the open set and joining the ends of the original path. We see, therefore, that given any path what-soever between two points in an open set, no matter how much the path twists and turns, there is an alternative path between the points in the open set which is just made up of a finite number of straight segments. We can even insist that the segments are parallel to the real or imaginary axes, giving a *step-path* in the open set (just by taking a step-path in each paving disc). (Fig. 2.2).

By techniques such as this we can use the Paving Lemma to illuminate

22

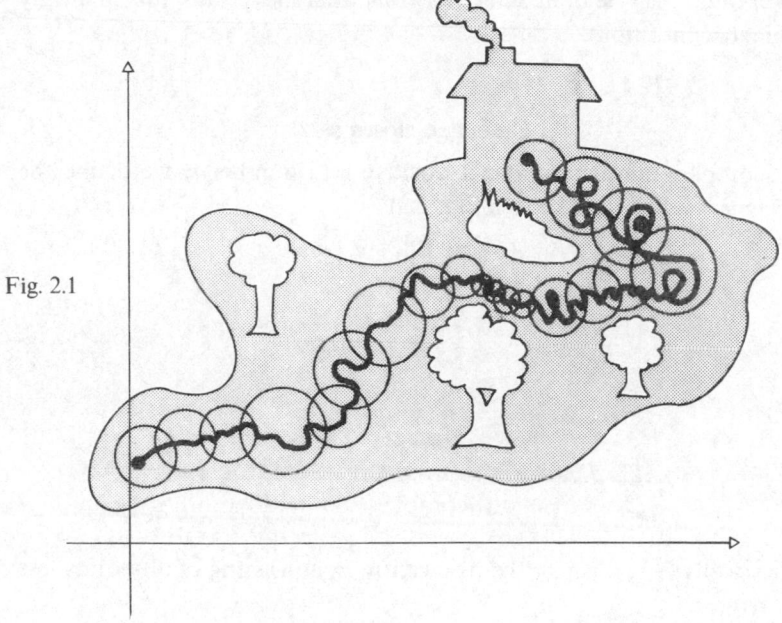

Fig. 2.1

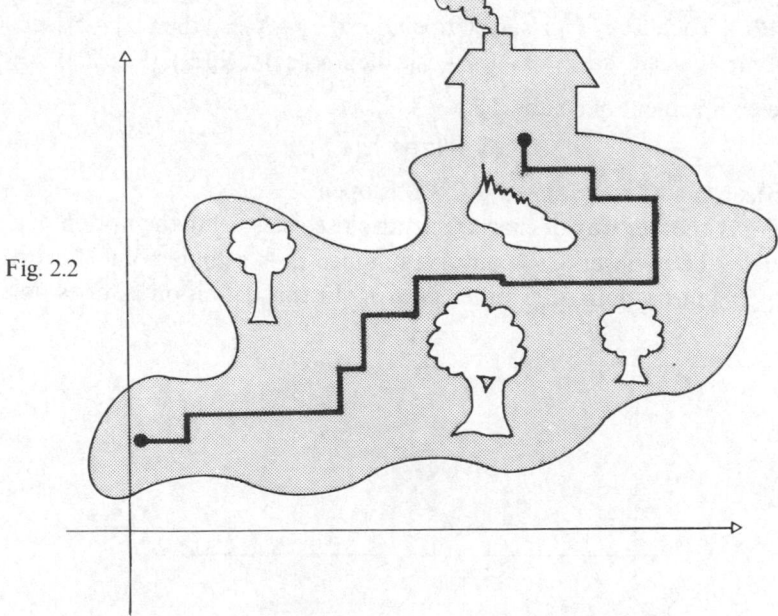

Fig. 2.2

complex analysis, yielding fully rigorous analytic proofs linked firmly to geometric intuition.

1. Open and closed sets

For a complex number z_0 and a positive real number ε, we define the ε-*neighbourhood* of z_0 to be (see Fig. 2.3)

$$N_\varepsilon(z_0) = \{z \in \mathbb{C} \,\|\, |z - z_0| < \varepsilon\}.$$

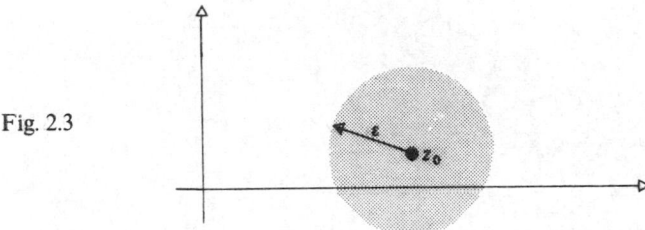

Fig. 2.3

Geometrically $N_\varepsilon(z_0)$ is just the disc centre z_0 consisting of all points less than ε from z_0.

A subset $S \subseteq \mathbb{C}$ is said to be *open* if every $z_0 \in S$ has a corresponding real number $\varepsilon > 0$ such that $N_\varepsilon(z_0) \subseteq S$. We emphasize the fact that ε may depend on z_0.

EXAMPLE. The disc $N_\varepsilon(z_0)$ is itself open, for if $z_1 \in N_\varepsilon(z_0)$ then $|z_1 - z_0| < \varepsilon$. Choose $\delta > 0$ such that $\delta \leqslant \varepsilon - |z_1 - z_0|$; then $N_\delta(z_1) \subseteq N_\varepsilon(z_0)$. (Fig. 2.4).

The complement of a subset S is

$$\mathbb{C} \backslash S = \{z \in \mathbb{C} : z \notin S\}.$$

A subset S is said to be *closed* if $\mathbb{C} \backslash S$ is open.

There is another way of characterizing closed sets using the notion of a *limit point* of a subset S. A complex number z_0 is a limit point if every $N_\varepsilon(z_0)$ contains a point of S other than z_0. In this definition z_0 does not

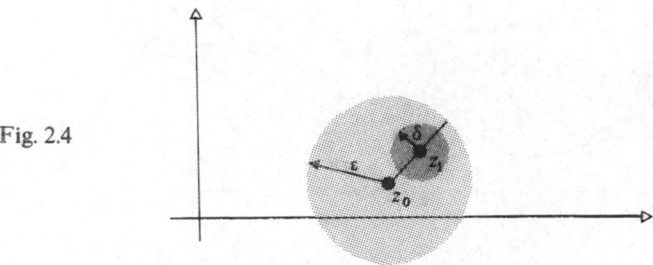

Fig. 2.4

have to be a member of S, though it may be. The essential feature of a limit point is that it has points of S arbitrarily close to it. In fact each $N_\varepsilon(z_0)$ must contain an *infinite* number of points of S (for if some $N_\varepsilon(z_0)$ contained only a finite number of elements $z_1, \ldots, z_n$ in S distinct from z_0, by taking ε_1 to be the smallest of the distances $|z_0 - z_r|$, then $N_{\varepsilon_1}(z_0)$ would contain *no* points of S).

As an alternative characterization of a closed set we have:

PROPOSITION 2.1. S is closed if and only if S contains all its limit points.

Proof. Suppose S is closed and z_0 is a limit point. If $z_0 \in \mathbb{C} \setminus S$, which is open, then $N_\varepsilon(z_0) \subseteq \mathbb{C} \setminus S$ for some $\varepsilon > 0$ and so $N_\varepsilon(z_0)$ would contain no points of S, contrary to the fact that z_0 is a limit point. Hence $z_0 \in S$.

Conversely, suppose that S contains all its limit points. Then $z_0 \in \mathbb{C} \setminus S$ is not a limit point of S, so there exists $N_\varepsilon(z_0)$ containing no point of S and $N_\varepsilon(z_0) \subseteq \mathbb{C} \setminus S$. Thus $\mathbb{C} \setminus S$ is open and S is closed. $\qquad\square$

Of course, not every point of a closed set need be a limit point of that set. For instance, if

$$T = \{z \in \mathbb{C} \mid z = 0 \text{ or } z = 1/n \text{ for a positive integer } n\},$$

then the only limit point of T is 0, which is in T, so T is closed.

Points in S which are not limit points are said to be *isolated points* of S. All the other points of T except 0 are therefore isolated points of T. An isolated point z_0 of S has a neighbourhood $N_\varepsilon(z_0)$ which contains no other points of S. For instance, in the set T just described, if $\varepsilon = 1/n - 1/(n+1)$, then $N_\varepsilon(1/n)$ contains no other points of T. On the other hand, it is clear from the definitions that *every* element of an *open* set is a limit point.

2. Limits of functions

The notion of a limit $\lim_{z \to z_0} f(z)$ is analogous to the real case and its properties follow by similar arguments.

DEFINITION. If $f : S \to \mathbb{C}$ is an arbitrary complex function and z_0 is a limit point of S, then $\lim_{z \to z_0} f(z) = l$ if, given $\varepsilon > 0$, there exists $\delta > 0$ such that

$$\text{for all } z \in S, \ 0 < |z - z_0| < \delta \quad \text{implies} \quad |f(z) - l| < \varepsilon. \tag{1}$$

Two points should be made about this definition:

(a) z_0 need not be a point of S and so $f(z_0)$ need not be defined; and even if $z_0 \in S$, we may have $f(z_0) \neq l$.

For example, if

$$f(z) = \begin{cases} 0 & (z \neq 0) \\ 1 & (z = 0) \end{cases}$$

then $\lim_{z \to 0} f(z) = 0 \neq f(0)$.

(b) It is essential that z_0 is a limit point of S, or else there would exist $\delta > 0$ such that $\{z | 0 < |z - z_0| < \delta\}$ contains no point of S. In this case the condition (1) would be vacuously true for any $l \in \mathbb{C}$.

As in the real case, however, if z_0 is a limit point of S and $\lim_{z \to z_0} f(z) = l$, then the limit is *unique*. For if $l' \neq l$ is another candidate for the limit, take $\varepsilon = \frac{1}{2} |l - l'|$ to find $\delta_1 > 0$, $\delta_2 > 0$ such that

$$z \in S, 0 < |z - z_0| < \delta_1 \quad \text{implies} \quad |f(z) - l| < \varepsilon,$$
$$z \in S, 0 < |z - z_0| < \delta_2 \quad \text{implies} \quad |f(z) - l'| < \varepsilon.$$

Because z_0 is a limit point, there exists $z^* \in S$, where $0 < |z_0 - z^*| < \min \{\delta_1, \delta_2\}$. Then

$$|l - l'| = |l - f(z^*) + f(z^*) - l'|$$
$$\leqslant |l - f(z^*)| + |f(z^*) - l'|$$
$$< \varepsilon + \varepsilon$$
$$= 2\varepsilon,$$

contradicting the choice of ε.

Standard properties of complex limits may be proved by using methods analogous to the real case:

PROPOSITION 2.2. if $\lim_{z \to z_0} f(z) = l$, $\lim_{z \to z_0} g(z) = k$, then

(i) $\lim_{z \to z_0} (f(z) + g(z)) = l + k$,

(ii) $\lim_{z \to z_0} (f(z) - g(z)) = l - k$

(iii) $\lim_{z \to z_0} (f(z)g(z)) = lk$

(iv) $\lim_{z \to z_0} (f(z)/g(z)) = l/k \quad$ (for $k \neq 0$).

Proof. Parts (i), (ii) are routine. So is part (iii), but a little trickier. We write

$$|f(z)g(z) - lk| = |f(z)g(z) - lg(z) + lg(z) - lk|$$
$$\leqslant |f(z) - l| \, |g(z)| + |l| \, |g(z) - k|.$$

Since $\lim_{z \to z_0} g(z) = k$, given $\varepsilon > 0$, there exists $\delta_0 > 0$ such that

$$z \in S \quad \text{and} \quad 0 < |z - z_0| < \delta_0 \quad \text{implies} \quad |g(z) - k| < \varepsilon$$

and so $|g(z)| < |k| + \varepsilon$, which means that $|g|$ is bounded above near z_0 by $M = |k| + \varepsilon$.

By continuity of f there exists $\delta_1 > 0$ such that

$$z \in S \text{ and } 0 < |z - z_0| < \delta_1 \quad \text{implies } |f(z) - l| < \varepsilon/(2M).$$

Let N be any real number exceeding $|l|$ (in particular, $N > 0$), then by the continuity of g there exists $\delta_2 > 0$ such that

$$z \in S \text{ and } 0 < |z - z_0| < \delta_2 \quad \text{implies } |g(z) - k| < \varepsilon/(2N).$$

For $\delta = \min \{\delta_0, \delta_1, \delta_2\}$, we then have

$z \in S \text{ and } 0 < |z - z_0| < \delta \quad$ implies

$$|f(z) - l||g(z)| + |l||g(z) - k| < \frac{\varepsilon}{2M} M + N \frac{\varepsilon}{2N} = \varepsilon.$$

Hence, for $z \in S$ and $0 < |z - z_0| < \delta$ we have $|f(z)g(z) - lk| < \varepsilon$.

The proof of (iv) only requires us to show that $\lim_{z \to z_0}(1/g(z)) = 1/k$, for we then may appeal to (iii) for the functions f, $1/g$ to give the full force of (iv). Because $g(z) \to k \neq 0$, we know that for some $\delta_1 > 0$,

$$z \in S \quad \text{and } 0 < |z - z_0| < \delta_1 \quad \text{implies } |g(z) - k| < \tfrac{1}{2}|k|,$$

$$\text{hence } |g(z)| > \tfrac{1}{2}|k|.$$

We now find $\delta_2 > 0$ such that

$$z \in S \quad \text{and } 0 < |z - z_0| < \delta_2 \quad \text{implies } |g(z) - k| < \tfrac{1}{2}|k|^2\varepsilon,$$

then for $\delta = \min \{\delta_1, \delta_2\}$,

$z \in S \quad \text{and } 0 < |z - z_0| < \delta \quad$ implies

$$\left| \frac{1}{g(z)} - \frac{1}{k} \right| = \frac{|k - g(z)|}{|g(z)| |k|} < \tfrac{1}{2}|k|^2 \varepsilon / (\tfrac{1}{2}|k|^2) = \varepsilon. \qquad \square$$

The real and imaginary parts of a complex function,

$$f(z) = \text{re } (f(z)) + i \text{ im } (f(z)),$$

may be considered separately. If

$$\lim_{z \to z_0} f(z) = l = \alpha + i\beta \qquad (\alpha, \beta \in \mathbb{R}) \tag{2}$$

then, because

$$|\text{re } (f(z)) - \alpha| = |\text{re } (f(z) - l)| \leqslant |f(z) - l|,$$

we deduce from the definition that

$$\lim_{z \to z_0} \text{re } (f(z)) = \alpha. \tag{3}$$

Similarly

$$\lim_{z \to z_0} \text{im } (f(z)) = \beta. \tag{4}$$

Conversely, if (3) and (4) both hold, from proposition 2.2(i) we deduce (2).

We can rephrase the preceding argument by recalling that the limit of a real function $\phi:S\to\mathbb{R}$ of two variables, $\phi(x, y)$ for $(x, y) \in S \subseteq \mathbb{R}^2$ is defined as follows:

$\phi(x, y)\to\lambda$ as $(x, y)\to(a, b)$ if, given $\varepsilon>0$, there exists $\delta>0$ such that

for all $(x, y) \in S$, $0<\sqrt{\{(x-a)^2+(y-b)^2\}}<\delta$ implies $|\phi(x, y)-\lambda|<\varepsilon$.

Identifying (x, y) with $x+iy$, as in Chapter 1, for $z=x+iy$, $z_0=a+ib$, this may be written as

$$z \in S \quad \text{and} \quad 0<|z-z_0|<\delta \quad \text{implies} \quad |\phi(x, y)-\lambda|<\varepsilon.$$

If we now write

$$f(z)=u(x, y)+iv(x, y)$$

where the real and imaginary parts of f are considered as real functions u and v of two real variables x, y, then we have proved:

PROPOSITION 2.3. $\lim\limits_{z\to z_0} f(z)=l=\alpha+i\beta$ if and only if

$$u(x, y)\to\alpha, v(x, y)\to\beta \quad \text{as } (x, y)\to(a, b)$$

where $f(z)=u(x, y)+iv(x, y)(z=x+iy)$ and $z_0=a+ib$.   □

In this way limits of complex functions are equivalent to limits of real functions of two real variables (but the notation in the complex case is usually much simpler).

3. Continuity

A function $f: S\to\mathbb{C}$ is said to be *continuous* at $z_0 \in S$ if, given $\varepsilon>0$ there exists $\delta>0$ such that

$$\text{for all } z \in S, |z-z_0|<\delta \quad \text{implies} \quad |f(z)-f(z_0)|<\varepsilon.$$

If z_0 is a limit point, this is equivalent to saying that $\lim_{z\to z_0} f(z)$ exists and

$$\lim\limits_{z\to z_0} f(z)=f(z_0).$$

If z_0 is an isolated point, then there is a neighbourhood $N_\delta(z_0)$ which contains no other points of S apart from z_0, so

$$\text{for all } z \in S, \quad |z-z_0|<\delta \quad \text{implies } z=z_0$$

which in turn implies

$$|f(z)-f(z_0)|=0,$$

so a complex function is always deemed to be continuous at an isolated point according to the given definition. Actually, this latter remark will not trouble us at all, for we shall consider only functions where all the points in S are limit points; in particular, S will usually be open. The remark about isolated points is simply made to tidy up a loose end.

We can rephrase the definition of continuity in terms of open discs: f is continuous at $z_0 \in S$ if, given $\varepsilon > 0$, there exists $\delta > 0$ such that

$$\text{for all } z \in S, \quad z \in N_\delta(z_0) \text{ implies } f(z) \in N_\varepsilon(f(z_0));$$

or, more succinctly,

$$f(N_\delta(z_0) \cap S) \subseteq N_\varepsilon(f(z_0)).$$

A function $f: S \to \mathbb{C}$ is said to be *continuous* if it is continuous at every point z_0 in S. We can develop an alternative way of defining continuous functions in terms of open sets. First we need a generalization: a subset $V \subseteq S$ is said to be *relatively open* in S, or just *open in S*, for short, if every $z_0 \in V$ has a corresponding $\sigma > 0$ such that $N_\sigma(z_0) \cap S \subseteq V$. A relatively open set need not be open.

EXAMPLE. The interval $(a, b) = \{x \in \mathbb{R} \mid a < x < b\}$ is open in $\mathbb{R}$, but not in $\mathbb{C}$. For $x \in (a, b)$, let $\sigma = \min \{x - a, b - x\}$, then

$$N_\sigma(x) \cap \mathbb{R} = \{y \in \mathbb{R} \mid x - \sigma < y < x + \sigma\} \subseteq (a, b),$$

but no open disc $N_\sigma(x)$ in $\mathbb{C}$ is contained in the real interval (a, b), so (a, b) is not open in $\mathbb{C}$. (Fig. 2.5)

Using the standard set-theoretical notation

$$f^{-1}(U) = \{z \in S \mid f(z) \in U\}$$

we get an alternative characterization of a continuous function:

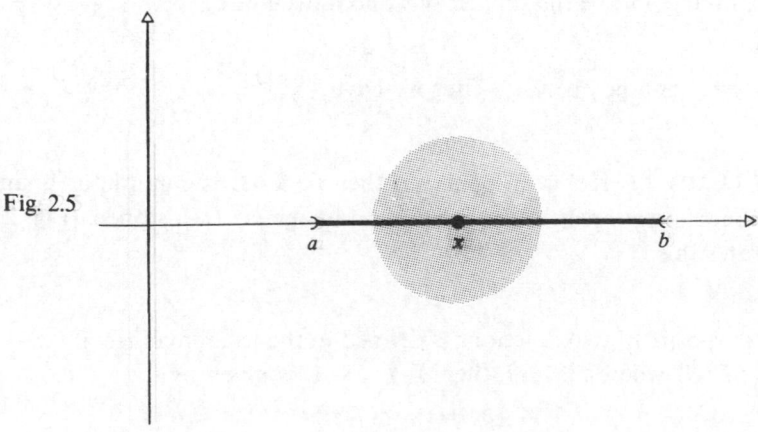

Fig. 2.5

PROPOSITION 2.4. A complex function $f : S \to \mathbb{C}$ is continuous if, and only if, for every open set U, the inverse image $f^{-1}(U)$ is open in S.

Proof. Suppose that f is continuous and U is open. Let $z_0 \in f^{-1}(U)$. Then $f(z_0) \in U$ and so there exists $\varepsilon > 0$ such that $N_\varepsilon(f(z_0)) \subseteq U$. By continuity of f there exists $\delta > 0$ such that

$$f(N_\delta(z_0) \cap S) \subseteq N_\varepsilon(f(z_0)) \subseteq U,$$

hence

$$N_\delta(z_0) \cap S \subseteq f^{-1}(U)$$

and $f^{-1}(U)$ is open.

Conversely, suppose that $f^{-1}(U)$ is open in S for every open set U. Given $z_0 \in S$ and $\varepsilon > 0$, then set $N_\varepsilon(f(z_0))$ is open, so $f^{-1}(N_\varepsilon(f(z_0)))$ is open in S and there exists $\delta > 0$ such that

$$N_\delta(z_0) \cap S \subseteq f^{-1}(N_\varepsilon(f(z_0))).$$

Hence

$$f(N_\delta(z_0) \cap S) \subseteq N_\varepsilon(f(z_0))$$

and f is continuous. $\square$

Proposition 2.4 is favoured by topologists as the *definition* of continuity in terms of open sets.

It is particularly useful when S is itself an open set. For then, given a subset $V \subseteq S$ open in S and $z_0 \in V$, we have

$$N_\varepsilon(z_0) \subseteq S \qquad \text{(because } S \text{ is open)}$$
$$N_\sigma(z_0) \cap S \subseteq V \qquad \text{(because } V \text{ is open in } S),$$

hence, taking δ to be the smaller of ε and σ, we find

$$N_\delta(z_0) \subseteq V$$

and V is a genuine open set. Thus we have:

COROLLARY 2.5. If S is an open set, then $f : S \to \mathbb{C}$ is continuous if, and only if, for every open set U, the inverse image $f^{-1}(U)$ is open. (Fig. 2.6 illustrates this.) $\square$

The composite of two functions is defined in the usual manner: if $f : S \to \mathbb{C}$ and $g : T \to \mathbb{C}$ where $f(S) \subseteq T$, then $g \circ f : S \to \mathbb{C}$ is given by

$$g \circ f(z) = g(f(z)).$$

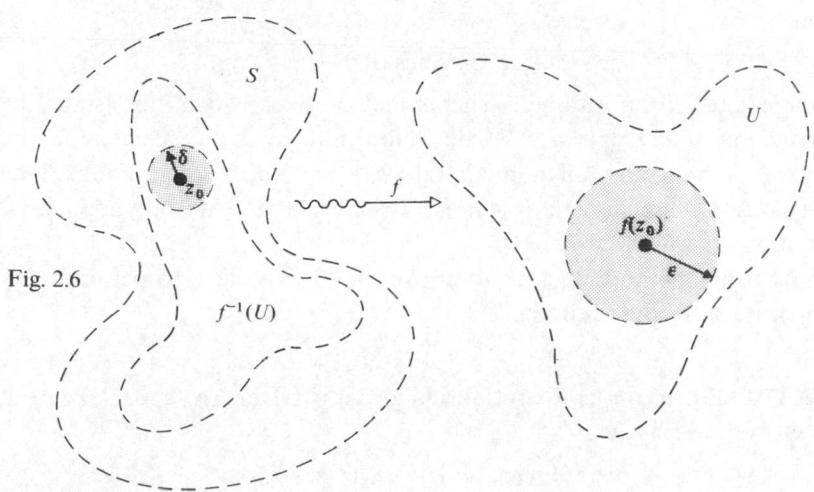

Fig. 2.6

PROPOSITION 2.6. If f is continuous at $z_0 \in S$ and g is continuous at $f(z_0)$ then $g \circ f$ is continuous at z_0.

Proof. Easy exercise. ☐

We add and multiply complex functions $f_1 : S \rightarrow \mathbb{C}$ and $f_2 : S \rightarrow \mathbb{C}$ on a given set S to get

$$f_1 + f_2 : S \rightarrow \mathbb{C} \quad \text{where} \quad (f_1 + f_2)(z) = f_1(z) + f_2(z) \quad (z \in S)$$
$$f_1 - f_2 : S \rightarrow \mathbb{C} \quad \text{where} \quad (f_1 - f_2)(z) = f_1(z) - f_2(z) \quad (z \in S)$$
$$f_1 \cdot f_2 : S \rightarrow \mathbb{C} \quad \text{where} \quad (f_1 \cdot f_2)(z) = f_1(z) f_2(z) \quad (z \in S)$$
$$f_1 / f_2 : S' \rightarrow \mathbb{C} \quad \text{where} \quad (f_1 / f_2)(z) = f_1(z) / f_2(z) \quad (z \in S').$$

Here $S' = \{z \in S \mid f_2(z) \neq 0\}$.

PROPOSITION 2.7. If f_1 and f_2 are continuous, then so are $f_1 f_2$, $f_1 - f_2$, $f_1 \cdot f_2$, f_1 / f_2.

Proof. This is a direct consequence of Proposition 2.2. ☐

This result allows us to show very quickly that certain functions built up from continuous functions are themselves continuous. For instance the constant function $k(z) = c$, and the identity function $I(z) = z$, are clearly continuous. (To show this let $\varepsilon > 0$. For a constant function take any positive δ, and for the identity function take $\delta = \varepsilon$, to obtain

$$|z - z_0| < \delta \quad \text{implies} \quad |k(z) - k(z_0)| = 0 < \varepsilon,$$

and

$$|z-z_0|<\varepsilon \quad \text{implies} \quad |I(z)-I(z_0)|<\varepsilon.)$$

Immediately Proposition 2.7 shows that $\phi(z)=cz$ is continuous and, by induction on n, $\phi_n(z)=a_n z^n$ is continuous for any $a_n \in \mathbb{C}$. Once more, by induction on n, any polynomial $p(z)=a_0+\cdots+a_n z^n$ is continuous; then any rational function $r(z)=p(z)/q(z)$ (where p and q are polynomials) is continuous wherever $q(z)\neq 0$.

By using methods such as these we shall rarely have to perform very intricate $\varepsilon-\delta$ computations.

EXAMPLE 1. $m(z)=|z|$ is continuous for all z. Given any $\varepsilon>0$, take $\delta=\varepsilon$, then, for $|z-z_0|<\delta$,

$$|m(z)-m(z_0)|=\big||z|-|z_0|\big|\leqslant|z-z_0|<\varepsilon.$$

EXAMPLE 2. $f(z)=(|z|^2+17z^3+1066z)/(1+z)$ is continuous for $z\neq -1$. We could show this by explicit computation, but far quicker is to note that $|z|^2=|z||z|$ is a product of continuous functions, hence continuous, $17z^3+1066z$ is a polynomial, hence continuous, thus the sum $|z|^2+17z^3+1066z$ is continuous. Finally $1+z$ is continuous and non-zero for $z\neq -1$, so the quotient $f(z)$ is continuous for $z\neq -1$.

Writing $f(z)=u(x,y)+iv(x,y)$ for $z=x+iy\in S$, where u and v are real functions of the real variables x, y, then Proposition 2.3 gives

PROPOSITION 2.8. $f(z)=u(x,y)+iv(x,y)$ is continuous at $z_0=x_0+iy_0$ if and only if u and v are continuous at (x_0,y_0). □

EXAMPLE 3. If $f(z)=z^2$, then $f(z)=(x+iy)^2=x^2-y^2+2ixy$. Hence $u(x,y)=x^2-y^2$ and $v(x,y)=2xy$. The function f is continuous for all $z\in\mathbb{C}$, just as the real functions u and v are continuous functions of x and y for all $(x,y)\in\mathbb{R}^2$.

An interesting case occurs when S is the real interval

$$[a,b]=\{x\in\mathbb{R}|a\leqslant x\leqslant b\},$$

considered as a subset of $\mathbb{C}$. For $z\in[a,b]$, $z=x+i0$, so here we may simplify our notation and write $f(z)=f(x)=u(x)+iv(x)$. Thus a function $f:[a,b]\to\mathbb{C}$ is continuous if and only if both u and v are continuous.

EXAMPLE 4. $f: [0, 1] \to \mathbb{C}$, where $f(x) = x^2 + ix^3$, is continuous, since $u(x) = x^2$ and $v(x) = x^3$ are both continuous.

4. Paths

A *path* in the complex plane is a continuous function $\gamma: [a, b] \to \mathbb{C}$. Its *initial point* is $\gamma(a)$ and *final point* is $\gamma(b)$. Sometimes we speak of a 'path in $\mathbb{C}$ from z_1 to z_2' if z_1 is the initial point and z_2 the final point. We shall also refer to the point $z = \gamma(t)$ as a 'point on the path γ', although, strictly speaking, z is on the *image* of the function γ. If we think of t as 'time' and imagine t increasing from a to b then the point $\gamma(t)$ traces out a curve in the plane from $\gamma(a)$ to $\gamma(b)$. In drawing a diagram we often indicate the direction of this motion by an arrow, though it is to be emphasized that this is a makeshift device since the curve may cross itself and the picture may get quite complicated. (Fig. 2.7).

In theory the path could be even more complicated: for instance it could be a 'space-filling' curve. We should therefore beware of representing the general case by a simplistic geometric picture. In even simple cases it is possible for two different paths to have the same image curve. For example

$$\gamma_1(t) = 2(t + it) \qquad (0 \leqslant t \leqslant \tfrac{1}{2})$$
$$\gamma_2(t) = t^2 + it^2 \qquad (0 \leqslant t \leqslant 1),$$

traverse the same set of points

$$\{x + iy \in \mathbb{C} \mid x = y, \ 0 \leqslant x \leqslant 1\}.$$

as in Figure 2.8.

To simplify matters, when we speak of line segments and circles, considered as paths, we shall assume them to be given by the following standard functions:

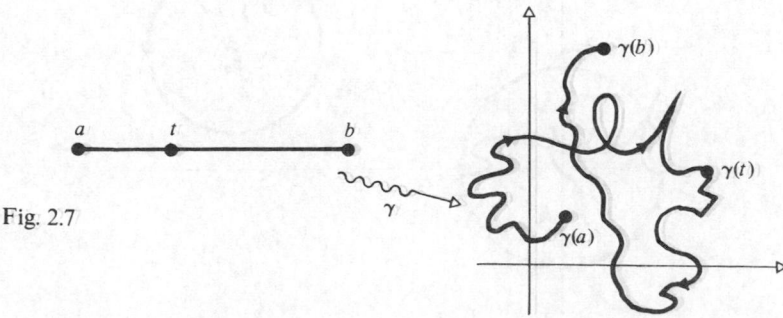

Fig. 2.7

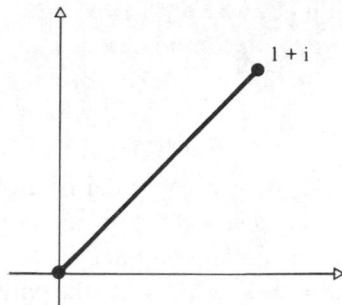

Fig. 2.8

(i) The line segment L from z_1 to $z_2 : L(t) = z_1(1-t) + z_2 t$ $(0 \leqslant t \leqslant 1)$. In this case we shall denote L by $[z_1, z_2]$. When $z_1 = a$, $z_2 = b$ are real numbers, then the image of the path $[z_1, z_2]$ is just the closed interval $[a, b]$.

(ii) The 'unit circle', $C(t) = \cos t + i \sin t$ $(0 \leqslant t \leqslant 2\pi)$. And, more generally,

(iii) The circle S, centre z_0, radius $r > 0$,

$$S(t) = z_0 + r(\cos t + i \sin t) \qquad (0 \leqslant t \leqslant 2\pi).$$

(Fig. 2.9).

For an arbitrary path $\gamma : [a, b] \to \mathbb{C}$, a *subpath* is found by restricting to a subinterval $[c, d]$ where $a \leqslant c \leqslant d \leqslant b$. If $a = x_0 < x_1 < \cdots < x_n = b$ and γ_r is the subpath given by restricting γ to $[x_{r-1}, x_r]$, we write

$$\gamma = \gamma_1 + \cdots + \gamma_n.$$

In doing so, we think of γ being made up by tracing along the subpaths $\gamma_1, \ldots, \gamma_n$ taken in order. (Fig. 2.10).

If the final point of a path γ_1 coincides with the initial point of γ_2, then

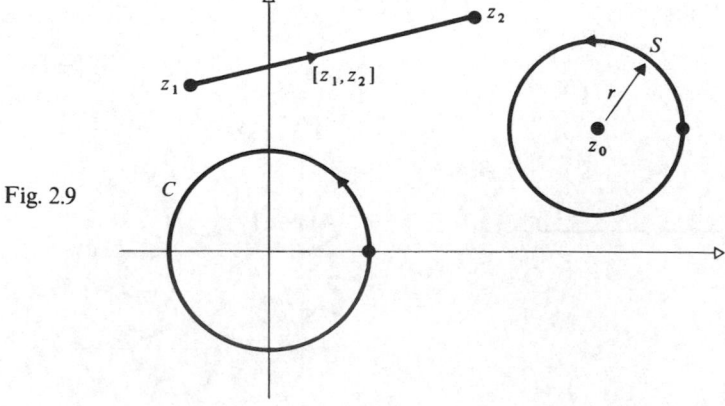

Fig. 2.9

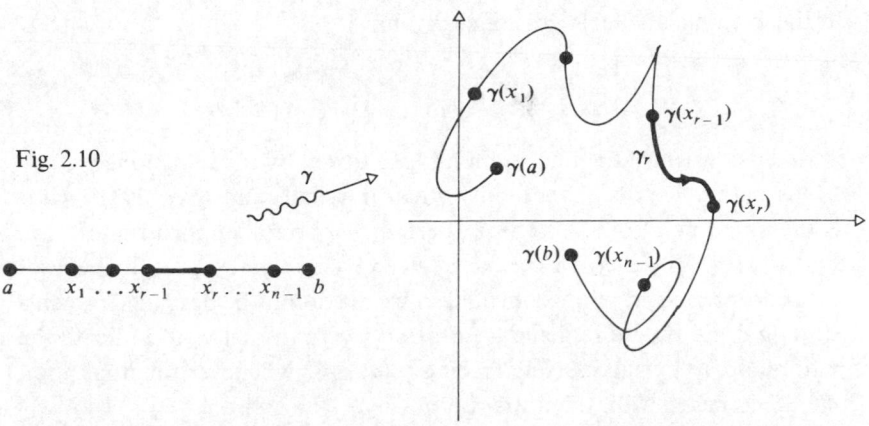

Fig. 2.10

it is useful on occasion to put them together to get a combined path found first by tracing along γ_1, then along γ_2. There is a minor technicality here, in that γ_1 may be defined on $[a, b]$, then γ_2 defined on $[c, d]$ where $b \neq c$.

EXAMPLE. $\gamma_1(t) = t \ (0 \leqslant t \leqslant 1), \gamma_2(t) = \cos t + i \sin t \ (0 \leqslant t \leqslant \pi).$ (Fig. 2.11).

In such a case we extend our notation a little and define the combination of $\gamma_1 : [a, b] \to \mathbb{C}$ with $\gamma_2 : [c, d] \to \mathbb{C}$ (provided $\gamma_1(b) = \gamma_2(c)$) to be $\gamma = \gamma_1 + \gamma_2$, where $\gamma : [a, b + d - c] \to \mathbb{C}$ is given by

$$\gamma(t) = \begin{cases} \gamma_1(t) & (a \leqslant t \leqslant b) \\ \gamma_2(t + c - b) & (b \leqslant t \leqslant d + b - c). \end{cases}$$

In effect, all we have done is to shift the parametric interval of the second part from $[c, d]$ to $[b, d + b - c]$, moving it along by adding $b - c$ to all points in $[c, d]$.

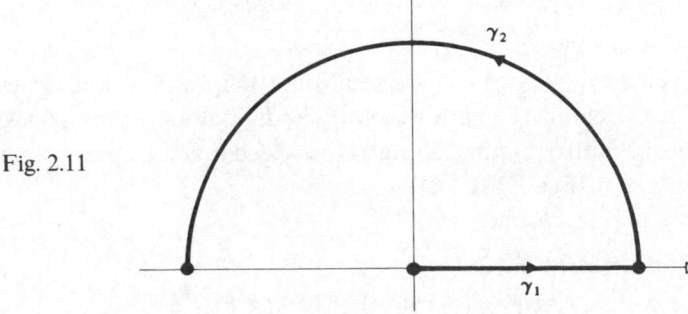

Fig. 2.11

In the example above, for instance, we have

$$\gamma(t) = \begin{cases} t & (0 \leqslant t \leqslant 1) \\ \cos(t-1) + i \sin(t-1) & (1 \leqslant t \leqslant 1+\pi), \end{cases}$$

so that γ consists of the line segment γ_1 followed by the semicircle γ_2.

The path $\gamma_1 + \gamma_2$ is defined only when the final point of γ_1 is the same as the initial point of γ_2, so this is, perhaps, not a fully appropriate use of the '+' sign. However, we do have $(\gamma_1 + \gamma_2) + \gamma_3 = \gamma_1 + (\gamma_2 + \gamma_3)$ whenever the appropriate endpoints coincide, so we may omit the brackets in such a 'sum' and use the more general notation $\gamma = \gamma_1 + \cdots + \gamma_n$ to indicate the path made up by successively tracing $\gamma_1, \ldots, \gamma_m$, whenever the final point of γ_{r-1} coincides with the initial point of γ_r $(1 < r \leqslant n)$.

Another useful concept, given a path $\gamma: [a, b] \to \mathbb{C}$, is the *opposite path* $-\gamma: [a, b] \to \mathbb{C}$ defined by

$$(-\gamma)(t) = \gamma(a + b - t) \qquad (a \leqslant t \leqslant b).$$

As t increases from a to b, so $-\gamma$ describes the same curve as γ, but it does so in the reverse order. If γ is a path from z_1 to z_2 then $-\gamma$ is a path from z_2 to z_1.

For example, if L is the line segment $[z_1, z_2]$ given by

$$L(t) = z_1(1 - t) + z_2 t \qquad (0 \leqslant t \leqslant 1)$$

then $-L$ is given by

$$(-L)(t) = z_1 t + z_2(1 - t) \qquad (0 \leqslant t \leqslant 1)$$

which is, of course, $[z_2, z_1]$.

If S is the circle centre z_0, radius $r > 0$, then

$$S(t) = z_0 + r(\cos t + i \sin t) \qquad (0 \leqslant t \leqslant 2\pi)$$

describes the point set of the circle once, anticlockwise, whilst the opposite path

$$(-S)(t) = z_0 + r(\cos(2\pi - t) + i \sin(2\pi - t))$$
$$= z_0 + r(\cos t - i \sin t) \qquad (0 \leqslant t \leqslant 2\pi)$$

describes it once, clockwise.

In a sum such as $\gamma_1 + (-\gamma_2) + \gamma_3$ we shall omit the brackets and write $\gamma_1 - \gamma_2 + \gamma_3$. This is simply the path which traces first along γ_1, then back along the opposite path of γ_2, then along γ_3 (provided that the appropriate endpoints match up). (Fig. 2.12).

For instance, if $0 < \varepsilon < R$ and

$$C_R(t) = R(\cos t + i \sin t) \qquad (0 \leqslant t \leqslant \pi)$$
$$C_\varepsilon(t) = \varepsilon(\cos t + i \sin t) \qquad (0 \leqslant t \leqslant \pi),$$

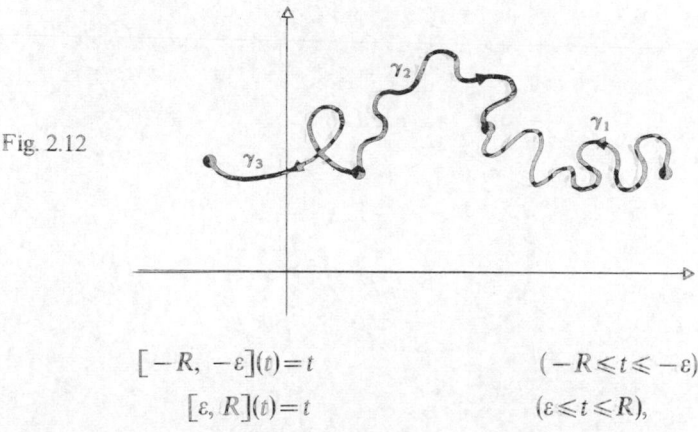

Fig. 2.12

$$[-R, -\varepsilon](t)=t \qquad\qquad (-R \leqslant t \leqslant -\varepsilon),$$
$$[\varepsilon, R](t)=t \qquad\qquad (\varepsilon \leqslant t \leqslant R),$$

then $C_R + [-R, -\varepsilon] - C_\varepsilon + [\varepsilon, R]$ is the path describing the curve in Figure 2.13 once, anticlockwise.

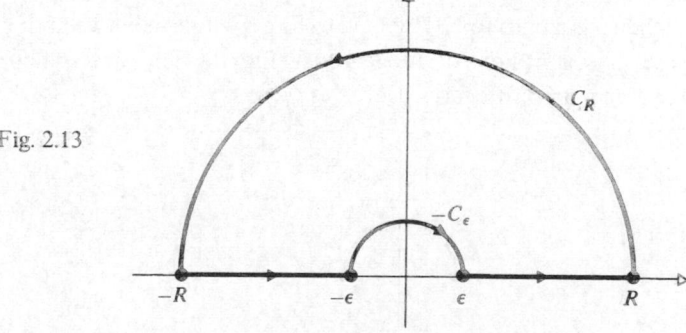

Fig. 2.13

5. The Paving Lemma

A path $\gamma: [a, b] \to \mathbb{C}$ is said to be *a path in S* if

$$\{\gamma(t) \mid a \leqslant t \leqslant b\} \subseteq S.$$

This will be signified by the notation $\gamma: [a, b] \to S$.

EXAMPLE 1. The unit circle $C(t) = \cos t + i \sin t$ $(0 \leqslant t \leqslant 2\pi)$ is a path in $S = \{z \in \mathbb{C} \mid \frac{1}{2} \leqslant |z| \leqslant 2\}$. (Fig. 2.14).

This example is a simple one. It must not lull us into a false sense of security, for much more intricate paths are possible.

EXAMPLE 2. If $V = \{z \in \mathbb{C} \mid z \neq 1/n$ for a non-zero integer $n\}$ and

$$\sigma(t) = t + it \cos(\pi/t) \quad (-1 \leqslant t \leqslant 1)$$

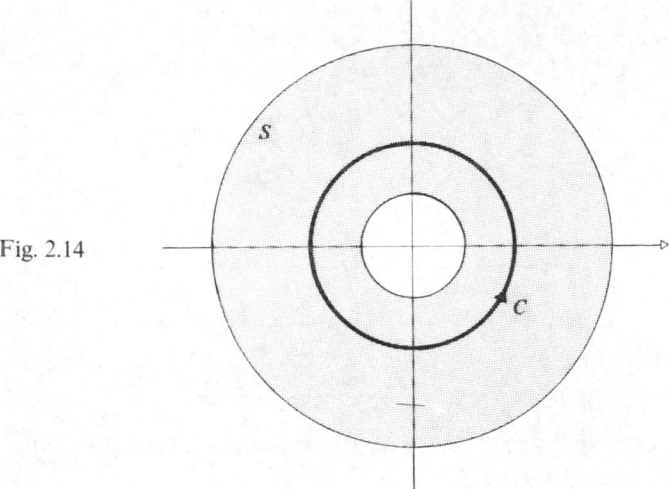

Fig. 2.14

(where $\sigma(0)=0$), then σ is a path in V. (Fig. 2.15). Here σ winds in between the points $1/n$ where n is a non-zero integer, crossing the real axis when $t=1/(n+\frac{1}{2})$ for n a non-zero integer.

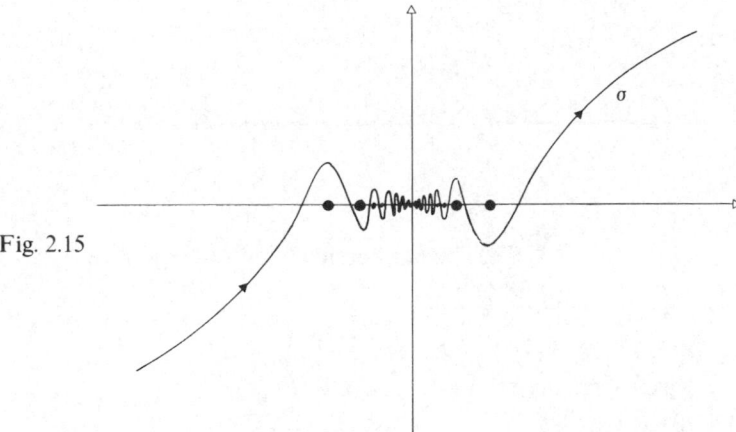

Fig. 2.15

It does not take much imagination to realize that an open set S can be very complicated and a path γ can be intertwined in S in a very intricate manner. Rather than trying to imagine all the possible intricacies, we side-step the issue, as follows. If S is an open set, we can cover γ with a finite number of open discs within S in the following fashion:

THE PAVING LEMMA (Lemma 2.9)

Let $\gamma: [a, b] \to S$ be a path in an open set S. Then there exists a subdivision $a = t_0 < t_1 < \ldots < t_n = b$ such that each subpath γ_r, given by restricting γ to $[t_{r-1}, t_r]$, lies inside an open disc $D_r \subseteq S$.

Proof. Because S is open, there exists $\varepsilon > 0$ such that $N_\varepsilon(\gamma(a)) \subseteq S$. The continuity of γ implies that there exists $\delta > 0$ such that

$$a \leqslant t < a + \delta \quad \text{implies} \quad \gamma(t) \in N_\varepsilon(\gamma(a)).$$

Let $t_1 = a + \frac{1}{2}\delta$, $D_1 = N_\varepsilon(\gamma(a))$, then the path $\gamma_1(t) = \gamma(t)$ $(a \leqslant t \leqslant t_1)$ lies in the disc $D_1 \subseteq S$.

We now try to move along the path, covering it bit by bit with discs, as in Figure 2.16. Our only fear is that we may not progress all the way; for instance the discs might decrease dramatically in size and we might never reach the end.

This fear is unjustified. Say that γ restricted to $[a, x]$ 'can be paved' if it can be subdivided into a finite number of subpaths $\gamma_1, \ldots, \gamma_m$ where each γ_r lies in a disc $D_r \subseteq S$ $(1 \leqslant r \leqslant m)$. Let C be the set of $x \in [a, b]$ such that γ restricted to $[a, x]$ can be paved. Then we know that $[a, a + \frac{1}{2}\delta] \in C$, so C is non-empty; also C is bounded above by b. This means that C has a least upper bound c where $a + \frac{1}{2}\delta \leqslant c \leqslant b$.

Because S is open, there is an open disc $N_\tau(\gamma(c)) \subseteq S$, and by the continuity of γ, for some $\kappa > 0$

$$t \in [a, b] \quad \text{and} \ldots c - \kappa < t < c + \kappa \quad \text{implies} \quad \gamma(t) \in N_\tau(\gamma(c)).$$

The least upper bound property gives $x \in C$ where $c - \kappa < x \leqslant c$ and, by covering $\gamma([x, c])$ with $D_{m+1} = N_\tau(\gamma(c))$, we have $c \in C$.

Fig. 2.16

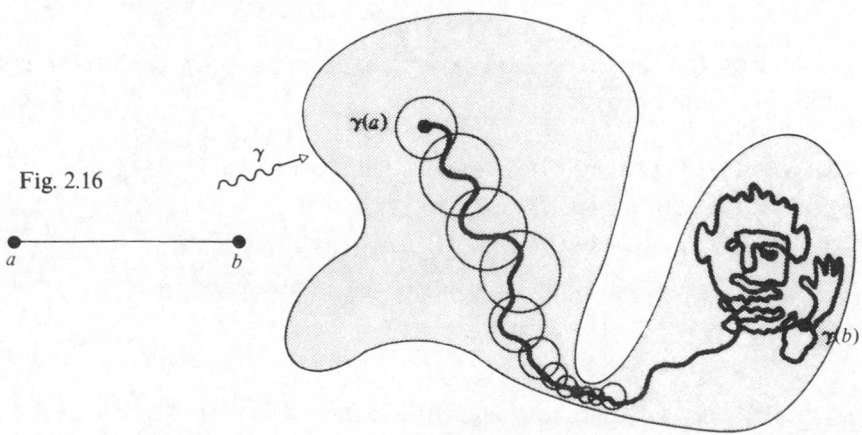

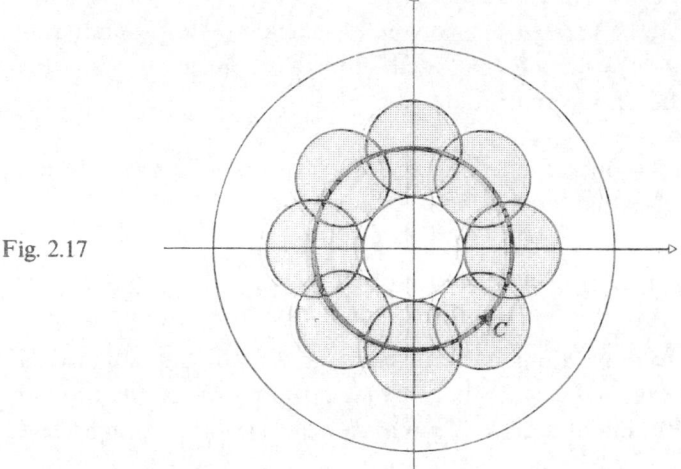

Fig. 2.17

We cannot have $c < b$, for in that case we could cover $\gamma([c, d])$ by $D_{m+1} = N_r(\gamma(c))$ for d in $c < d < c + \kappa$, giving $d \in C$, contradicting the fact that c is the least upper bound. Hence $c = b$, so $b \in C$ and the whole of $[a, b]$ can be paved in the desired manner. $\qquad\square$

In Example 1 (p. 37), for any point z on the unit circle C, we have $N_{1/2}(z) \subseteq S$. Clearly a finite number of such circles will pave C within S. (Fig. 2.17).

The path σ in Example 2 cannot be paved by a finite number of discs in V. Any disc covering the origin (which lies on σ) includes points of the form $1/n$ lying outside V. But this also means that V is not open, so the conditions necessary for the Paving Lemma are not satisfied.

6. Connectedness

A subset $S \subseteq \mathbb{C}$ is said to be *path connected* if given $z_1, z_2 \in S$ there exists a path γ in S from z_1 to z_2.

EXAMPLE 1. Any open disc $N_r(z_0)$ is path connected. For, given z_1, $z_2 \in N_r(z_0)$, let $L(t) = z_1(1-t) + z_2 t$ $(0 \leqslant t \leqslant 1)$; then

$$
\begin{aligned}
|L(t) - z_0| &= |\{z_1(1-t) - z_0(1-t)\} + \{z_t t - z_0 t\}| \\
&\leqslant (1-t)|z_1 - z_0| + t|z_2 - z_0| \quad \text{(for } 0 \leqslant t \leqslant 1) \\
&< (1-t)r + tr \\
&= r,
\end{aligned}
$$

so the line segment L lies in $N_r(z_0)$. (Fig. 2.18).

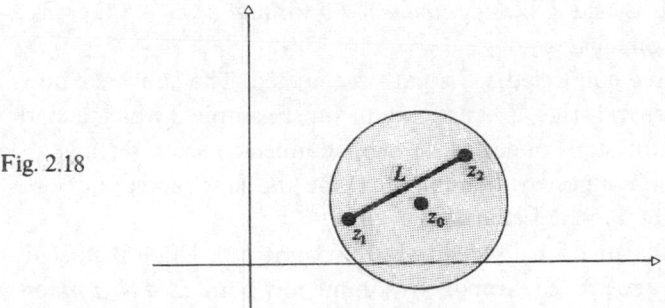

Fig. 2.18

EXAMPLE 2. $S = \{z \in \mathbb{C} \mid z = t + it^2,\ t \in \mathbb{R}\}$ is path connected, for if $t_1 + it_1^2$, $t_2 + it_2^2 \in S$ where $t_1 < t_2$, then $\gamma(t) = t + it^2\ (t_1 \leqslant t \leqslant t_2)$ is a path in S between the given points. (Fig. 2.19).

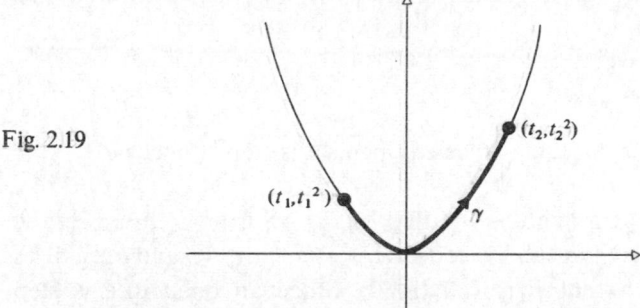

Fig. 2.19

Sometimes it is more convenient to use a simpler type of path. A *step path* in S is a path $\gamma : [a, b] \to S$ together with a subdivision $a = t_0 < t_1 \ldots < t_n = b$ such that in each subinterval $[t_{r-1}, t_r]$ either re γ or im γ is constant. This means that the image of γ consists of a finite number of straight line segments, each parallel to the real or imaginary axis. (Fig. 2.20).

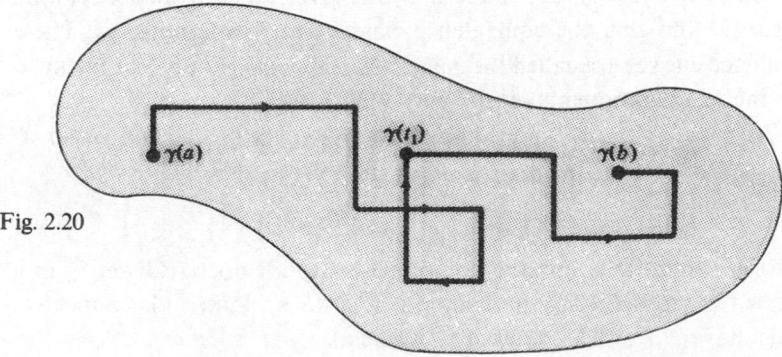

Fig. 2.20

A subset $S \subseteq \mathbb{C}$ is said to be *step connected* if for any $z_1, z_2 \in S$ there is a step path in S from z_1 to z_2.

Evidently a step connected set is path connected. The converse, however, is not in general true, as can be seen from Example 2 which is path connected but not step-connected. A path connected *open* set may be shown to be step connected; this proves to be the first success (albeit a minor one) for the Paving Lemma.

We first note that an open disc $N_r(z_0)$ is step connected. This is pictorially obvious and may be proved formally by joining any point $z_1 \in N_r(z_0)$ stepwise to z_0 (from whence another stepwise path can take us to any other $z_2 \in N_r(z_0)$):

If $z_0 = x_0 + iy_0$, $z_1 = x_1 + iy_1$, we let $w = x_1 + iy_0$, then

$$|w - z_0| = \sqrt{\{(x_1 - x_0)^2 + (y_0 - y_0)^2\}} = |x_1 - x_0| \leqslant |z_1 - z_0| < r$$

and $w \in N_r(z_0)$. The line segments $[z_1, w]$, $[w, z_0]$ then provide a step connection from z_1 to z_0.

We can now prove:

PROPOSITION 2.10. A path connected open set is step connected.

Proof. Given $z_1, z_2 \in S$, there is a path $\gamma : [a, b] \to S$ from z_1 to z_2. By the Paving Lemma there is a subdivision $a = t_0 < t_1 \ldots < t_n = b$ and open discs $D_1, \ldots, D_n \subseteq S$ such that $\gamma([t_{r-1}, t_r]) \subseteq D_r$. Since an open disc is step connected, we have a step connection π_r from $\gamma(t_{r-1})$ to $\gamma(t_r)$ in D_r, and hence in S. Piecing together these step connections gives a step path $\pi = \pi_1 + \cdots + \pi_n$ from z_1 to z_2 in S. $\square$

Arbitrary steps need not be connected. However, for any set S whatsoever we may define the relation $z_1 \sim z_2$ ($z_1, z_2 \in S$) to mean that there is a path in S from z_1 to z_2. It is easy to see that this gives an equivalence relation (Exercise 7) and that the equivalence classes are path connected. These equivalence classes are called the *connected components* of S. For instance, the connected components of

$$A = \{z \in \mathbb{C} \mid |z| \neq 1\}$$

are clearly

$$A_1 = \{z \in \mathbb{C} \mid |z| < 1\}, \quad A_2 = \{z \in \mathbb{C} \mid |z| > 1\}.$$

If S is *open* then its connected components are all open. (Given z_0 in a connected component S_0, since some $N_r(z_0) \subseteq S$ and $N_r(z_0)$ is connected, we must have $N_r(z_0) \subseteq S_0$, whence S_0 is open.)

A particularly interesting case is the complement of a path $\gamma: [a, b] \to \mathbb{C}$, by which we mean

$$\{z \in \mathbb{C} \mid z \neq \gamma(t) \text{ for any } t \in [a, b]\}.$$

The complement of γ may be connected, for instance

$$\gamma(t) = t \qquad (0 \leqslant t \leqslant 1).$$

On the other hand it may have two or more components. (Fig. 2.21).

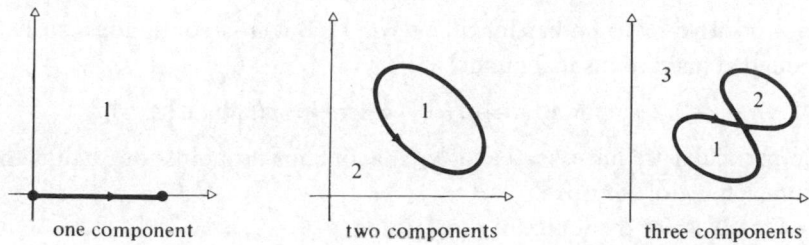

one component   two components   three components

Fig. 2.21

The complement of $\gamma = \gamma_1 - \gamma_2$, where

$$\gamma_1(0) = 0, \ \gamma_1(t) = t + it \sin(\pi/t) \quad (0 < t \leqslant 1)$$
$$\gamma_2(0) = 0, \ \gamma_2(t) = t - it \sin(\pi/t) \quad (0 < t \leqslant 1),$$

has an *infinite* number of components. (Fig. 2.22).

Fig. 2.22

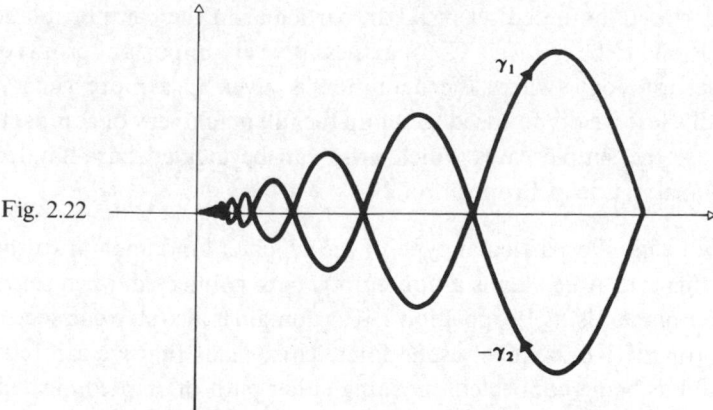

However complicated a path is, we may show:

PROPOSITION 2.11. (i) The image of a path in the complex plane is closed and bounded.

(ii) The complement of a path is open and has precisely one unbounded component.

Proof. (i) Let $\gamma: [a, b] \to \mathbb{C}$ be a path. Then $m(t) = |\gamma(t)|$, being the composite of continuous functions, gives a continuous function $m: [a, b] \to \mathbb{R}$. From real analysis it is bounded, say $m(t) \leqslant K$, whence $|\gamma(t)| \leqslant K$ for all $t \in [a, b]$ and the image of γ lies inside the disc centre the origin, radius K.

It only remains to prove (ii), since this implies the image of the path must be closed. Suppose that z_0 lies in the complement of γ. Then

$$\mu(t) = |\gamma(t) - z_0|$$

is a positive real number for all $t \in [a, b]$. But μ is continuous, so it is bounded and attains its bounds:

$$\mu(t) \geqslant k \quad \text{for all } t \in [a, b], \quad \text{where } k = \mu(t_0)(t_0 \in [a, b]).$$

In particular we have $k > 0$ and $N_k(z_0)$ contains no points on γ; thus the complement of γ is open.

Finally, since the image of γ lies inside

$$A = \{z \in \mathbb{C} | |z| \leqslant K\}$$

and since

$$B = \{z \in \mathbb{C} | |z| > K\}$$

is clearly connected, the only unbounded component must be the one containing B; all other components are bounded within A. □

Remark. A closed, bounded set in $\mathbb{R}^n$, in particular in the complex plane (let $n = 2$), is said to be *compact*. Compactness is a very important property in point-set topology (where the definition is given in a more general form). In this text we do not need to set up the full machinery of compactness because the simple cases which arise can be tackled bare-handed using real analysis, as in Proposition 2.11.

We now define the particular type of set which is fundamental to the theory in this text. A *domain* is a non-empty, path connected, open set in the complex plane. Using Proposition 2.10, a domain is also step connected and this property will be most useful later. This means that we can refer to a domain as being connected, meaning either path or step connected, as appropriate.

As the theory of complex analysis unfolds, the reader will see the immense importance of this definition. Complex functions f will be restricted to those of the form $f: D \to \mathbb{C}$ where D is a domain. The fact that D is open will allow us to deal neatly with limits, continuity and differentiability because $z_0 \in D$ implies $N_\varepsilon(z_0) \subseteq D$ (for some positive ε) and so $f(z)$ is defined for all z near z_0. The connectedness of D guarantees a (step) path between any two points in D and this in turn will allow us to define the integral from one point to another along such a path.

However, restricting complex functions to those defined on domains has far more subtle consequences than merely providing a platform for the appropriate definition. Those who have studied the intricacies of real analysis in depth will find untold riches quite unlike the real case. For instance, if two differentiable complex functions are defined on the same domain D and they happen to be equal in a small disc in D, then they are equal throughout D! No such result holds for general differentiable functions in the real case, and this result is just one of many to come that illustrate the beauty and simplicity of complex analysis. We shall not reach it until Chapter 10; we mention it here to underline the fundamental importance of establishing the correct topological foundations for the subject.

Exercises 2

1. Let $z_0 \in \mathbb{C}$ and $a, b \in \mathbb{R}$ be arbitrary. Draw the set consisting of all $z \in \mathbb{C}$ satisfying the following conditions. In each case, state whether the set is open, closed, or neither.
   (i) $1 < |z| < 2$
   (ii) $\operatorname{re} z \geqslant 0$ and $\operatorname{im} z \leqslant \operatorname{re} z$
   (iii) $\operatorname{im} z \geqslant 0$ and $1 < |z| < 2$
   (iv) $\operatorname{re}(z z_0) > 0$
   (v) $a < \arg(z - z_0) < b$ (where $-\pi < a < b \leqslant \pi$)
   (vi) $|z - \bar{z}_0| = |\bar{z} - z_0|$
   (vii) $|z - \bar{z}_0| < |\bar{z} - z_0|$
   (viii) $|z - \bar{z}_0| \leqslant |\bar{z} - z_0|$
   (ix) $\operatorname{re}(z^2) > 0$
   (x) $\operatorname{re}(z^2) \leqslant 0$
   (xi) $\operatorname{re}(z^2) > 1$
   (xii) $\operatorname{re}(z^2) \leqslant 1$.
2. Prove Proposition 2.2 parts (a) and (b).
3. Find the following limits, if they exist:
   (i) $\lim_{z \to 0} |z|/z$
   (ii) $\lim_{z \to 0} |z|^2/z$
   (iii) $\lim_{z \to 0} z/|z|^2$
   (iv) $\lim_{z \to 0} (z - \operatorname{re} z)/\operatorname{im} z$.
4. Prove, from first principles, that the following functions are continuous:
   (i) $\operatorname{re} z$ (ii) $\operatorname{im} z$
   (iii) $z + |z|$ (iv) $1/z \ (z \neq 0)$
   (v) $|z|^2$.

5. Prove Proposition 2.6.
6. For each of the following sets and pairs of points, define (if possible) (i) a path in the set joining the two points, (ii) a step path in the set joining the two points.
    (i) $|z| < 2$; $1+i$, $1-i$.
    (ii) $|z| = 1$; $-i$, i.
    (iii) $1 \leqslant |z| < 2$; $\sqrt{2}$, $-\sqrt{2}$.
    (iv) $|\text{re } z| > 5$; $-9+37i$, $1066 + i(\pi + \sqrt{5})/17$.
    (v) $|1 - |z|| > \frac{1}{2}$; $i/3$, $49(i+1)$.
7. Let S be a subset of $\mathbb{C}$. If z, $w \in S$ define $z \sim w$ if and only if there is a path in S from z to w. Show that $\sim$ is an equivalence relation. The equivalence classes are called *components*. If S is open and non-empty, show that each component of S is a domain.
8. Let S be a (path) connected subset of $\mathbb{C}$, and let $f : S \to \mathbb{C}$ be a continuous function. Prove that $f(S)$ is connected.
9. Give explicit functions for the paths which describe the curves of Figure 2.23 in the given direction. (All the subpaths are parts of circles or line segments; $0 < \varepsilon < R$ and x_1, x_2, y_1, y_2 are positive reals.)

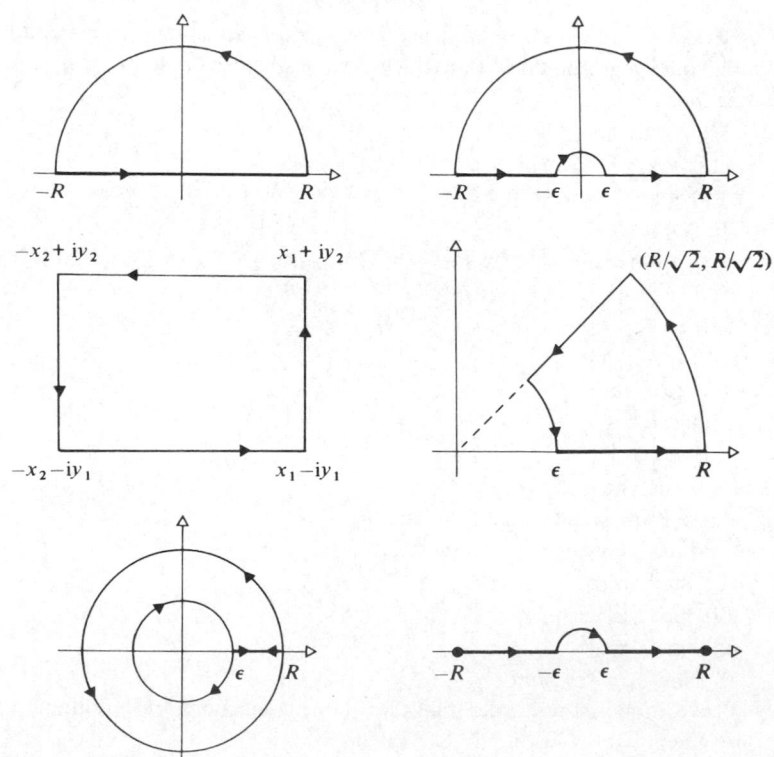

Fig. 2.23

10. Let S be a subset of $\mathbb{C}$. A point $z \in \mathbb{C}$ is a *boundary point* of S if z is a limit point of S and also a limit point of the complement $\mathbb{C} \setminus S$. The *boundary* ∂S of S is the set of all boundary points of S. In the following cases describe ∂S, and state whether ∂S is (path) connected. Draw a picture in each case.

   (i) $S = \{z \in \mathbb{C} | 1 < |z| < 2\}$

   (ii) $S = \{z \in \mathbb{C} | z \neq 0\}$

   (iii) $S = \{z \in \mathbb{C} | z = x + iy$ where x and y are rational numbers$\}$

   (iv) $S = \{z \in \mathbb{C} | 0 \leqslant \operatorname{re} z \leqslant 1, 0 \leqslant \operatorname{im} z \leqslant 1\}$

   (v) $S =$ the intersection of the sets S in (iii) and (iv)

   (vi) $S = \{z \in \mathbb{C} | z \neq iy$ where $y \in R, y \leqslant 0\}$

   (vii) $S =$ the intersection of the sets S in (vi) and (ii).

11. In the following cases the boundary of S (Exercise 10) can be described as the image of a path. Draw a picture of S and specify a function giving the path.

   (i) $S = \{z \in \mathbb{C} | |z| \leqslant 1, \operatorname{im} z \geqslant 0\}$

   (ii) $S = \{z \in \mathbb{C} | 1 \leqslant |z| \leqslant 2, \operatorname{im} z \geqslant 0\}$

   (iii) $S = \{z \in \mathbb{C} | 0 \leqslant \operatorname{re} z \leqslant 1, 0 \leqslant \operatorname{im} z \leqslant 1\}$

   (iv) $S = \{z \in \mathbb{C} | 1 \leqslant |z| \leqslant 2, 0 \leqslant \operatorname{im} z \leqslant \operatorname{re} z\}$.

3

Power series

Many of the more important functions studied in real analysis, such as the exponential or trigonometric functions, are most conveniently defined using power series. The familiar power series for the sine and cosine go back at least to Newton in 1676: they were derived again by de Moivre in 1698, by James Bernoulli in 1702, and were widely used by Euler in the 1730s.

Power series are, if anything, even more important for the study of complex functions. In the 1820s, Cauchy made considerable use of power series $\Sigma\, a_n z^n$ in a complex variable z. In particular any real function having a power series development automatically gives rise to a complex function with the corresponding complex power series, which provides a natural method for generalizing functions from the real to the complex case. And in the 1840s Weierstrass showed how to build up the whole of complex function theory by using power series methods.

In this chapter we shall develop some elementary properties of sequences and series of complex numbers, mostly by direct analogy with the real case, and then specialize to a deeper study of power series.

1. Sequences

For our purposes, sequences are required only as a stepping-stone towards series. A (complex) *sequence* is a function

$$f : \mathbb{N} \to \mathbb{C}$$

where $\mathbb{N}$ denotes the natural numbers $\{0, 1, 2, \ldots\}$. It is convenient to write

$$z_n = f(n)$$

and to arrange the numbers z_n, called the *terms* of the sequence, in a line

$$z_0, z_1, z_2, \ldots .$$

48

Alternative notation

$$\{z_n\}(n \geqslant 0)$$

or just

$$\{z_n\}$$

are often useful for brevity, and it is sometimes helpful to start the sequence at 1 instead of 0, thus:

$$z_1, z_2, z_3, \ldots .$$

We say that a sequence $\{z_n\}$ *tends to the limit z as n tends to* ∞ if, given any real $\varepsilon > 0$ there exists a natural number $N(\varepsilon)$ such that

$$n > N(\varepsilon) \quad \text{implies} \quad |z_n - z| < \varepsilon.$$

Note that this is identical with the usual definition for real sequences, except that z_n and z are allowed to be complex, and the absolute value $|z_n - z|$ is defined as in §1.5.

We write

$$\lim_{n \to \infty} z_n = z$$

or

$$z_n \to z \quad \text{as } n \to \infty.$$

The geometric content of this definition is that for sufficiently large n, the terms z_n all lie inside arbitrarily small circles centred on z, as in Figure 3.1. This again is reminiscent of the real case.

A sequence that tends to a limit is said to be *convergent*.

The problem of finding limits of complex sequences can be reduced directly to the real case:

LEMMA 3.1. Let $\{z_n\}$ be a sequence, and let $z_n = x_n + iy_n$ where $x_n, y_n \in \mathbb{R}$. Let $z = x + iy$ where $x, y \in \mathbb{R}$. Then

$$\lim_{n \to \infty} z_n = z$$

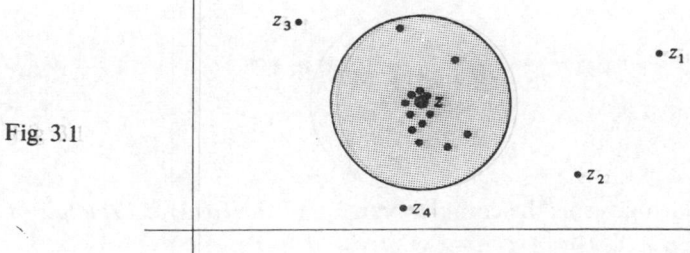

Fig. 3.1

if and only if

$$\lim_{n\to\infty} x_n = x$$

$$\lim_{n\to\infty} y_n = y.$$

Proof. Suppose that $\lim_{n\to\infty} z_n = z$. Let $\varepsilon > 0$. There exists $N \in \mathbb{N}$ such that $|z_n - z| < \varepsilon$ if $n > N$. For any $w = u + iv \in \mathbb{C}$ we have

$$|u| \leqslant \sqrt{u^2 + v^2} = |w|, \quad |v| \leqslant \sqrt{u^2 + v^2} = |w|,$$

so, taking $w = z_n - z$, for $n > N$ we have

$$|x_n - x| \leqslant |z_n - z| < \varepsilon,$$

$$|y_n - y| \leqslant |z_n - z| < \varepsilon.$$

Therefore $x_n \to x$ and $y_n \to y$.

Conversely, suppose that $x_n \to x$, $y_n \to y$. Given $\varepsilon > 0$ there exist $M, N \in \mathbb{N}$ such that

$$|x_m - x| < \varepsilon/2 \quad \text{for all } m > M$$

$$|y_n - y| < \varepsilon/2 \quad \text{for all } n > N.$$

Let $R = \max(M, N)$. If $n > R$ then

$$|z_n - z| \leqslant |x_n - x| + |y_n - y| < \varepsilon/2 + \varepsilon/2 = \varepsilon.$$

Hence $z_n \to z$. $\qquad\square$

Note that $\{x_n\}$ and $\{y_n\}$ are *real* sequences, so their limits may be found using the techniques of real analysis.

EXAMPLE. Let $z_n = (n + in^2 + 1)^{-1}$. Does z_n converge? If so, to what?
Separate into real and imaginary parts, thus:

$$z_n = ((n+1) + in^2)^{-1} = (n+1)[(n+1)^2 + n^4]^{-1} - in^2[(n+1)^2 + n^4]$$

using equation (1.8). Then

$$x_n = \frac{n+1}{(n+1)^2 + n^4} \to 0 \quad \text{as } n \to \infty,$$

and

$$y_n = \frac{-n^2}{(n+1)^2 + n^4} \to 0 \quad \text{as } n \to \infty.$$

So

$$z_n \to 0 + i0 = 0.$$

A similar idea gives us the complex version of the *General Principle of Convergence*:

THEOREM 3.2. A sequence $\{z_n\}$ tends to a limit z as $n \to \infty$ if and only if for all $\varepsilon > 0$ there exists $N(\varepsilon) \in N$ such that

$$m, n > N(\varepsilon) \quad \text{implies} \quad |z_m - z_n| < \varepsilon. \tag{1}$$

Proof. First let $z_n \to z$. There exists $N = N(\varepsilon)$ such that $|z_n - z| < \varepsilon/2$ for all $n > N$. If $m, n > N$ then

$$|z_m - z_n| \leqslant |z_m - z| + |z - z_n| < \varepsilon/2 + \varepsilon/2 = \varepsilon.$$

So (1) holds.

Conversely, assume (1) holds. Let $x_n = \mathrm{re}(z_n)$, $y_n = \mathrm{im}(z_n)$. If $m, n > N(\varepsilon)$ then

$$|x_m - x_n| \leqslant |z_m - z_n| < \varepsilon,$$
$$|y_m - y_n| \leqslant |z_m - z_n| < \varepsilon.$$

By the General Principle of Convergence for real series (that is, the fact that Cauchy sequences are convergent), it follows that $x_n \to x$ and $y_n \to y$ for some $x, y \in \mathbb{R}$. By Lemma 3.1, we have $z_n \to z = x + iy$. $\qquad\square$

EXAMPLE. Use the General Principle of Convergence to show that the sequence defined by

$$z_n = i\sqrt{2} + \left(\frac{3 - 4i}{6} \right)^n$$

converges.

We compute

$$
\begin{aligned}
|z_m - z_n| &= \left| \left(\frac{3-4i}{6} \right)^m - \left(\frac{3-4i}{6} \right)^n \right| \\
&\leqslant \left| \left(\frac{3-4i}{6} \right)^m \right| + \left| \left(\frac{3-4i}{6} \right)^n \right| \\
&= \left(\tfrac{5}{6} \right)^m + \left(\tfrac{5}{6} \right)^n \\
&\leqslant 2 \cdot \left(\tfrac{5}{6} \right)^r
\end{aligned}
$$

where $r = \min(m, n)$. Since $\frac{5}{6} < 1$ we can make this less than any given ε by taking r large enough.

Note that trying to tackle this question using Lemma 3.1 directly does not work out very easily (although the intrepid reader who converts $(3 - 4i)/6$ to polar coordinates stands a better chance). It may of course be shown from the definition that $z_n \to i\sqrt{2}$.

2. Series

Given a sequence $\{z_n\}$ we can form another sequence of *partial sums* defined by

$$s_n = \sum_{r=0}^{n} z_r = z_0 + z_1 + \cdots + z_n.$$

If s_n tends to a limit $s \in C$ then we define

$$\sum_{r=0}^{\infty} z_r = s = \lim_{n \to \infty} \sum_{r=0}^{n} z_r,$$

and say that the *series*

$$\sum_{r=0}^{\infty} z_r$$

converges to s. A series that converges to some s is called *convergent*. We may also write

$$s = z_0 + z_1 + \ldots$$

where this is convenient (but note that this is a *definition* of the expression $z_0 + z_1 + \ldots$ since infinite additions do not have any meaning until we give them one).

A series that is not convergent is called *divergent*.

By way of this definition, any question about the convergence of a series can be turned into one about the sequence $\{s_n\}$ of partial sums. For example, Theorem 3.2 applied to $\{s_n\}$ yields:

LEMMA 3.3. A series $\Sigma_{r=0}^{\infty} z_r$ converges if and only if, for all $\varepsilon > 0$, there exists $N(\varepsilon) \in \mathbb{N}$ such that

$$m, n > N(\varepsilon) \quad \text{implies} \quad \left| \sum_{r=m+1}^{n} z_r \right| < \varepsilon.$$

Proof. Note that $\Sigma_{r=m+1}^{n} z_n = s_n - s_m$. Now use Theorem 3.2.   □

EXAMPLE. Let $z_n = \left(\dfrac{3-4i}{6} \right)^n$. Does $\Sigma_{r=1}^{\infty} z_r$ converge?

We have

$$\left| \sum_{r=m}^{n} \left(\frac{3-4i}{6} \right)^r \right| \leq \sum_{r=m}^{n} \left| \left(\frac{3-4i}{6} \right)^r \right| \quad \text{(triangle inequality!)}$$

$$= \sum_{r=m}^{n} \left(\frac{5}{6} \right)^r.$$

Now this series is a (finite) geometric progression, whose sum is known to be

$$[(\tfrac{5}{6})^m - (\tfrac{5}{6})^{n+1}](1 - \tfrac{5}{6})^{-1}$$

which obviously may be made less than a given ε by taking m, n large enough, because $\tfrac{5}{6} < 1$.

WARNING. There is an ancient and venerable joke about two country yokels who each bought horses and wanted to distinguish them. So they cut the tail off one. Next day, someone had cut the tail off the other. So they tried again with an ear . . . with the same result. Finally, in desperation, one said to the other: 'I'll tell you what; you take the black one and I'll take the white one.'

For some reason, many students become confused between sequences and series. There is no reason to fall into this trap, because:

$$\text{series are the ones that begin with } \sum_{r=0}^{\infty}.$$

Actually this observation does not eliminate *all* sources of confusion, but it's a good start.

As with sequences, it is sometimes preferable to start at 1 rather than 0. This gives series of the form

$$\sum_{r=1}^{\infty} z_r$$

or

$$z_1 + z_2 + \ldots$$

whose precise definition is left to the reader. Variations, such as

$$\sum_{r=3}^{\infty} z_r,$$

are also feasible. However, this is no more than $\sum_{r=0}^{\infty} z_r$ where $z_0 = z_1 = z_2 = 0$. To simplify notation we shall often write

$$\sum z_r$$

instead of $\sum_{r=0}^{\infty} z_r$.

The summation of complex series may, if desired, be reduced to equivalent real series. Applying Lemma 3.1 to the partial sums, we obtain an immediate proof of:

LEMMA 3.4. Let $z_r = x_r + iy_r$ where x_r, $y_r \in \mathbb{R}$. Then Σz_r converges if and only if both the real series Σx_r, Σy_r converge, in which case

$$\sum z_r = \sum x_r + i \sum y_r. \qquad \qquad \square$$

EXAMPLE. Let $z_n = (i)^n/n^2$. Does Σz_r converge?
We have

$$x_n = \mathrm{re}\,(z_n) = \begin{cases} 0 & \text{if } n \text{ is odd}, \\ \dfrac{(-1)^{n/2}}{n^2} & \text{if } n \text{ is even}. \end{cases}$$

Now Σx_r is an alternating series (ignoring the zero terms) whose nth term tends to zero, so it converges.

Similarly $y_n = \mathrm{im}\,(z_n)$ leads to a convergent series Σy_r. Hence Σz_r converges.

In the same way, considering real and imaginary parts, or by mimicking the proof for real series directly, we get:

LEMMA 3.5. Let Σa_r and Σb_r be convergent series, and let c be a complex number. Then $\Sigma (a_r + b_r)$ and $\Sigma (ca_r)$ converge and their sums are given by

$$\sum (a_r + b_r) = \sum a_r + \sum b_r,$$

$$\sum (ca_r) = c \sum a_r. \qquad \qquad \square$$

When verifying that a complex series converges without computing its actual sum, it is often better to concentrate on the modulus rather than the real and imaginary parts. By analogy with the real case, we say that a series Σz_r is *absolutely convergent* if and only if the series $\Sigma |z_r|$ is convergent.

THEOREM 3.6. An absolutely convergent series is convergent.

Proof. Let Σz_r be absolutely convergent and let $z_r = x_r + iy_r$. Then $|x_r| \leqslant |z_r|$ and $|y_r| \leqslant |z_r|$. But $\Sigma |z_r|$ is convergent, so the comparison test for real series implies $\Sigma |x_r|$ and $\Sigma |y_r|$ are convergent. Since absolute convergence implies convergence in the real case, Σx_r and Σy_r are convergent. The result follows by Lemma 3.4. $\qquad \square$

EXAMPLE. $\Sigma z_n = \Sigma (i^n/n^2)$ converges because $\Sigma |z_n| = \Sigma 1/n^2$ is convergent. (Compare this proof with the previous example.)

Theorem 3.6 is useful because it allows us to verify the convergence of many *complex* series $\Sigma \, z_r$ by reference to the *real* series $\Sigma \, |z_r|$. Since the latter has *positive* terms there is a complex version of:

THE COMPARISON TEST. Let $\Sigma \, a_r$, $\Sigma \, b_r$ be complex series with $\Sigma \, a_r$ absolutely convergent. If

$$|b_r| < K|a_r| \quad (r > N)$$

for some positive K and integer N, then $\Sigma \, b_r$ is absolutely convergent, and hence convergent. $\qquad\qquad\square$

Proof. $\Sigma \, |a_r|$ converges so, by the real comparison test, $\Sigma \, |b_r|$ converges and, by Theorem 3.6, $\Sigma \, b_r$ is convergent.

There is also a complex version of:

THE RATIO TEST. Let $\Sigma \, a_r$ be a complex series with non-zero terms such that

$$\lim_{r \to \infty} \frac{|a_r|}{|a_{r-1}|} = \lambda.$$

If $\lambda < 1$, the series $\Sigma \, a_r$ is absolutely convergent; if $\lambda > 1$, it is divergent; and for $\lambda = 1$, it may be convergent or divergent.

Proof. For $\lambda < 1$, let $\rho = \frac{1}{2}(\lambda + 1)$, then $\lambda < \rho < 1$ and there exists N such that

$$|a_r|/|a_{r-1}| < \rho \quad \text{for } r > N.$$

Hence

$$|a_r| < \rho|a_{r-1}| < \rho^2|a_{r-2}| < \cdots < \rho^{r-N}|a_N| \quad (r > N)$$

and $\Sigma \, a_r$ converges by comparison with $\Sigma \, \rho^r$ for real $\rho < 1$.

In the case $\lambda > 1$, we see that for some N:

$$|a_r|/|a_{r-1}| > 1 \quad (r > N)$$

so

$$|a_r| > |a_{r-1}| > |a_{r-2}| \ldots > |a_N| \quad (r > N)$$

and $\Sigma \, a_r$ cannot converge because the terms $\{a_r\}$ do not tend to zero.

The proof is completed by noting that $\Sigma \, 1/n$ and $\Sigma \, 1/n^2$ both have $\lambda = 1$, but the former diverges and the latter converges. $\qquad\square$

3. Power series

Let $z_0 \in \mathbb{C}$. A series of the form

$$\sum_{n=0}^{\infty} a_n (z - z_0)^n$$

with *coefficients* $a_n \in \mathbb{C}$ is called a *power series about* z_0. By the change of variable $z' = z - z_0$ we may usually consider only the case $z_0 = 0$. Here, convergence is governed by the following results:

LEMMA 3.7. (i) If a power series $\Sigma a_n z^n$ converges for $z = z_1 \neq 0$, then it converges absolutely for all z with $|z| < |z_1|$. (ii) If $\Sigma a_n z^n$ diverges at $z = z_2$ then it diverges for all z such that $|z| > |z_2|$.

Proof. (i) If $\Sigma a_n z_1^n$ converges then $|a_n z_1^n| \to 0$ as $n \to \infty$, by the General Principle of Convergence. Thus there exists $K \in \mathbb{R}$ such that $|a_n z_1^n| < K$ for all n. If $|z| < |z_1|$ then $q = |z/z_1| < 1$. Now

$$|a_n z^n| = |a_n z_1^n| \, |z/z_1|^n < K q^n$$

so, by the comparison test, $\Sigma |a_n z^n|$ converges. (ii) If $|z| > |z_2|$ and $\Sigma a_n z^n$ converges then by (i), $\Sigma a_n z_2^n$ also converges, a contradiction. So $\Sigma a_n z^n$ is divergent. $\qquad\qquad\square$

These results lead to an important concept. If we let $R = \sup\{|z|| \text{ there exists } z \text{ such that } \Sigma a_n z^n \text{ converges}\}$ (allowing $R = \infty$ if no real supremum exists) then it follows at once that

$$\sum a_n z^n \quad \begin{array}{l} \textit{converges for } |z| < R \\ \textit{diverges for } |z| > R. \end{array}$$

(We cannot say yet what happens when $|z| = R$.) We call R the *radius of convergence* of the series, and the set

$$\{z \in \mathbb{C} | \, |z| < R\}$$

is its *disc of convergence*. Geometrically this is the interior of a circle (which may be just the origin, or all of $\mathbb{C}$, in extreme cases).

EXAMPLE. The series $1 + z + z^2 + \ldots$ is convergent for $|z| < 1$, since it is absolutely convergent in that case; it diverges for $z = 1$ and hence for all z such that $|z| > 1$. Therefore its radius of convergence is 1.

When $|a_r|/|a_{r-1}|$ tends to a limit as r tends to infinity, then the radius of convergence may be computed as follows. Let

$$\lim_{r \to \infty} |a_r|/|a_{r-1}| = l.$$

Then

$$\lim_{r \to \infty} |a_r z^r| / |a_{r-1} z^{r-1}| = |z| \, l.$$

By the ratio test we have the series convergent for $|z| l < 1$ and divergent for $|z| l > 1$. Hence the radius of convergence is $1/l$. Put another way, the radius of convergence of $\Sigma \, a_r z^r$ is

$$\lim_{r \to \infty} |a_{r-1} / a_r|,$$

provided that the latter exists.

EXAMPLE. If $\Sigma \, a_n z^n = \Sigma \, z^n / n$, then

$$\lim_{r \to \infty} |a_{r-1} / a_r| = 1,$$

so the radius of convergence is 1.

In general the ratio $|a_{r-1} / a_r|$ may not tend to a limit. Sometimes it is possible to spot the radius of convergence by native wit, but when all else fails we may use a powerful technique which works in *all* cases:

THEOREM 3.8. The radius of convergence of $\Sigma \, a_n z^n$ is given by

$$1/R = \lim \sup |a_n|^{1/n}. \tag{2}$$

Here by convention $1/0 = \infty$, $1/\infty = 0$.

Proof. Define R by the formula (2).

Suppose first that $|z| < R$. Then we may choose ρ such that $|z| < \rho < R$. By the definition of lim sup, we have $1/\rho > |a_n|^{1/n}$ for all n larger than some N. Hence $|a_n| < 1/\rho^n$, so

$$|a_n z^n| = |a_n| \rho^n |z/\rho|^n < |z/\rho|^n.$$

Now $|z/\rho| < 1$ so $\Sigma_{n=N}^{\infty} |z/\rho|^n$ converges. By the comparison test, $\Sigma_{n=N}^{\infty} a_n z^n$ converges, hence $\Sigma_{n=0}^{\infty} a_n z^n$ also converges.

Next suppose $|z| > R$. Choose ρ such that $|z| > \rho > R$. Then $1/\rho < |a_n|^{1/n}$ for all n larger than some N. As above, it follows by comparison with $\Sigma \, |z/\rho|^n$ that $\Sigma \, a_n z^n$ diverges. $\qquad \square$

EXAMPLE. The series $\sum \dfrac{z^n}{n!}$ has radius of convergence R where

$$1/R = \lim \sup (1/n!)^{1/n} = 0.$$

so $R = \infty$ and the series is absolutely convergent for all z. Its sum is defined to be the complex exponential function $\exp(z)$ or e^z.

The series $\sum (-1)^n \dfrac{z^{2n}}{(2n)!}$ and $\sum (-1)^n \dfrac{z^{2n+1}}{(2n+1)!}$ similarly have $R = \infty$.
They define the cosine and sine functions (see Chapter 5 for more details).

The series $\sum \dfrac{z^n}{n}$ has radius of convergence given by

$$1/R = \limsup \left(\frac{1}{n}\right)^{1/n} = 1,$$

so $R = 1$. It converges for $|z| < 1$, diverges for $|z| > 1$. When $|z| = 1$, further
analysis is required for which we do not yet have the technique.   □

In the next chapter we shall use power series to define complex exponential
and trigonometric functions. In order to derive some of their basic properties rapidly, we shall make use of a theorem on the multiplication of series
which says, roughly, that if we multiply 'term-by-term' and collect terms
together in the right way, we get a series converging to the product. More
precisely:

THEOREM 3.9. Suppose that $\Sigma \, a_n$ and $\Sigma \, b_n$ are absolutely convergent, and
that their sums are a and b respectively. Let

$$c_r = a_0 b_r + a_1 b_{r-1} + a_2 b_{r-2} + \cdots + a_r b_0.$$

Then $\Sigma \, c_n$ is convergent, and its sum is ab.

Proof. This is a direct extension of the corresponding result in real analysis.
For students unfamiliar with this, we give details in the appendix to this
chapter, §5.   □

We use Theorem 3.9 once only, in the next section, to show by a com-
binatorial argument that $\exp(z+w) = \exp(z)\exp(w)$. A more elegant
proof of the latter, not depending on results established by its use (that is,
avoiding circular reasoning) is given in §5.1.

4. Manipulating power series

Our results so far permit us to calculate with power series 'as if they are
infinite polynomials', *provided they are absolutely convergent*. To see this,
let $\Sigma \, a_n z^n$ and $\Sigma \, b_n z^n$ be power series with radii of convergence R_a, R_b
respectively, and let $|z| < \min(R_a, R_b)$. Then, by Lemma 3.5 we have

$$\sum a_n z^n + \sum b_n z^n = \sum (a_n + b_n) z^n.$$

And by Theorem 3.9,

$$\left(\sum a_n z^n\right)\left(\sum b_n z^n\right) = \sum (a_0 b_n + a_1 b_{n-1} + \cdots + a_n b_0) z^n.$$

Note that if we replace the sum to infinity $\sum$ by a finite sum, say $\sum_1^r$, these formulae become the usual ones for adding and multiplying polynomials.

It is this feature that makes power series so useful: it is (relatively) easy to *calculate* with them.

We shall take advantage of this to exhibit some important features of the complex exponential and trigonometric functions. These are *defined* by the power series

$$\exp(z) = \sum \frac{1}{n!} z^n$$

$$\cos(z) = \sum \frac{(-1)^n}{(2n)!} z^{2n}$$

$$\sin(z) = \sum \frac{(-1)^n}{(2n+1)!} z^{2n+1}$$

These particular series are of course motivated by the real case. We know (§3.3) that they are absolutely convergent for all $z \in \mathbb{C}$.

We begin with the expression $\cos\theta + i\sin\theta$ of §1.7. By adding the power series, we have

$$\cos\theta + i\sin\theta = \sum c_r \theta^r$$

where

$$c_r = \frac{(-1)^{r/2}}{r!} \quad \text{if } r \text{ is even}$$

$$c_r = \frac{i(-1)^{(r-1)/2}}{r!} \quad \text{if } r \text{ is odd.}$$

Now $i^2 = -1$, $i^3 = -i$, $i^4 = 1$, so we have

$$(-1)^{r/2} = i^r \quad (r \text{ even})$$
$$i(-1)^{(r-1)/2} = i^r \quad (r \text{ odd}).$$

Hence

$$\sum c_r \theta^r = \sum \frac{i^r \theta^r}{r!} = \sum \frac{(i\theta)^r}{r!} = \exp(i\theta).$$

We thus have the important formula

$$\cos\theta + i\sin\theta = \exp(i\theta).$$

Hence the polar coordinate form $r(\cos\theta + i\sin\theta)$ for a complex number may be written as $r \cdot \exp(i\theta)$, or, more briefly, $re^{i\theta}$.

Now we apply the formula for the product of two series to evaluate
$$\exp(z)\exp(w)$$
where $z, w \in \mathbb{C}$. We find that
$$\exp(z)\exp(w) = \left(\sum \frac{z^n}{n!}\right)\left(\sum \frac{w^n}{n!}\right)$$
$$\sum_{n=0}^{\infty}\left(\sum_{r=0}^{n}\frac{1}{r!}\frac{1}{(n-r)!}z^r w^{n-r}\right)$$
$$= \sum_{n=0}^{\infty}\frac{1}{n!}\left(\sum_{r=0}^{n}\binom{n}{r}z^r w^{n-r}\right).$$
By the Binomial Theorem this is
$$\sum_{n=0}^{\infty}\frac{1}{n!}(z+w)^n = \exp(z+w).$$
Therefore
$$\exp(z)\exp(w) = \exp(z+w).$$

5. Appendix

We sketch a proof of:

THEOREM 3.9. Suppose that $\Sigma\, a_n$ and $\Sigma\, b_n$ are absolutely convergent, and that their sums are a and b respectively. Let
$$c_r = a_0 b_r + a_1 b_{r-1} + a_2 b_{r-2} + \cdots + a_r b_0.$$
Then $\Sigma\, c_n$ is convergent, and its sum is ab.

Proof. The proof may be easier to follow using Figure 3.2, which represents all possible cross-products $a_s b_t$ on a square grid. The idea is that a partial sum of $\Sigma\, c_n$ is the sum of terms in a right-angled triangular region; whereas partial sums of $\Sigma\, a_n$ or $\Sigma\, b_n$, when multiplied together, give rectangular regions. Our main task is to estimate the sum of terms in certain triangular regions by approximating with rectangles. Here are the details:

Let $\Sigma\, |a_n| = A$, $\Sigma\, |b_n| = B$. Given $\varepsilon > 0$ we can choose N large enough so that all of the following conditions hold. (To do this, choose an N for each condition, then take the maximum of the three Ns chosen.)

$$\text{(i)} \quad \left|\sum_{n=0}^{N} a_n \sum_{n=0}^{N} b_n - ab\right| < \varepsilon/(A+B+1).$$

$$\text{(ii)} \quad \sum_{n=N+1}^{2N} |a_n| < \varepsilon/(A+B+1).$$

$$(iii) \quad \sum_{n=N+1}^{2N} |b_n| < \varepsilon/(A+B+1).$$

Then

$$\left| \sum_{n=0}^{2N} c_n - ab \right| \leqslant \left| \sum_{n=0}^{2N} c_n - \sum_{n=0}^{N} a_n \sum_{n=0}^{N} b_n \right| + \left| \sum_{n=0}^{N} a_n \sum_{n=0}^{N} b_n - ab \right|$$

which, from Figure 3.2, is less than or equal to

$$\sum_{n=N+1}^{2N} |a_n| \sum_{n=0}^{N} |b_n| + \sum_{n=0}^{N} |a_n| \sum_{n=N+1}^{2N} |b_n| + \varepsilon/(A+B+1)$$

$$< \frac{\varepsilon B}{A+B+1} + \frac{A\varepsilon}{A+B+1} + \frac{\varepsilon}{A+B+1}$$

$$= \varepsilon.$$

Therefore $\Sigma\, c_n$ converges to ab. $\qquad\qquad\qquad\qquad\qquad\qquad\qquad\qquad$ □

Fig. 3.2

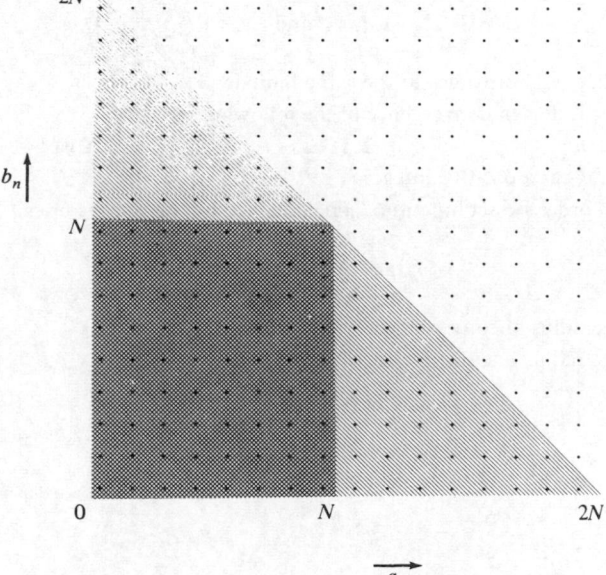

Exercises 3

1. Determine whether the following sequences converge, and find the limits of those which are convergent:

   (i) $\{(1+i)^n\}$ (ii) $\{(1+i)^n/n\}$ (iii) $\{(1+i)^n/n!\}$

   (iv) $\{(1/(1+i)^n\}$ (v) $\{n/(1+i)^n\}$ (vi) $\{n!/(1+i)^n\}$.

2. For what values of $z \in \mathbb{C}$ do each of the following sequences converge?

   (i) $\{z^n\}$ (ii) $\{z^n/n\}$ (iii) $\{n!z^n\}$

   (iv) $\{z^n/n!\}$ (v) $\{z^n/n^k\}$ (where k is a positive integer)

   (vi) $\{a(a-1)\ldots(a-n+1)z^n/n!\}$ (where a is a fixed complex number).

3. Let $a \in \mathbb{R}$ have the decimal expansion
   $$a = a_0.a_1a_2\ldots a_n \ldots$$
   where each a_i is an integer and $0 \leqslant a_n \leqslant 9$ for $n \geqslant 1$. Find all values of a for which the sequence $\{a_n\}$ converges.

4. Let $z_n = \frac{1}{2}(i^n + (-i)^n)$. Write down the first few terms of the sequence $\{z_n\}$. Derive similar expressions for the nth term of sequences which begin as follows:

   (i) $1, 0, 0, 0, 1, 0, 0, 0, 1, \ldots$

   (ii) $1, 0, 0, 0, -1, 0, 0, 0, 1, 0, 0, 0, -1, \ldots$

   (iii) $1, 0, -\frac{1}{2}, 0, \frac{1}{3}, 0, -\frac{1}{4}, 0, \ldots$

   (iv) $1, 0, 1\frac{1}{2}, 0, \frac{2}{3}, 0, 1\frac{1}{4}, 0, \frac{4}{5}, 0, 1\frac{1}{6}, \ldots$

5. Let $\{u_n\}$ be a convergent sequence in $\mathbb{C}$. Let $v_n = (\sum_{r=1}^{n} u_r)/n$. By writing $v_n = v'_n + v''_n$ where
   $$v'_n = \left(\sum_{r \leqslant \sqrt{n}} u_n\right)\Big/ n \quad \text{and} \quad v''_n = \left(\sum_{\sqrt{n} < r \leqslant n} u_r\right)\Big/ n,$$
   show that $\{v_n\}$ converges to the same limit as $\{u_n\}$.

6. Find the radius of convergence of the following series:

   (i) $\sum z^n/n$ (ii) $\sum z^n/n!$ (iii) $\sum n!z^n$

   (iv) $\sum n^k z^n$ for a positive integer k (v) $\sum z^{n!}$.

7. The 0th order Bessel function $J_0(z)$ is defined by the power series
   $$J_0(z) = \sum_{n=0}^{\infty} (-1)^n \frac{1}{(n!)^2} \frac{z^n}{2^n}.$$
   Find its radius of convergence.

8. Find the radius of convergence of the following:

   (i) $z - \dfrac{z^3}{3!} + \dfrac{z^5}{5!} - \dfrac{z^7}{7!} + \ldots$

   (ii) $1 - \dfrac{z^2}{2!} + \dfrac{z^4}{4!} - \dfrac{z^6}{6!} + \ldots$

   (iii) $z - \dfrac{z^2}{2} + \dfrac{z^3}{3} - \dfrac{z^4}{4} + \ldots$

   (iv) $1 + az + \dfrac{a(a-1)}{2!} z^2 + \cdots + \dfrac{a(a-1)\ldots(a-n+1)}{n!} z^n + \ldots$ (for $a \in \mathbb{C}$).

   (Remark: in part (iv), the radius of convergence is different for certain values of a.)

9. Show that $\sum_{n=1}^{\infty} z^{n!}$ converges for all $|z| < 1$, but diverges for infinitely many z with $|z| = 1$.

10. Suppose that $\sum a_n z^n$ has radius of convergence R and let C be the circle

$\{z \in C | |z| = R\}$. Prove or disprove the following (which may or may not be true):

(i) If $\Sigma a_n z^n$ converges at some point z on C then it converges everywhere on C.

(ii) If $\Sigma a_n z^n$ converges absolutely at one point z on C then it converges absolutely everywhere on C.

(iii) If $\Sigma a_n z^n$ converges at every point z on C, except possibly one, then it converges at all points on C.

(Hint: The series $\Sigma z^n/n$ could prove useful in this question.)

11. If $\Sigma a_n z^n$ has radius of convergence R, use the formula $1/R = \limsup \{|a_n|^{1/n}\}$ to find the radius of convergence of

(i) $\Sigma n^3 a_n z^n$ (ii) $\Sigma a_n z^{3n}$ (iii) $\Sigma a_n^3 z^n$.

12. Prove that if each of the series $\Sigma a_n z^n$, $\Sigma b_n z^n$ and $\Sigma a_n b_n z^n$ has radius of convergence equal to 1, then so have the series $\Sigma a_n b_n^2 z^n$ and $\Sigma a_n^2 b_n z^n$.

13. For $|a_n| \leqslant 1$, show that $\Sigma a_n z^n$ is absolutely convergent for all $|z| < 1$. If $\Sigma a_n z^n = f(z)$ for $|a_n| \leqslant 1$, $|z| < 1$, show that

$$|f(z)| \leqslant 1/(1 - |z|).$$

14. Prove that, for $z \neq 1$,

$$\sum_{n=1}^{k} z^n/n = \frac{z}{1-z} \left(\sum_{n=1}^{k-1} 1/(n(n+1)) - \sum_{n=1}^{k-1} z^n/(n(n+1)) + \frac{1-z^k}{k} \right).$$

Show that the series $\Sigma_{n=1}^{\infty} z^n/n$ and $\Sigma_{n=1}^{\infty} z^n/(n(n+1))$ have radius of convergence 1; that the latter series converges everywhere on $|z| = 1$ whilst the former converges everywhere on $|z| = 1$ except $z = 1$.

15. Suppose that the power series $\Sigma_{n=0}^{\infty} a_n z^n$ has a recurring sequence of coefficients, say $a_n = a_{n+k}$ for all n, where k is a fixed positive integer. Prove that the series converges for $|z| < 1$ to a rational function $p(z)/q(z)$ where p, q are polynomials, and that the roots of q are all on the unit circle.

What happens if $a_{n+k} = a_n/k$ instead?

4

Differentiation

The derivative of a real function is defined by a limiting process, which generalizes without difficulty to complex functions now that we have developed the necessary concept of limits. There are few surprises in this chapter: the results on differentiation of sums, products, composites of functions, and power series parallel the real case – and even the proofs are essentially unchanged. One minor surprise is that the condition of differentiability implies certain relations between the real and imaginary parts of a complex function, known as the *Cauchy–Riemann Equations*. We apply these immediately to prove that if a function with connected domain has zero derivative, it must be constant. While the *result* is analogous to the real case, the proof is not. The idea of differentiation can also be extended to *hybrid* functions $\mathbb{R} \to \mathbb{C}$ or $\mathbb{C} \to \mathbb{R}$, still without surprises.

To counter the impression that the results of real analysis go through without change, the final section gives a preview of a dramatic difference between the two theories: every differentiable complex function can be differentiated arbitrarily many times.

1. Basic results

If f is a complex function defined on an open set S, then, by analogy with the real case, f is said to be *differentiable at a point* $z_0 \in S$ with *derivative* $f'(z_0) \in \mathbb{C}$ if

$$\lim_{z \to z_0} \frac{f(z) - f(z_0)}{z - z_0} = f'(z_0).$$

If f is differentiable at every point of S then f is said to be a *differentiable function*. In this case the derivative may also be considered as a function $f' : S \to \mathbb{C}$. If

$$\lim_{z \to z_0} \frac{f'(z) - f'(z_0)}{z - z_0}$$

64

exists then we define this limit to be $f''(z_0)$, and by a repetition of this process we obtain the usual notion of higher derivatives $f''(z_0), f'''(z_0), \ldots,$ $f^{(n)}(z_0), \ldots$ where $f^{(n)}(z_0)$ denotes the nth derivative of f at z_0.

EXAMPLE. $f(z) = z^2$.

$$\lim_{z \to z_0} \frac{f(z) - f(z_0)}{z - z_0} = \lim_{z \to z_0} \frac{z^2 - z_0^2}{z - z_0} = 2z_0.$$

Hence $f'(z_0) = 2z_0$ for all $z_0 \in \mathbb{C}$. Similarly $f''(z_0) = 2$ and $f^{(n)}(z_0) = 0$ for $n \geqslant 3$.

Several alternative notations are used for differentiation, the two most popular being $Df(z)$ or $df(z)/dz$ in place of $f'(z)$. The second derivatives are then denoted by $D^2 f(z)$ and $d^2 f(z)/dz^2$ respectively. Sometimes $f(z)$ is denoted by w and its derivative by dw/dz. In this text the notations most frequently used will be $f'(z)$ and $Df(z)$ because, when we think of the derivative as a function, we can denote it by f' or Df.

In many ways results concerning complex differentiation follow naturally by analogy with the real case:

PROPOSITION 4.1. If f is differentiable at z_0, then f is continuous at z_0.

Proof.

$$\lim_{z \to z_0} (f(z) - f(z_0)) = \lim_{z \to z_0} \frac{f(z) - f(z_0)}{z - z_0} (z - z_0)$$
$$= f'(z_0) \cdot 0$$
$$= 0,$$

hence differentiability at z_0 implies

$$\lim_{z \to z_0} f(z) = f(z_0). \qquad \square$$

Recall (from Chapter 2 section 4) that the sum $f + g$, difference $f - g$, product $f \cdot g$ and quotient f/g of two functions $f: S \to \mathbb{C}$ and $g: S \to \mathbb{C}$ are defined in the usual manner:

$$(f + g)(z) = f(z) + g(z) \quad z \in S$$
$$(f - g)(z) = f(z) - g(z) \quad z \in S$$
$$(f \cdot g)(z) = f(z)g(z) \quad z \in S$$
$$(f/g)(z) = f(z)/g(z) \quad z \in S, g(z) \neq 0.$$

We obtain the expected rules for differentiation:

PROPOSITION 4.2. If f and g are differentiable at z_0, then so are $f+g$, $f-g$, $f \cdot g$ and f/g (in the latter case provided that $g(z_0) \neq 0$) and the derivatives are

$$\text{(i)} \quad (f+g)' = f' + g'$$
$$\text{(ii)} \quad (f-g)' = f' - g'$$
$$\text{(iii)} \quad (f \cdot g)' = f \cdot g' + g \cdot f'$$
$$\text{(iv)} \quad (f/g)' = (gf' - fg')/g^2.$$

Proof. The computations are analogous to the real case; for instance,

$$(f \cdot g)'(z_0) = \lim_{z \to z_0} \frac{f(z)g(z) - f(z_0)g(z_0)}{z - z_0}$$

$$= \lim_{z \to z_0} \frac{f(z)g(z) - f(z)g(z_0) + f(z)g(z_0) - f(z_0)g(z_0)}{z - z_0}$$

$$= \lim_{z \to z_0} f(z) \lim_{z \to z_0} \frac{g(z) - g(z_0)}{z - z_0} + g(z_0) \lim_{z \to z_0} \frac{f(z) - f(z_0)}{z - z_0}$$

(using the algebra of limits)

$$= f(z_0)g'(z_0) + g(z_0)f'(z_0)$$

(since f differentiable implies f continuous, by the previous proposition). The other cases follow after the same fashion. $\qquad\square$

If as usual we denote the composite of $f:S \to \mathbb{C}$ and $g:T \to \mathbb{C}$ where $f(S) \subseteq T$ by $g \circ f$,

$$(g \circ f)(z) = g(f(z)),$$

then we obtain ...

PROPOSITION 4.3 (The Chain Rule).   If f is differentiable at z_0 and g is differentiable at $f(z_0)$, then $g \circ f$ is differentiable at z_0 and

$$(g \circ f)'(z_0) = g'(f(z_0))f'(z_0).$$

Proof. A common method of attempting to demonstrate the chain rule is to write

$$\frac{g(g(z)) - g(g(z_0))}{z - z_0} = \frac{g(g(z)) - g(g(z_0))}{g(z) - g(z_0)} \frac{g(z) - g(z_0)}{z - z_0} \cdots \quad (1)$$

(provided that $g(z) \neq g(z_0)$). Since g is differentiable at z_0, it is continuous (by Proposition 3.1), so $z \to z_0$ implies $g(z) \to g(z_0)$ and this gives

$$\lim_{z \to z_0} \frac{g(g(z)) - g(g(z_0))}{g(z) - g(z_0)} = g'(g(z_0)).$$

Letting $z \to z_0$ in (1) then gives $(g \circ f)'(z_0) = g'(f(z_0))f'(z_0)$.

Unfortunately, this demonstration has a nasty gap in it, because $f(z) - f(z_0)$ might be zero. If it were known to be non-zero in some neighbourhood of z_0, for $z \neq z_0$, we could work in that neighbourhood and the proof would then be valid. In fact Theorem 10.11 below shows that in the *complex* case such a neighbourhood always exists; an interesting example of the extra simplicity of complex analysis, since in the *real* case no such statement holds.

However, the proof commonly used in the real case to patch up the hole also works in the complex case, and is more elementary than Theorem 10.11. It goes like this.

Let $u = f(z_0)$. Define

$$h(w) = \frac{g(w) - g(u)}{w - u} - g'(u) \quad \text{if } w \neq u$$

$$h(u) = 0.$$

Clearly h is continuous and defined near u. Also, as $z \to z_0$ we see that $h \circ f(z) \to h(f(z_0)) = h(u) = 0$. Now the definition of h, applied when $w = f(z)$, can be written in the shape

$$g(f(z)) - g(u) = (h(f(z)) + g'(u))(f(z) - u)$$

when $f(z) \neq u$. It *is also obviously true when* $f(z) = u$. Let $z \neq z_0$, divide both sides by $z - z_0$, let $z \to z_0 \ldots$ Voilà! $\square$

2. The Cauchy–Riemann equations

If we write a complex function f in terms of two real functions u, v of two real variables

$$f(z) = u(x, y) + iv(x, y)$$

where $z = x + iy$, then the differentiability of f imposes restrictions on the partial derivatives of u and v. We will use the notation

$$\frac{\partial u}{\partial x}(x, y) = \lim_{h \to 0} \frac{u(x + h, y) - u(x, y)}{h}$$

$$\frac{\partial u}{\partial y}(x, y) = \lim_{k \to 0} \frac{u(x, y + k) - u(x, y)}{k}$$

and abbreviate these to $\partial u / \partial x$ and $\partial u / \partial y$ where no confusion can occur. Then we have

PROPOSITION 4.4. If f is differentiable at $z = x + iy$, then $\partial u / \partial x$, $\partial v / \partial x$, $\partial u / \partial y$, $\partial v / \partial y$ all exist at (x, y) and

$$\frac{\partial u}{\partial x} = \frac{\partial v}{\partial y}, \frac{\partial v}{\partial x} = -\frac{\partial u}{\partial y}.$$

Proof. We calculate $f'(z)$ in two different ways. First we take a point near z in the form $z+h=(x+h)+iy$ where h is real and compute

$$f'(z)=\lim_{h\to 0}\frac{f(z+h)-f(z)}{h}$$

$$=\lim_{h\to 0}\frac{u(x+h, y)+iv(x+h, y)-u(x, y)-iv(x, y)}{h}$$

$$=\lim_{h\to 0}\frac{u(x+h, y)-u(x, y)}{h}+\lim_{h\to 0}\frac{i(v(x+h, y)-v(x, y))}{h}$$

$$=\frac{\partial u}{\partial x}+i\frac{\partial v}{\partial x}.$$

Next we take a point in the form $z+ik=x+i(y+k)$ where k is real; as this point tends to z, k tends to zero, so that

$$f'(z)=\lim_{k\to 0}\frac{u(x, y+k)+iv(x, y+k)-u(x, y)-iv(x, y)}{ik}$$

$$=\frac{\partial v}{\partial y}-i\frac{\partial u.}{\partial y}$$

The proof is completed by equating real and imaginary parts.   □

The equations

$$\frac{\partial u}{\partial x}=\frac{\partial v}{\partial y},\quad \frac{\partial v}{\partial x}=-\frac{\partial u}{\partial y}$$

are called the *Cauchy–Riemann equations* after Cauchy (1789–1852) and Riemann (1826–66). They were known to D'Alembert who noted them in 1752.

The converse of Proposition 4.4 is false, as the following example shows:

Let $f(x+iy)=0$ if one or both of x, y is zero,

$f(x+iy)=1$ if neither of x, y is zero.

For this function the partial derivatives of u and v all exist at the origin and all are zero, so the Cauchy–Riemann equations are certainly satisfied. But f is not even continuous at the origin, which means that it cannot be differentiable there.

Once again we find that the introduction of real analysis leads to complications. However, in this case it is a relatively easy matter to patch things up. It is rather like starting a well-tuned vintage car on an icy morning. Once it is going, it will run smoothly. Complex analysis will prove to be a well-oiled machine; but taking real analysis as a point of departure requires adequate starting conditions. In this piece of mathematics suitable conditions are that the partial derivatives be continuous. With these conditions satisfied, the machine runs well, but to verify this, we have to take a very close look at the mechanics. We begin with a technical lemma.

LEMMA 4.5. If $\partial u/\partial x$, $\partial u/\partial y$ exist at (x, y) and $\partial u/\partial x$ is continuous there, then

$$u(x+h, y+k) - u(x, y) = h\left(\frac{\partial u}{\partial x}(x, y) + \varepsilon(h, k)\right) + k\left(\frac{\partial u}{\partial y}(x, y) + \eta(h, k)\right)$$

where $\varepsilon, \eta \to 0$ as $h, k \to 0$.

Proof. Write $u(x+h, y+k) - u(x, y)$ as

$$u(x+h, y+k) - u(x, y+k) + u(x, y+k) - u(x, y).$$

By the Mean Value Theorem for one real variable applied to $\phi(t) = u(x+t, y+k)$, there exists θ in $0 < \theta < 1$ such that

$$u(x+h, y+k) - u(x, y+k) = h\frac{\partial u}{\partial x}(x+\theta h, y+k).$$

By continuity of $\partial u/\partial x$,

$$\frac{\partial u}{\partial x}(x+\theta h, y+k) - \frac{\partial u}{\partial x}(x, y) = \varepsilon(h, k)$$

where $\varepsilon(h, k) \to 0$ as $h, k \to 0$. Hence

$$u(x+h, y+k) - u(x, y+k) = h\left(\frac{\partial u}{\partial x}(x, y) + \varepsilon(h, k)\right). \qquad (2)$$

It is an easier matter to use the fact that $k \to 0$ implies

$$\frac{u(x, y+k) - u(x, y)}{k} \to \frac{\partial u}{\partial y}(x, y)$$

to deduce that if

$$\eta(h, k) = \frac{u(x, y+k) - u(x, y)}{k} - \frac{\partial u}{\partial y}(x, y)$$

then

$$u(x, y+k) - u(x, y) = k\left(\frac{\partial u}{\partial y} + \eta(h, k)\right) \qquad (3)$$

where $\eta(h, k) \to 0$ as $h, k \to 0$. (In fact η depends only on k.)
Adding (2) and (3) gives the required result. $\qquad \square$

THEOREM 4.6. If $f(z)=u(x, y)+iv(x, y)$ where f is a complex function defined on an open set S and at some point $z_0=x_0+iy_0 \in S$ the partial derivatives $\partial u/\partial x$, $\partial u/\partial y$, $\partial v/\partial x$, $\partial v/\partial y$ all exist, are continuous and satisfy the Cauchy–Riemann equations

$$\frac{\partial u}{\partial x}=\frac{\partial v}{\partial y}, \quad \frac{\partial v}{\partial x}=-\frac{\partial u}{\partial y},$$

then f is differentiable at z_0.

Proof. Using Lemma 4.5, we can write

$$f(z)-f(z_0)=u(x_0+h, y_0+k)+iv(x_0+h, y_0+k)-u(x_0, y_0)-iv(x_0, y_0)$$

$$=h\left(\frac{\partial u}{\partial x}+\varepsilon_1\right)+k\left(\frac{\partial u}{\partial y}+\eta_1\right)+ih\left(\frac{\partial v}{\partial x}+\varepsilon_2\right)+ik\left(\frac{\partial v}{\partial y}+\eta_2\right)$$

where $\varepsilon_1, \varepsilon_2, \eta_1, \eta_2 \to 0$ as $h, k \to 0$.

Using the Cauchy–Riemann equations we have

$$f(z)-f(z_0)=(h+ik)\left(\frac{\partial u}{\partial x}+i\frac{\partial v}{\partial x}\right)+h\varepsilon_1+k\eta_1+h\varepsilon_2+k\eta_2$$

$$=(z-z_0)\left(\frac{\partial u}{\partial x}+i\frac{\partial v}{\partial x}\right)+\rho$$

where $\rho=h\varepsilon_1+k\eta_1+h\varepsilon_2+k\eta_2$, and so

$$\frac{f(z)-f(z_0)}{z-z_0}=\frac{\partial u}{\partial x}+i\frac{\partial v}{\partial x}+\rho/(z-z_0).$$

But

$$\left|\frac{\rho}{z-z_0}\right|=\frac{|\rho|}{\sqrt{(h^2+k^2)}}\leqslant\frac{|h||\varepsilon_1|+|k||\eta_1|+|h||\varepsilon_2|+|k||\eta_2|}{\sqrt{(h^2+k^2)}}$$

$$\leqslant|\varepsilon_1|+|\eta_1|+|\varepsilon_2|+|\eta_2|.$$

Let $h, k \to 0$, then $|\rho/(z-z_0)| \to 0$, and so

$$\lim_{z \to z_0}\frac{f(z)-f(z_0)}{z-z_0}=\frac{\partial u}{\partial x}+i\frac{\partial v}{\partial x}$$

as required.   □

EXAMPLE. The function $f(z)=|z|^2$ is differentiable at the origin and nowhere else, since

$$u(x, y)=x^2+y^2, \quad v(x, y)=0$$

Hence $\partial u/\partial x=2x$, $\partial u/\partial y=2y$, $\partial v/\partial x=0=\partial v/\partial y$, and the Cauchy–Riemann equations are satisfied only at $x=y=0$, at which point the partial derivatives are all continuous.

3. Connected sets and differentiability

If $f(z)=$ constant, then $f'(z)=0$, but what of the converse? When the derivative is zero, does this imply that the function is constant? The answer is in the affirmative when f is defined on a *connected* set. We recall that a connected, open set is called a domain and now prove:

THEOREM 4.7. If f is differentiable in a domain D and $f'(z)=0$ throughout D, then f is constant on D.

Proof. $f'(z)=\partial u/\partial x+i\partial v/\partial x=\partial v/\partial y-i\partial u/\partial y$, and so $f'(z)=0$ implies that all the partial derivatives of u and v are zero.

From real analysis, if $\phi'=0$ on a closed interval $[a, b]$, then ϕ is constant on $[a, b]$.

If $L=\{t+iy_0|a\leqslant t\leqslant b\}$ is a line segment in D, let $\phi(t)=u(t, y_0)$, then $\partial u/\partial x=\phi'=0$ and so u is constant on L. By a similar argument, u and v are both constant on any horizontal or vertical line segment in D. Hence $f(z)=u(x, y)+iv(x, y)$ is constant on any step path in D. But any two points in a connected set can be joined by a step path, so f is constant throughout D. $\qquad\square$

The same technique proves

PROPOSITION 4.8. If f is differentiable in a domain D and any one of re f, im f or $|f|$ is constant, then f is constant.

Proof. If $f=u+iv$ and re $f=u$ is constant, then $\partial u/\partial x=\partial u/\partial y=0$; the Cauchy–Riemann equations give $\partial v/\partial x=\partial v/\partial y=0$ and by the argument in the previous proof, $f=u+iv$ is constant in the domain D. The case when im f is constant is similar.

If $|f|$ is constant, then $u^2+v^2=c$. For $c=0$, we have $f=0$, so we may suppose that $c\neq 0$. Differentiating we have

$$2u\frac{\partial u}{\partial x}+2v\frac{\partial v}{\partial x}=0$$

$$2u\frac{\partial u}{\partial y}+2v\frac{\partial v}{\partial y}=0$$

and the Cauchy–Riemann equations give

$$u\frac{\partial u}{\partial x}-v\frac{\partial u}{\partial y}=0$$

$$u\frac{\partial u}{\partial y}+v\frac{\partial u}{\partial x}=0.$$

Adding u times the first equation to v times the second, we get

$$(u^2 + v^2)\frac{\partial u}{\partial x} = 0$$

and, because $u^2 + v^2 = c \neq 0$, we have $\partial u/\partial x = 0$. Similarly the other partial derivatives of u and v are zero and it follows that $f = u + iv$ is constant in the domain D.   □

4. Hybrid functions

At this point we may briefly consider hybrid functions, by which we mean real-valued functions of a complex variable or complex-valued functions of a real variable. There are evident notions of differentiation in both cases. For instance, a real-valued function of a complex variable $f:D \to \mathbb{R}$ where D is an open subset of the complex plane may be regarded merely as a complex function with imaginary part zero. Such a hybrid function is a very dull fellow, for if it is differentiable, its constant imaginary part implies that it must be constant (by Proposition 4.8).

We fare a little better with complex functions of a real variable. The most interesting case is $f:[a, b] \to \mathbb{C}$, which (when continuous) is a path in the complex plane. Defining the derivative at $t_0 \in [a, b]$ to be

$$\lim_{t \to t_0} \frac{f(t) - f(t_0)}{t - t_0}$$

in the obvious way (and allowing appropriate one-sided derivatives at a and b) we obtain the expected generalizations of the properties of differentiation:

PROPOSITION 4.9. If $f:[a, b] \to \mathbb{C}$, $g:[a, b] \to \mathbb{C}$ are differentiable at $t \in [a, b]$, then

$$(f \pm g)'(t) = f'(t) \pm g'(t)$$
$$(f \cdot g)'(t) = f(t)g'(t) + g(t)f'(t)$$
$$(f/g)'(t) = (g(t)f'(t) - g(t)g'(t))/(g(t))^2 \quad (g(t) \neq 0).$$ □

The chain rule involving a complex function f of a real variable appears in two guises. We may either precede f by a real function h, or follow it by a complex function g to obtain

PROPOSITION 4.10. If $h:[c, d] \to [a, b]$, $f:[a, b] \to D$ and $g:D \to \mathbb{C}$, then

$$(f \circ h)'(s) = f'(h(s))h'(s)$$
$$(g \circ f)'(t) = g'(f(t))f'(t)$$

wherever the derivatives on the right-hand side of the equations are defined.   □

The proofs of 4.9 and 4.10 follow the same patterns as the real or complex case. These results will prove to be of great value later when we consider paths in the domain of a complex function and we get a blend of the hybrid function (the path) and the complex function itself.

5. Power series

More exciting creatures are power series, for they will later prove to be the foundation for all differentiable complex functions. The derivative of a polynomial

$$p(z) = a_0 + a_1 z + \cdots + a_n z^n$$

is known to be

$$p'(z) = a_1 + 2a_2 z + \cdots na_n z^{n-1}.$$

This suggests that for a power series

$$f(z) = \sum a_n z^n$$

we should have

$$f'(z) = \sum na_n z^{n-1}.$$

If this is the case we say that $f(z)$ *may be differentiated term by term.* When *is* this the case? Certainly $f(z)$ must be convergent. The next two results show that this is almost sufficient; in fact term by term differentiation is always possible within the disc of convergence of $f(z)$.

LEMMA 4.11. Let $f(z) = \sum a_n z^n$ converge absolutely for $|z| < R$; then

$$g(z) = \sum na_n z^{n-1}$$

converges for $|z| < R$.

Proof. For $|z| < R$, choose r such that $|z| < r < R$. Then $\sum a_n r^n$ converges absolutely, so, as in Lemma 3.7, there exists $K \in \mathbb{R}$ such that

$$|a_n r^n| < K$$

for all n. Now $q = |z|/r$ is less than 1, so

$$|na_n z^{n-1}| = n|a_n||z/r|^{n-1} r^{n-1}$$

$$< \frac{nK}{r} q^{n-1}.$$

But for $0 \leqslant q < 1$ the real series

$$\sum n K q^{n-1}$$

converges to $K(1-q)^{-2}$. By the comparison test $\sum |na_n z^{n-1}|$ converges, so (by Theorem 3.6) the series $\sum na_n z^{n-1}$ is convergent. $\square$

THEOREM 4.12. A power series $f(z) = \sum a_n z^n$ may be differentiated term by term within its disc of convergence, so that

$$f'(z) = \sum na_n z^{n-1}.$$

Proof. From Lemma 4.11 we know that

$$g(z) = \sum na_n z^{n-1}$$

is absolutely convergent for $|z| < R$. We must show that for $|z_0| < R$,

$$f'(z) = \lim_{z \to z_0} \left\{ \frac{f(z) - f(z_0)}{z - z_0} \right\} = g(z_0)$$

or, equivalently, that

$$\lim_{z \to z_0} \left(\frac{f(z) - f(z_0)}{z - z_0} - g(z_0) \right) = 0$$

Taking our courage in both hands we compute

$$\frac{f(z) - f(z_0)}{z - z_0} - g(z_0) = \sum_{n=1}^{\infty} \left(a_n \frac{z^n - z_0^n}{z - z_0} - na_n z_0^{n-1} \right)$$

(since power series may be added and subtracted term by term, by Lemma 3.5)

$$= \sum_{n=1}^{\infty} a_n \{ z^{n-1} + z_0 z^{n-2} + \cdots + z_0^{n-1} - nz_0^{n-1} \}$$

$$= \sum_{n=1}^{N} a_n \{ z^{n-1} + z_0 z^{n-2} + \cdots + z_0^{n-1} - nz_0^{n-1} \}$$

$$+ \sum_{n=N+1}^{\infty} a_n \{ z^{n-1} + z_0 z^{n-1} + \cdots + z_0^{n-1} - nz_0^{n-1} \}$$

$$= \sum_1 + \sum_2, \quad \text{say.}$$

Given any $\varepsilon > 0$, we first choose any r such that $|z_0| < r < R$. Then $\sum na_n r^{n-1}$ is convergent and (as in Lemma 3.3) there must exist $N = N(\varepsilon)$ such that

$$\sum_{n=N+1}^{\infty} |na_n r^{n-1}| < \varepsilon/4.$$

Since $|z_0| < r$, if z is close enough to z_0 to ensure that $|z| < r$ also, then

$$\left|\sum_2\right| \leq \sum_{n=N+1}^{\infty} 2n|a_n|r^{n-1} < \varepsilon/2. \tag{4}$$

Furthermore, Σ_1 is a polynomial in z and so $\Sigma_1 \to 0$ as $z \to z_0$. We can therefore find $\delta > 0$ such that

$$|z - z_0| < \delta \quad \text{implies} \quad \left|\sum_1\right| < \varepsilon/2. \tag{5}$$

We now ensure that z is close enough to z_0 to ensure (4) and (5) both hold and then

$$\left|\frac{f(z) - f(z_0)}{z - z_0} - g(z_0)\right| = \left|\sum_1 + \sum_2\right| \leq \left|\sum_1\right| + \left|\sum_2\right| < \frac{\varepsilon}{2} + \frac{\varepsilon}{2} = \varepsilon,$$

and so $f'(z_0) = g(z_0)$, as claimed. $\square$

Theorem 4.12 is immensely important, for it tells us not only about first derivatives, but about higher derivatives as well; we just repeat it as often as we want to get:

COROLLARY 4.13. All the higher derivatives f', f'', f''', ..., $f^{(n)}$, ... of a power series $f(z) = \Sigma\, a_n z^n$ exist for z within the disc of convergence and

$$f^{(k)}(z) = \sum_{n=k}^{\infty} n(n-1) \ldots (n-k+1)a_n z^{n-k}$$

$$= \sum_{n=k}^{\infty} \frac{n!}{(n-k)!} a_n z^{n-k}.$$

Proof. Use induction on k. $\square$

Replacing z by $z - z_0$ in Corollary 4.13, we find that if the power series $f(z) = \Sigma\, a_n(z - z_0)^n$ has disc of convergence $|z - z_0| < R$, then inside this disc of convergence, all the higher derivatives of f exist and

$$f^{(k)}(z) = \sum_{n=k}^{\infty} \frac{n!}{(n-k)!} a_n (z - z_0)^{n-k}.$$

Putting $z = z_0$ in this series we get

$$f^{(k)}(z_0) = k!\, a_k$$

which gives yet another important corollary:

COROLLARY 4.14. If $f(z) = \Sigma\, a_n(z - z_0)^n$ for $|z - z_0| < R$, then

$$a_k = f^{(k)}(z_0)/k!$$

and we can express f as a *Taylor series*

$$f(z) = \sum \frac{f^{(n)}(z_0)}{n!}(z - z_0)^n \quad (|z - z_0| < R).$$ □

EXAMPLE. $f(z) = 1/(1 + z) = 1 + z + z^2 + \cdots + z^n + \cdots$ for $|z| < 1$.
Hence we know that

$$f'(z) = 1/(1 - z)^2 = 1 + 2z + \cdots + nz^{n-1} + \cdots$$
$$f''(z) = 2/(1 - z)^3 = 2 + 6z + \cdots + n(n-1)z^{n-2} + \cdots.$$

and so on. We also have $f^{(n)}(0) = n!$ and $f(z) = \Sigma f^{(n)}(0)z^n/n!$.

6. A glimpse into the future

In the real case there exist functions which are differentiable n times but
not $n + 1$ times. A simple example in which $n = 1$ is given by

$$\phi(x) = \begin{cases} 0 & x \leqslant 0 \\ x^2 & x \geqslant 0. \end{cases}$$

Trivially

$$\phi'(x) = \begin{cases} 0 & x < 0 \\ 2x & x > 0, \end{cases}$$

and an easy calculation gives

$$\phi'(0) = \lim_{x \to 0} \frac{\phi(x) - \phi(0)}{x} = 0$$

Hence ϕ' exists and is even continuous at 0. But $\phi''(0)$ does not exist
because

$$\frac{\phi'(x) - \phi'(0)}{x} = \begin{cases} 0 & \text{for } x < 0 \\ 2 & \text{for } x > 0. \end{cases}$$

More generally, the function $\phi(x) = x^{n+1}$ for $x \geqslant 0$ and $\phi(x) = 0$ for
$x \leqslant 0$ is differentiable n times everywhere, but not $n + 1$ times at the origin.

We shall see later that there is no way in which we can piece together
functions in the complex case to yield this type of behaviour. A real
function only has two ways in which a limit point x_0 can be approached,
from the left and from the right. We can piece real functions together quite
happily, provided that we see that the right and left derivatives are the
same to whatever extent we deem necessary. As well as the above cases,
where we piece together the zero function on the left to the function
x^{n+1} on the right to give a patched up function that can be differentiated
n times, we can even find functions like $F(x) = 0$ for $x \leqslant 0$ and $F(x) = e^{-1/x}$
for $x > 0$ which have *all* derivatives equal to the right and left at the origin

(and all equal to zero at that: see exercise 16). This Frankenstein-like creation is patched up well, but there is something unnatural about it: because all its derivatives are zero, its Taylor series at the origin is

$$\sum_{n=0}^{\infty} F^{(n)}(0)x^n/n! = 0 + 0x + \cdots + 0x^n \cdots = 0$$

which is clearly convergent for all real x. But this Taylor series *does not equal* $F(x)$, because the latter is non-zero for positive x. In the real case we can find functions whose Taylor series exist but which do not equal the function. (There is nothing mysterious about this; it simply means that the remainder term $R_n(x)$ in

$$F(x) = a_0 + a_1 x + \cdots + a_n x^n + R_n(x)$$

does not tend to zero. In the case just mentioned, we always have $R_n(x) = F(x)$.)

Real analysis has even more grey areas, inhabited by functions which are continuous everywhere yet differentiable nowhere. Let $G(x)$ be the distance from the real number x to the nearest integer. This has a graph like the teeth of a saw; $G(x) = x(0 \leqslant \leqslant \frac{1}{2})$, $G(x) = 1 - x(\frac{1}{2} \leqslant x \leqslant 1)$ and G is periodic with $G(n+x) = G(x)$ for any integer n. It is not differentiable at $x = \frac{1}{2}n$ for any integer n, and $0 \leqslant G(x) \leqslant \frac{1}{2}$ for all real x. The function $G_n(x) = (\frac{1}{4})^n G(4^n x)$ is not differentiable at $x = \frac{1}{2}(\frac{1}{4})^n m$ for any integer m and satisfies $0 \leqslant G_n(x) \leqslant \frac{1}{2}(\frac{1}{4})^n$. If we form the sum

$$b(x) = \sum_{n=0}^{\infty} G_n(x),$$

then we get a very bad function indeed. We find

$$0 \leqslant b(x) \leqslant \sum_{n=0}^{\infty} \frac{1}{2}(\frac{1}{4})^n = \frac{\frac{1}{2}}{1 - \frac{1}{4}} = \frac{2}{3},$$

and it may be shown that b is continuous everywhere (relatively easy) but is differentiable nowhere (a little more intricate). A sketch of the latter is as follows. If $\gamma(x) = (b(x) - b(\alpha))/(x - \alpha)$, then for b to be differentiable at α, as x tends to α, $\gamma(x)$ tends to the derivative. We construct a sequence α_n tending to α such that $\gamma(\alpha_n)$ cannot tend to any limit. To do this, note that under each straight line segment (a half-tooth), there are two identical teeth of G_{n+1} of length $(\frac{1}{4})^n$, so it is possible to find $\alpha_n = \alpha \pm (\frac{1}{4})^n$ such that

$$G_m(\alpha_n) = G_m(\alpha) \quad \text{(for } m \geqslant n+1),$$

but the gradient of the straight bit of the tooth of G_n and larger teeth is

$$\frac{G_m(\alpha_n) - G_m(\alpha)}{\alpha_n - \alpha} = 1 \quad \text{for } m \leqslant n.$$

Hence

$$\gamma(\alpha_n) = \frac{b(\alpha_n) - b(\alpha)}{\alpha_n - \alpha} = \sum_{m=0}^{\infty} \frac{G_m(\alpha_n) - G_m(\alpha)}{\alpha_n - \alpha}$$

is a sum of $n+1$ terms, each ± 1. So $\gamma(\alpha_n)$ is an *odd integer* when n is even and an *even integer* when n is odd: the sequence $\gamma(\alpha_n)$ cannot tend to a finite limit as n tends to infinity.

Having created the bad function b, its antiderivative

$$b_1(x) = \int_0^x b(t)\, dt$$

is such that b_1 is differentiable once everywhere (with derivative b), but is not twice differentiable anywhere. By induction on n, the function b_n given by $b_n(x) = \int_0^x b_{n-1}(t)\, dt$, is differentiable everywhere precisely n times, but differentiable nowhere $n+1$ times.

Real analysis is a very hairy subject indeed. But what is the relevance of such bizarre functions in complex analysis? The answer is: NONE WHATSOEVER They have been mentioned once only to be dismissed. We shall find that no such animals live in the complex world; as we have said in Chapter 0, complex analysis is *simple* (relative to the real case). In the complex case there are very easy functions which are continuous everywhere but differentiable nowhere. An example is $f(z) = i|z|$, where continuity is easy to prove and non-differentiability follows from a consideration of the Cauchy–Riemann equations. There are also functions, such as $f(z) = |z|^2$ at the origin, which are differentiable only at isolated points, and are therefore not differentiable twice. But that is the end of the line. If a complex function is differentiable in a domain, then, as we demonstrate in a later chapter, it is differentiable again and again, it has a Taylor series, and it is equal to its Taylor series.

In computing the derivative

$$f'(z_0) = \lim_{z \to z_0} \frac{f(z) - f(z_0)}{z - z_0}$$

z can approach z_0 from any direction whatsoever. The existence of such a limit is such a strong condition that it precludes any possibility of patching together functions as in the real case.

The key to the whole theory is the fact that every differentiable complex function can be shown to have a power series expansion, whence it is equal to its own Taylor series as in Corollary 4.14. This will be established by a roundabout route in Chapter 10, but it is worth waiting for, and it underlines our emphasis on power series. They are not just good examples of differentiable functions; in a very genuine sense they will prove to be the *only* examples.

Exercises 4

1. From first principles differentiate:
   (i) $f(z) = z^2 + 2z$ (ii) $f(z) = 1/z$ $(z \neq 0)$ (iii) $f(z) = z^3 + z^2$.
2. Show that $f(z) = |z|$ is continuous everywhere and differentiable nowhere. Show that $f(z) = |z|^2$ is differentiable at the origin but nowhere else.
3. Differentiate (i) $(z^2 + 3)/(4z^3 + 5)^2$ (ii) $(z^2 + 3)^5(z^4 + 26z)^{12}$.
4. Let $\mathbb{C}_\pi = \{z \in \mathbb{C} | z \neq x \text{ for } x \in \mathbb{R}, x \leqslant 0\}$ be the 'cut plane' with the negative real axis removed. Define $r : \mathbb{C}_\pi \to \mathbb{C}$ by
   (i) $(r(z))^2 = z$,   (ii) re $(r(z)) > 0$.
   Prove that r is continuous in $\mathbb{C}_\pi$ and hence show from first principles that $r'(z) = \frac{1}{2}/r(z)$.
5. Let $f(z)$ be a polynomial in $z \in \mathbb{C}$. Prove that the function given by $g(z) = \overline{f(\bar{z})}$ is differentiable everywhere, but that $h(z) = \overline{f(z)}$ is differentiable at 0 if, and only if, $f'(0) = 0$.
6. In each of the following cases, for f defined on the domain D, find explicit formulae for $u(x, y)$, $v(x, y)$ where $f(z) = u(x, y) + iv(x, y)$, $z = x + iy$, and u, v, x, y are all real,
   (i) $f(z) = 1/z$, $D = \{z \in \mathbb{C} | z \neq 0\}$,
   (ii) $f(z) = |z|$, $D = \mathbb{C}$,
   (iii) $f(z) = \bar{z}$, $D = \mathbb{C}$.
   Show that u, v satisfy the Cauchy–Riemann equations everywhere in (i) and nowhere in (ii), (iii).
7. Verify the Cauchy–Riemann equations for the functions $u(x, y)$, $v(x, y)$ defined in the given domains by:
   (i) $u(x, y) = x^3 - 3xy^2$, $v(x, y) = 3x^2 y - y^3$,
   (ii) $u(x, y) = \sin x \cosh y$, $v(x, y) = \cos x \sinh y$,
   (iii) $u(x, y) = x/(x^2 + y^2)$, $v(x, y) = -y/(x^2 + y^2)$ $(x^2 + y^2 \neq 0)$,
   (iv) $u(x, y) = \frac{1}{2}\log(x^2 + y^2)$, $v(x, y) = \sin^{-1}(y/(x^2 + y^2)^{\frac{1}{2}})$ $(x > 0)$.
   In each case, verify that $u(x, y)$ and $v(x, y)$ are the real and imaginary parts of a differentiable complex function.
8. For $z = x + iy$, let
$$f(z) = \frac{x^3(1+i) - y^3(1-i)}{x^2 + y^2} \quad (z \neq 0), \ f(0) = 0.$$
   Show that f is continuous at the origin, the Cauchy–Riemann equations are satisfied there, yet $f'(0)$ does not exist. Why does this not contradict Theorem 4.6?
9. Let $f(z) = \sqrt{|xy|}$ for $z = x + iy$. Show that the Cauchy–Riemann equations are satisfied at the origin yet $f'(0)$ does not exist.
10. Consider $f(z) = \dfrac{xy^2(x + iy)}{x^2 + y^2}$ $(z = x + iy \neq 0)$, $f(0) = 0$.

Verify that $\lim (f(z)-f(0))/z=0$ as $z\to 0$ along any straight line $z=(a+ib)t$, $t\in\mathbb{R}$. This does not prove that $f'(0)=0$, however; by considering $z\to 0$ along the path $z(t)=t+it$, show that f is not differentiable at 0.
(This shows that in computing f', it is not sufficient to consider the limit taken along certain paths.)

11. For each of the following, compute $f'(\gamma(t))$, $\gamma'(t)$, $(f\gamma)'(t)$ and verify that
$$(f\gamma)'(t)=f'(\gamma(t))\gamma'(t).$$
  (i) $f(z)=z^2$, $\gamma(t)=t^3+it^4$ $(z\in\mathbb{C}, t\in[0,1])$,
  (ii) $f(z)=1/z$, $\gamma(t)=\cos t+i\sin t$ $(z\neq 0, t\in[0,2\pi])$,
  (iii) $f(z)=1+z+z^2+z^3+\cdots+z^n+\cdots$, $\gamma(t)=t+it^2$ $(|z|<1, t\in[0,\tfrac{1}{2}])$.

12. Suppose that $f(z)=\Sigma_{n=0}^{\infty} a_n z^n$ is convergent for all $z\in\mathbb{C}$ and satisfies $f'=f$ and $f(0)=1$. Find a_n for all $n\geqslant 0$. Consider the derivative of g where
$$g(z)=f(c-z)f(z)$$
for $c\in\mathbb{C}$ and deduce that
$$f(a+b)=f(a)f(b)$$
for all $a, b\in\mathbb{C}$. Compute $f(1)$ to 5 decimal places. (A calculator may be used, but isn't necessary!)

13. Show that the power series
$$f_\alpha(z)=1+\sum_{n=1}^{\infty}\frac{\alpha(\alpha-1)\cdots(\alpha-n+1)}{n!}z^n$$
is convergent (for all α) for $|z|<1$ and that in this domain its derivative is $\alpha f_\alpha(z)/(1+z)$.
Is the radius of convergence 1 for all $\alpha\in\mathbb{C}$?
By differentiating $f_\alpha(z)f_\beta(z)/f_{\alpha+\beta}(z)$, or otherwise, show that
$$f_{\alpha+\beta}(z)=f_\alpha(z)f_\beta(z).$$
Hence deduce that for $|z|<1$,
$$f_n(z)=(1+z)^n$$
for every integer (positive or negative) and
$$(f_{1/n}(z))^n=1+z$$
for a positive integer n.
(The power series for $f_\alpha(z)$ may be used as a definition of $(1+z)^\alpha$ for complex α and any z in the domain $|z|<1$.)

14. The two power series $s(z)=\Sigma_{n=0}^{\infty} a_n z^n$ and $c(z)=\Sigma_{n=0}^{\infty} b_n z^n$ are given to be convergent for all complex z. They satisfy the relations $s'(z)=c(z)$, $c'(z)=-s(z)$. Deduce the identities
$$a_n=-a_{n-2}/(n(n-1)),\quad b_n=-b_{n-2}/(n(n-1)).$$
Given that $s(0)=0$, $c(0)=1$, determine $s(z)$ and $c(z)$ completely. By differentiation, or otherwise, prove that
$$(s(z))^2+(c(z))^2=1.$$

15. For a positive integer n the Bessel function of order n is defined by

$$J_n(z)=\sum_{r=0}^{\infty}\frac{(-1)^r(\tfrac{1}{2}z)^{n+2r}}{r!(n+r)!}.$$

Show that this converges for all complex z and satisfies the differential equation

$$z^2\frac{d^2y}{dz^2}+z\frac{dy}{dz}+(z^2-n^2)y=0.$$

Verify the following:

(i) $J_{n-1}(z)+J_{n+1}(z)=\dfrac{2n}{z}J_n(z),$

(ii) $J_n'(z)=\dfrac{n}{z}J_n(z)-J_{n+1}(z),$

(iii) $J_n'(z)=\tfrac{1}{2}(J_{n-1}(z)-J_{n+1}(z)),$

(iv) $J_n'(z)=J_{n-1}(z)-\dfrac{n}{z}J_n(z),$

(v) $\dfrac{d}{dz}(z^nJ_n(z))=z^nJ_{n-1}(z),$

(vi) $J_2(z)-J_0(z)=2J_0''(z).$

16. (Frankenstein's Monster.) Define $f:\mathbb{R}\to\mathbb{R}$ by

$$f(x)=0 \qquad (x\leqslant0)$$
$$f(x)=e^{-1/x} \quad (x>0).$$

Show that f is differentiable arbitrarily many times, and that $f^{(n)}(0)=0$ for all n.

5

The exponential function

An abstract theory of functions may be intellectually diverting, but it pays its way by its more concrete applications: studies of certain 'special' functions of particular interest. As a step in this direction we now study the complex versions of the usual (real) exponential and trigonometric functions exp, sin, and cos (together with simple variants such as tan and cosec), defining them as power series. From this definition we develop some of the more basic properties of these functions.

Euler's famous formula

$$e^{i\theta} = \cos\theta + i\sin\theta$$

follows at once from these definitions, and shows that the fundamental function here is exp: all of the others may be defined in terms of it.

Most of the material generalizes directly from the real case, and will be presented in a compact form. In addition to deriving the standard formulae, we place some emphasis on convincing the reader that the power series do indeed represent the *usual* functions, and that in particular the geometric interpretations of sin and cos hold good.

The complex version of the logarithm requires a deeper analysis, and is held over to Chapter 7.

1. The exponential function

We have already defined this by the power series

$$\exp(z) = \sum_{n=0}^{\infty} \frac{z^n}{n!} \tag{1}$$

which is absolutely convergent for all $z \in \mathbb{C}$. We may therefore differentiate term by term: the result is the identical series, so we know that

$$\frac{d}{dz}\exp(z) = \exp(z). \tag{2}$$

82

With a little ingenuity we can use (2) to prove the formula

$$\exp(z_1 + z_2) = \exp(z_1)\exp(z_2) \tag{3}$$

obtained in a cumbersome way in Chapter 3. Consider

$$f(z) = \exp(z)\exp(c - z)$$

where $c \in \mathbb{C}$. Differentiating, we get

$$f'(z) = \exp'(z)\exp(c - z) + \exp(z)\exp'(c - z)\cdot(-1) = 0.$$

By Theorem 4.7 this implies that $f(z)$ is constant: the constant must equal $f(0) = c$. So $\exp(z)\exp(z - c) = \exp(c)$. Putting $c = z_1 + z_2$, $z = z_1$, we obtain (3).

We wish to use the customary notation e^z for $\exp(z)$. To avoid ambiguity we must show that for rational numbers z this agrees with the usual real exponential $e^{m/n} = \sqrt[n]{e^m}$. We do this as follows.

Define the real number

$$e = \exp(1) = 2.7182818\ldots. \tag{4}$$

Using (3) and induction on n we find that

$$\exp(nz) = (\exp(z))^n$$

for any positive integer n. Therefore

$$\exp(n) = (\exp(1))^n = e^n.$$

Clearly

$$\exp(0) = 1.$$

Then

$$\exp(n)\exp(-n) = \exp(n - n) = \exp(0) = 1$$

so that

$$\exp(-n) = (\exp(n))^{-1}.$$

Now, for a rational number m/n ($n > 0$) we have

$$(\exp(m/n))^n = \exp(nm/n) = \exp(m) = e^m$$

so that

$$\exp(m/n) = (e^m)^{1/n} = e^{m/n}.$$

Thus the notation

$$e^z = \exp(z)$$

does not conflict with the standard notation for powers of e, so we may (and do) use it henceforth. Note that (3) now takes the form

$$e^{z_1 + z_2} = e^{z_1}e^{z_2}. \tag{5}$$

If $z = x + iy$ then it follows that

$$e^z = e^x e^{iy}.$$

Now e^x is just the *real* exponential function; so we know how e^z behaves if we understand e^x, and the complex function e^{iy} of a real y. We study these in the next two sections.

2. Real exponentials and logarithms

We briefly recall the standard properties.

Clearly $e^x > 1 + x$ for $x > 0$, which implies that $e^x > 0$ for $x \geqslant 0$ and $e^x \to \infty$ as $x \to \infty$. Also $e^{-x} = 1/e^x$ so $e^x > 0$ for all x and $e^x \to 0$ as $x \to -\infty$. The derivative of e^x is $e^x > 0$, so e^x is monotonic increasing for all x.

By the Intermediate Value Theorem e^x defines a strictly increasing continuous function from $\mathbb{R}$ onto $\mathbb{R}^+ = \{x \in \mathbb{R} | x > 0\}$. It therefore has a continuous strictly increasing inverse function called the *natural logarithm*

$$\log : \mathbb{R}^+ \to \mathbb{R}$$

with the property

$$y = \log x \Leftrightarrow x = e^y \quad (x > 0).$$

From (5) we obtain

$$\log(x_1 x_2) = \log x_1 + \log x_2 \quad (x_1, x_2 > 0). \tag{6}$$

Let $y = \log x$, $y_0 = \log x_0$ ($x, x_0 \in \mathbb{R}^+$). Since $\log$ is continuous,

$$\lim_{x \to x_0} \frac{\log x - \log x_0}{x - x_0} = \lim_{y \to y_0} \frac{y - y_0}{e^y - e^{y_0}} = 1/e^{y_0} = 1/x_0.$$

$$\text{Therefore } \frac{d}{dx} \log x = \frac{1}{x}.$$

3. Trigonometric functions

We have already defined the sine and cosine for complex numbers by the power series

$$\cos z = \sum_{n=0}^{\infty} (-1)^n \frac{z^{2n}}{2n!} \tag{7}$$

$$\sin z = \sum_{n=0}^{\infty} (-1)^n \frac{z^{2n+1}}{(2n+1)!}. \tag{8}$$

We know these are absolutely convergent for all $z \in \mathbb{C}$.

Putting $-z$ for z in (7), (8), we see that $\cos$ is an even function and $\sin$ an odd function, that is,

$$\cos(-z) = \cos(z)$$
$$\sin(-z) = -\sin(z).$$

Also

$$\cos 0 = 1$$
$$\sin 0 = 0.$$

Differentiating term by term,

$$\frac{d}{dz} \cos z = -\sin z \tag{9}$$

$$\frac{d}{dz} \sin z = \cos z. \tag{10}$$

By term-by-term addition, as in Chapter 3, we have *Euler's Formula*

$$e^{iz} = \cos z + i \sin z. \tag{11}$$

Since $(e^{iz})^n = e^{inz}$ for any integer n, (11) implies *De Moivre's Formula*

$$(\cos z + i \sin z)^n = \cos nz + i \sin nz. \tag{12}$$

This may be used to obtain rapid derivations of formulae for $\cos n\theta$ and $\sin n\theta$ in terms of $\cos \theta$ and $\sin \theta$ when $\theta \in \mathbb{R}$, by equating real and imaginary parts of both sides (exercise 6).

Replacing z by $-z$ in (11) gives

$$e^{-iz} = \cos(-z) + i \sin(-z) = \cos z - i \sin z. \tag{13}$$

From (11) and (13) we find

$$\cos z = \tfrac{1}{2}(e^{iz} + e^{-iz}) \tag{14}$$

$$\sin z = \frac{1}{2i}(e^{iz} - e^{-iz}). \tag{15}$$

From (14), (15), and (5) we obtain the usual addition formulae for sin and cos, now valid for all complex numbers z_1, z_2:

$$\sin(z_1 + z_2) = \sin z_1 \cos z_2 + \cos z_1 \sin z_2 \tag{16}$$

$$\sin(z_1 - z_2) = \sin z_1 \cos z_2 - \cos z_1 \sin z_2 \tag{17}$$

$$\cos(z_1 + z_2) = \cos z_1 \cos z_2 - \sin z_1 \sin z_2 \tag{18}$$

$$\cos(z_1 - z_2) = \cos z_1 \cos z_2 + \sin z_1 \sin z_2. \tag{19}$$

Putting $z_1 = z_2 = z$ in (19) gives

$$\cos^2 z + \sin^2 z = 1. \tag{20}$$

4. The analytic definition of π

Historically the real number π was defined as the ratio of the circumference of a circle to its diameter, and only later did its importance for trigonometric functions emerge. We shall reverse the process, defining π analytically and eventually showing (§7.1) that our definition agrees with the geometric one.

The idea is to define $\pi/2$ as the first positive solution of the equation $\cos x = 0$: the problem is to show that there *is* such a thing.

We know that cos and sin are continuous functions. Also,

$$\cos(2) = 1 - \frac{2^2}{2!} + \frac{2^4}{4!} - \cdots - \frac{2^{4n-2}}{(4n-2)!} + \frac{2^{4n}}{(4n)!} - \cdots$$

$$= 1 - 2 + \frac{2}{3} - \cdots - \frac{2^{4n-2}}{(4n)!} \left[4n(4n-1) - 4) \right] - \cdots$$

$$< 1 - 2 + \frac{2}{3}$$

$$< 0.$$

But $\cos 0 = 1$. By the intermediate value theorem, $\cos t_0 = 0$ for some $t_0 \in (0, 2)$. Let k be the greatest lower bound of $\{t \in \mathbb{R} \mid t > 0, \cos t = 0\}$. By continuity,

$$\cos k = 0.$$

By the definition of k, if $0 \leqslant x < k$ then $\cos x > 0$.

We define

$$\pi = 2k.$$

Then π has been defined by the property $\cos(\pi/2) = 0$, and $0 \leqslant x < \pi/2$ implies $\cos x > 0$. Since $\cos(2) < 0$ it follows that $0 < \pi < 4$. This is a crude estimate: we shall improve it in exercises 17 and 18.

5. The behaviour of real trigonometric functions

We know that $\cos x$ is positive for $0 \leqslant x < \pi/2$. Since $D \sin x = \cos x$, it follows that sin is strictly increasing in $[0, \pi/2]$. Since

$$\sin^2 \frac{\pi}{2} + \cos^2 \frac{\pi}{2} = 1$$

and $\cos \pi/2 = 0$, we must have $\sin \pi/2 = \pm 1$; but since sin is increasing from 0 in $[0, \pi/2]$ it follows that

$$\sin \frac{\pi}{2} = 1.$$

By (17) it now follows that

$$\sin\left(\frac{\pi}{2}-x\right)=\cos x. \tag{21}$$

Hence cos decreases monotonically from 1 to 0 in $[0, \pi/2]$. Using (16) and (18) repeatedly, we can deduce the behaviour of sin and cos in the intervals $[\pi/2, \pi]$, $[\pi, 3\pi/2]$, and $[3\pi/2, 2\pi]$. We have

$$\cos\left(\frac{\pi}{2}+x\right)=-\sin x$$

$$\sin\left(\frac{\pi}{2}+x\right)=\cos x$$

$$\sin(\pi+x)=-\sin x$$

and so on. We can tabulate the results as follows, where $a \nearrow b$ means 'strictly increasing from a to b' and $a \searrow b$ means 'strictly decreasing from a to b'.

interval	cos	sin
$\left[0, \dfrac{\pi}{2}\right]$	$1 \searrow 0$	$0 \nearrow 1$
$\left[\dfrac{\pi}{2}, \pi\right]$	$0 \searrow -1$	$1 \searrow 0$
$\left[\pi, \dfrac{3\pi}{2}\right]$	$-1 \nearrow 0$	$0 \searrow -1$
$\left[\dfrac{3\pi}{2}, 2\pi\right]$	$0 \nearrow 1$	$-1 \nearrow 0$

From the table we have

$$\sin 2\pi = 0$$
$$\cos 2\pi = 1.$$

Therefore

$$\cos(x+2\pi)=\cos x \cos 2\pi - \sin x \sin 2\pi = \cos x$$
$$\sin(x+2\pi)=\sin x \cos 2\pi + \cos x \sin 2\pi = \sin x.$$

It follows that

$$\cos(x+2n\pi)=\cos x$$
$$\sin(x+2n\pi)=\sin x$$

for all integers. So the behaviour shown in the table repeats in each interval $[2n\pi, (2n+2)\pi]$.

In this way purely formal considerations show that $\sin x$, $\cos x$ have the usual geometric properties for real x, at least in outline. More precise computations of the values of these functions may be performed to any desired degree of accuracy. By inspection it may be seen that for real x (positive or negative) the power series for $\sin x$ and $\cos x$ always have terms which alternate in sign. From the standard theory for alternating series, the sum of the first n terms alternately overestimates and under-estimates the actual limit, allowing us to make very precise estimates of the trigonometric functions.

EXAMPLE. $e^i = \sin 1 + i \cos 1$
$$= (1 - 1/2! + 1/4! - 1/6! + \cdots)$$
$$+ i(1/1! - 1/3! + 1/5! - 1/7! + \cdots).$$

Considering partial sums gives

$$\cos 1 < 1$$
$$\cos 1 > 1 - 1/2! \qquad\qquad\qquad = 0.5$$
$$\cos 1 < 1 - 1/2! + 1/4! \qquad\qquad = 0.54166\ldots$$
$$\cos 1 > 1 - 1/2! + 1/4! - 1/6! \qquad = 0.54027\ldots$$
$$\cos 1 < 1 - 1/2! + 1/4! - 1/6! + 1/8! = 0.54030\ldots$$
$$\ldots$$

and so

$$\cos 1 = 0.5403 \text{ (to 4 decimal places)}.$$

A similar argument gives

$$\sin 1 = 0.8415 \text{ (to 4 decimal places)}$$

and so

$$e^i \simeq 0.5430 + 0.8415\, i.$$

6. Complex exponential and trigonometric functions are periodic

A complex function $f: S \to \mathbb{C}$ is said to have *period* $\rho \in \mathbb{C}$ if

$$f(z + \rho) = f(z) \quad \text{for all } z \in S.$$

(This requires that $z + \rho \in S$ whenever $z \in S$.) Obviously if ρ is a period of f, so is $n\rho$ for any positive integer n. (If, further, $z \in S$ implies $z - \rho \in S$, then $n\rho$ is a period for *any* integer n.)

For the complex exponential, we find that

$$e^{2\pi i} = \cos 2\pi + i \sin 2\pi = 1$$

so that

$$e^{z + 2\pi i} = e^z e^{2\pi i} = e^z \cdot 1 = e^z.$$

Therefore $2\pi i$ is a period for exp. So is $2n\pi i$ for any integer n.

PROPOSITION 5.1. The complex number ρ is a period for exp if and only if $\rho = 2n\pi i$ ($n \in \mathbb{Z}$).

Proof. If ρ is a period then $e^\rho = e^{z+\rho}/e^z = 1$. Therefore if $\rho = u + iv$ we have

$$1 = e^{u+iv} = e^u (\cos v + i \sin v).$$

Taking the modulus, we get

$$1 = |e^u| \, |\cos v + i \sin v| = e^u.$$

By properties of the real exponential function established in section 2 this implies $u = 0$. So now

$$\cos v + i \sin v = 1.$$

Taking real and imaginary parts, $\cos v = 1$, $\sin v = 0$. By the table in section 5, these imply $v = 2n\pi$ ($n \in \mathbb{Z}$). □

We now investigate sin and cos for periodicity. Certainly $2n\pi$ is a period for both sin and cos: the calculation at the end of section 5 works if we replace x by any complex z.

PROPOSITION 5.2. The complex number ρ is a period for sin or cos if and only if $\rho = 2n\pi$ ($n \in \mathbb{Z}$).

Proof. Since $\sin(\pi/2 + z) = \cos z$ by (16) it follows that ρ is a period for sin if and only if ρ is a period for cos. Then

$$\cos(z + \rho) = \cos z$$
$$\sin(z + \rho) = \sin z$$

for all complex z. But then

$$e^{i(z+\rho)} = \cos(z + \rho) + i \sin(z + \rho) = \cos z + i \sin z = e^{iz}$$

so that

$$e^{w+i\rho} = e^{i(-iw+\rho)} = e^{i(-iw)} = e^w.$$

Therefore $i\rho$ is a period for exp. By Proposition 5.1, $-i\rho = 2n\pi$, so $\rho = 2n\pi i$. □

We can easily find the zeros of the functions sin and cos.

PROPOSITION 5.3. Let $z \in \mathbb{C}$. Then

$$\cos z = 0 \quad \text{if and only if } z = (n + \tfrac{1}{2})\pi$$
$$\sin z = 0 \quad \text{if and only if } z = n\pi,$$

where $n \in \mathbb{Z}$.

Proof. Since $\cos z = \sin (\pi/2 + z)$ the second implies the first. Now $\sin z = 0$ if and only if $(e^{iz} - e^{-iz})/(2i) = 0$. Multiplying by $2ie^{iz}$ (which is non-zero) we find that this holds if and only if $e^{2iz} = 1$, so $2iz = 2n\pi i$. Therefore $z = n\pi$ as claimed.   □

7. Other trigonometric functions

If $z \neq (n + \tfrac{1}{2})\pi$ we have $\cos z \neq 0$, so we may define

$$\tan z = \frac{\sin z}{\cos z}.$$

If $S = \{z \in \mathbb{C} | z \neq (n + \tfrac{1}{2})\pi, n \in \mathbb{Z}\}$ then S is a domain, and

$$\tan : S \to \mathbb{C}$$

is a differentiable function. Its derivative is given by

$$D \tan z = \frac{\cos z \, D \sin z - \sin z \, D \cos z}{\cos^2 z}$$

$$= \frac{\cos^2 z + \sin^2 z}{\cos^2 z}$$

$$= 1 + \tan^2 z.$$

Similarly we define

$$\cot z = \frac{\cos z}{\sin z} \quad (z \neq n\pi) \tag{22}$$

$$\sec z = \frac{1}{\cos z} \quad (z \neq (n + \tfrac{1}{2})\pi) \tag{23}$$

$$\operatorname{cosec} z = \frac{1}{\sin z} \quad (z \neq n\pi). \tag{24}$$

These are all differentiable functions (on the obvious domains) whose derivative may be calculated in the usual manner. All of the standard formulae relating trigonometric functions may be deduced from properties of sin and cos, and hence may also be derived for complex values of the variables. For example, using (16) and (18) we find that

$$\tan{(z_1 + z_2)} = \frac{\tan z_1 + \tan z_2}{1 - \tan z_1 \tan z_2}$$

(provided that z_1, z_2, $z_1 + z_2 \in S$). This implies that $\tan{(z + \pi)} = \tan z$, so π is a period for tan. It is easy to see that the only periods for tan are the numbers $n\pi$ ($n \in \mathbb{Z}$).

The reader is encouraged to develop all of his favourite trigonometric formulae for complex functions, including the basic properties of cot, sec, and cosec.

8. Hyperbolic functions

As in the real case, we define

$$\sinh z = \tfrac{1}{2}(e^z - e^{-z})$$
$$\cosh z = \tfrac{1}{2}(e^z + e^{-z})$$

for $z \in \mathbb{C}$. Differentiating, we find

$$D \sinh z = \cosh z$$
$$D \cosh z = \sinh z.$$

Properties of the hyperbolic functions, analogous to those of trigonometric functions (such as addition formulae for $\sinh{(z_1 + z_2)}$) follow either by direct computation or by using the identities

$$\sin iz = i \sinh z$$
$$\cos iz = \cosh z.$$

For example,

$$\cosh^2 z - \sinh^2 z = \cos^2 iz - (-i \sin iz)^2$$
$$= \cos^2 iz + \sin^2 iz$$
$$= 1.$$

The functions tanh, coth, sech, cosech are defined in the obvious way: we leave it to the reader to discover their properties, including zeros and periods.

The hyperbolic functions appear in expressions for the real and imaginary parts of $\sin z$ and $\cos z$. Thus, let $z = x + iy$. Then

$$\sin z = \sin{(x + iy)}$$
$$= \sin x \cos iy + \cos x \sin iy$$
$$= \sin x \cosh y + i \cos x \sinh y. \qquad (25)$$

Similarly

$$\cos z = \cos x \cosh y - i \sin x \sinh y. \qquad (26)$$

Exercises 5

1. Express the following in the form $a + ib$ for real a, b:
   (i) $\exp(i)$ (ii) $e^{2+i\pi}$ (iii) $1/\exp(2+i\pi)$

2. Express the following in the form $a + ib$ for real a, b:
   (i) $\sin i$ (ii) $\cos i$ (iii) $\sinh i$ (iv) $\cosh i$ (v) $\cos(\pi/4 - i)$ (vi) $\tan(1 + i)$.

3. Differentiate the functions given by the following formulae:
   (i) $\exp(z^2 + 2z)$ (ii) $1/\exp(z)$ (iii) $\exp(z^2)/\exp(z+1)$.

4. Differentiate the functions given by the following formulae:
   (i) $\tan(z^2)$ (ii) $\sinh(z+2)/\exp(z^3)$ (iii) $\sin z \cosh z \exp z$.

5. Use the identity $e^{i\theta} e^{i\phi} = e^{i(\theta + \phi)}$ to derive the usual formulae for $\sin(\theta + \phi)$ and $\cos(\theta + \phi)$. By a similar method, show that

   $$1/(\cos\theta + i\sin\theta) = \cos\theta - i\sin\theta.$$

6. Use the identity $(e^{i\theta})^3 = e^{i3\theta}$ to give formulae for $\cos 3\theta$, $\sin 3\theta$ in terms of $\cos\theta$, $\sin\theta$. Derive similar formulae for $\cos 4\theta$, $\sin 4\theta$, $\cos 5\theta$, $\sin 5\theta$.

7. Draw the following paths:
   (i) $z(t) = e^{-it}$ $(0 \leqslant t \leqslant \pi)$,
   (ii) $z(t) = 1 + i + 2e^{it}$ $(0 \leqslant t \leqslant 2\pi)$,
   (iii) $z(t) = z_0 + r e^{it}$ $(0 \leqslant t \leqslant 2\pi)$, where $z_0 \in \mathbb{C}$ and $r > 0$,
   (iv) $z(t) = t + i\cosh t$ $(-1 \leqslant t \leqslant 1)$,
   (v) $z(t) = \cosh t + i\sinh t$ $(-1 \leqslant t \leqslant 1)$.

8. 'Osborne's Rule' states that any formula involving sine and cosine has an analogous formula involving sinh and cosh which is the same in every way except that a product of two sines must be replaced by *minus* the product of corresponding sinhs. For each of the following formulae, write down the corresponding formula using Osborne's Rule and verify it from first principles.
   (i) $\sin^2 A + \cos^2 B = 1$,
   (ii) $\sin(A - B) = \sin A \cos B - \cos A \sin B$,
   (iii) $\cos(A + B) = \cos A \cos B - \sin A \sin B$.
   Comment on the basis of Osborne's Rule in the light of the formulae

   $$\cos iz = \cosh z, \quad \sin iz = i\sinh z.$$

9. Show that the complex conjugate of $\cos z$ is $\cos \bar{z}$ and of $\sin z$ is $\sin \bar{z}$. Verify the relations

   $$|\sin z|^2 = \tfrac{1}{2}(\cosh 2y - \cos 2x) = \sinh^2 y + \sin^2 x = \cosh^2 y - \cos^2 x$$
   $$|\cos z|^2 = \tfrac{1}{2}(\cosh 2y + \cos 2x) = \sinh^2 y + \cos^2 x = \cosh^2 y - \sin^2 x.$$

10. Show that $|\cos z|^2 + |\sin z|^2 = 1$ if and only if z is real; and that $\cos z$ is unbounded on $\mathbb{C}$ (i.e. no K exists such that $|\cos z| < K$).

11. Derive formulae for the real and imaginary parts of the following and check that they satisfy the Cauchy–Riemann equations:

    (i) $\sin z$　　(ii) $\cos z$　　(iii) $\exp z$　　(iv) $\sinh z$　　(v) $\cosh z$.

12. Derive the formulae for the real and imaginary parts of the following, specifying the domain on which they are defined and check that they satisfy the Cauchy–Riemann equations:

    (i) $\tan z$　(ii) $\tanh z$　(iii) $\operatorname{cosec} z$　(iv) $\operatorname{cosech} z$　(v) $\cot z$　(vi) $\coth z$.

13. Write $\tanh(x+iy)$ as real and imaginary parts and show that if $\tanh(x+iy)$ is real, then $y = n\pi/2$.

14. For each of the functions exp, cos, sin, tan, cosh, sinh, tanh, find the set of points on which it assumes

    (i) real values,　　(ii) purely imaginary values.

15. By considering the real and imaginary parts of

    $$1 + z + z^2 + \cdots + z^n = (1 - z^{n+1})/(1 - z),$$

    find the sums:

    (i) $1 + \cos x + \cos 2x + \cdots + \cos nx$,

    (ii) $\sin x + \sin 2x + \cdots + \sin nx$.

    By similar methods, find

    (iii) $\cos x + \cos 3x + \cdots + \cos(2n-1)x$,

    (iv) $\sin x + \sin 3x + \cdots + \sin(2n-1)x$,

    (v) $\sin x - \sin 2x + \cdots + (-1)^{n-1}\sin nx$,

    (vi) $\cos\theta + \cos(\theta+\phi) + \cdots + \cos(\theta+n\phi)$,

    (vii) $\sin\theta + \sin(\theta+\phi) + \cdots + \sin(\theta+n\phi)$.

16. If $z(t) = x(t) + iy(t)$ is a solution of

    $$\frac{d^2 z}{dt^2} + \lambda z = k_0 e^{i\omega t} \quad (\lambda, k_0, \omega \in \mathbb{R}), \tag{1}$$

    show that $x = x(t) = \operatorname{re}(z(t))$ is a solution of

    $$\frac{d^2 x}{dt^2} + \lambda x = k_0 \cos t. \tag{2}$$

    By considering solutions of (1) of the form $z = k e^{i\omega t}$, find a solution of (2).

    If k_0 is *complex*, say $k_0 = k_1 e^{i\varepsilon}(k_1, \varepsilon \in \mathbb{R})$, and λ, ω are real, write down the real part of equation (1) to obtain

    $$\frac{d^2 x}{dt^2} + \lambda x = k_0 \cos(\omega t + \varepsilon). \tag{3}$$

    Show that a solution of (3) may be taken in the form

    $$x(t) = k_1 \cos(\omega t + \varepsilon)/(\lambda - \omega^2).$$

17. By using the sum of the geometric progression

    $$1 + z + z^2 + \cdots + z^n = (1 - z^{n+1})/(1 - z),$$

find $A_n(x)$ such that
$$1/(1+x^2)=1-x^2+x^4-\cdots+(-1)^n x^{2n}+A_n(x).$$

Verify that the derivative of $\tan^{-1}x$ is $1/(1+x^2)$ where $\tan^{-1}:\mathbb{R}\to(-\pi/2, \pi/2)$ is the inverse function of $\tan:(-\pi/2, \pi/2)\to\mathbb{R}$. Hence deduce that

$$\tan^{-1}t=\int_0^t 1/(1+x^2)\,dx=t-t^3/3+t^5/5-\cdots+\int_0^t A_n(x)\,dx.$$

By estimating the size of the last named integral, show that the power series
$$t-t^3/3+t^5/5-\cdots$$
converges to $\tan^{-1}t$ for $|t|\leqslant 1$.

Deduce *Gregory's Series:*
$$\frac{\pi}{4}=1-\frac{1}{3}+\frac{1}{5}-\frac{1}{7}+\cdots.$$

18. Gregory's series converges very slowly. Better methods for calculating π may be obtained using the identities:

(1) $\dfrac{\pi}{4}=\tan^{-1}\dfrac{1}{2}+\tan^{-1}\dfrac{1}{3}$,

(2) $\dfrac{\pi}{4}=4\tan^{-1}\dfrac{1}{5}-\tan^{-1}\dfrac{1}{239}$.

Verify (1) and (2) using the addition formula for tan. Use (2) to compute π correct to 5 decimal places. (This only requires one term of the expansion of $\tan^{-1}$ $(1/239)$ and five terms of $\tan^{-1}$ $(1/5)$.)

6
Integration

The next part of the grand plan is to define complex integration by analogy with the real case and establish the inverse relation between differentiation and integration.

Consider a complex function $f: D \to \mathbb{C}$ in a domain D and let $z_0, z_1 \in D$. The real integral $\int_a^b f(t)\, dt$ does not generalize immediately to the complex case $\int_{z_0}^{z_1} f(z)\, dz$ because we must specify how to get from z_0 to z_1. We do so by selecting a path γ between them. A sensible notation for the integral of f along γ is then $\int_\gamma f(z)\, dz$, or $\int_\gamma f$ for short.

To define $\int_\gamma f$ the reader has a choice of two approaches. The first is to build up the theory of complex Riemann sums by mimicking the real case. This will be done in §§1 and 2. It leads to the result

$$\int_\gamma f = \int_a^b f(\gamma(t))\gamma'(t)\, dt \qquad (1)$$

when $\gamma: [a, b] \to D$ has a continuous derivative γ'. In §3 the length L of such a path is computed to be

$$L = \int_a^b |\gamma'(t)|\, dt. \qquad (2)$$

These equations lead to a second possible approach, in which the reader is assumed to be familiar with real integration. Why not use (1) and (2) to *define* $\int_\gamma f$ and L? The integrand $f(\gamma(t))\gamma'(t)$ in (1) happens to be complex, but by writing it as $U(t) + iV(t)$ the integral $\int_\gamma f$ can be found from two real integrals:

$$\int_\gamma f = \int_a^b U(t)\, dt + i \int_a^b V(t)\, dt.$$

This method is adopted in §4 as an alternative, allowing §1–3 to be omitted. Such a short cut has its price. It means that a couple of proofs later in the chapter must be given in a slightly more technical and less intuitive manner. But this price is not very great and it leaves the reader

95

with a genuine choice: work through the theory of complex Riemann integration to see the full analogy with the real case, or bypass the next three sections and start at §4.

1. The real case

For the reader who has chosen to build up the analogy between the real and complex integral, we begin by recalling the real case.

The Riemann integral $\int_a^b \phi(t)\, dt$ of a real function $\phi:[a, b]\to\mathbb{R}$ is defined in stages. First take a partition P of $[a, b]$ given by $a=t_0<t_1<\cdots<t_n=b$, and choose intermediate points s_r in each subinterval $t_{r-1}\leqslant s_r\leqslant t_r$. Then form the sum

$$S(P, \phi)= \sum_{r=1}^{n} \phi(s_r)(t_r-t_{r-1}).$$

The points $t_0,\ldots,t_n$ are called the *division points* of P and another partition Q is said to be *finer* than P if the division points of P are all included in those of Q.

The following result is well known from real analysis; we quote it without proof.

LEMMA 6.1. Let $\phi:[a, b]\to\mathbb{R}$ be continuous. Then there exists a real number A such that for any $\varepsilon>0$ there is a partition P_ε of $[a, b]$ with every partition P finer than P_ε giving

$$|S(P, \phi)-A|<\varepsilon.$$

The real number A is denoted by $\int_a^b \phi(t)\, dt$.   □

The actual computation of $\int_a^b \phi(t)\, dt$ is usually performed by antidifferentiation using:

LEMMA 6.2 (The Fundamental Theorem of Calculus)
(i) If ϕ is continuous on $[a, b]$ and $F'=\phi$, then

$$\int_a^b \phi(t)\, dt = F(b)-F(a).$$

(ii) If $I(x)=\int_a^x \phi(t)\, dt$ $(a\leqslant x\leqslant b)$, then $I'=\phi$.   □
(Part (ii) also requires ϕ continuous on $[a, b]$, of course.)

EXAMPLE. To compute $\int_a^b t^5 \, dt$, we do not need to calculate sums $S(P, \phi)$ where $\phi(t) = t^5$. Because $F(t) = \frac{1}{6} t^6$ satisfies $F' = \phi$, Lemma 6.2(i) immediately gives

$$\int_a^b t^5 \, dt = \frac{1}{6} b^6 - \frac{1}{6} a^6.$$

In general we compute $\int_a^b \phi(t) \, dt$ by seeking an antiderivative F of ϕ and using Lemma 6.2(i).

Before passing to the complex case, it will be helpful to consider a slight generalization: the Riemann–Stieltjes integral $\int_a^b \phi(t) \, d\theta$ where θ is a second real function. Here, given a partition P of $[a, b]$, as above, we form the sum

$$S(P, \phi, \theta) = \sum_{r=1}^n \phi(s_r)(\theta(t_r) - \theta(t_{r-1})).$$

From real analysis we have a generalization of Lemma 6.1:

LEMMA 6.3. Let ϕ, θ be real functions defined on $[a, b]$ such that ϕ is continuous and θ has continuous derivative θ'. Then the real number $B = \int_a^b \phi(t)\theta'(t) \, dt$ satisfies the following condition:

Given $\varepsilon > 0$, there exists a partition P_ε such that for any finer partition P,

$$|S(P, \phi, \theta) - B| < \varepsilon. \qquad \square$$

The limit of the sum $S(P, \phi, \theta)$ in Lemma 6.3 is also denoted by $\int_a^b \phi(t) \, d\theta$. What Lemma 6.3 tells us is that (for differentiable θ) the Riemann–Stieltjes integral $\int_a^b \phi(t) \, d\theta$ is equal to the Riemann integral $\int_a^b \phi(t)\theta'(t) \, dt$. The conditions of the lemma merely ensure that the integrand $\phi(t)\theta'(t)$ exists and is continuous. Of course the Riemann–Stieltjes integral (and the Riemann integral) exist under far more general conditions on ϕ, θ than we have mentioned, but those in Lemma 6.3 will prove to be all that is necessary for our generalization to the complex case.

2. Complex integration along smooth paths

The Riemann integral of a complex function f is defined by analogy with the limit of the sum $\Sigma \, \phi(s_r)(t_r - t_{r-1})$. We simply consider $\Sigma \, f(\zeta_r)(z_r - z_{r-1})$ where ζ_r and z_r are complex. We select ζ_r and z_r along a path γ in the domain of f as in Figure 6.1.

For these purposes we suppose that
(i) $f : D \to \mathbb{C}$ is continuous,
(ii) $\gamma : [a, b] \to D$ is a path such that γ' exists and is continuous throughout

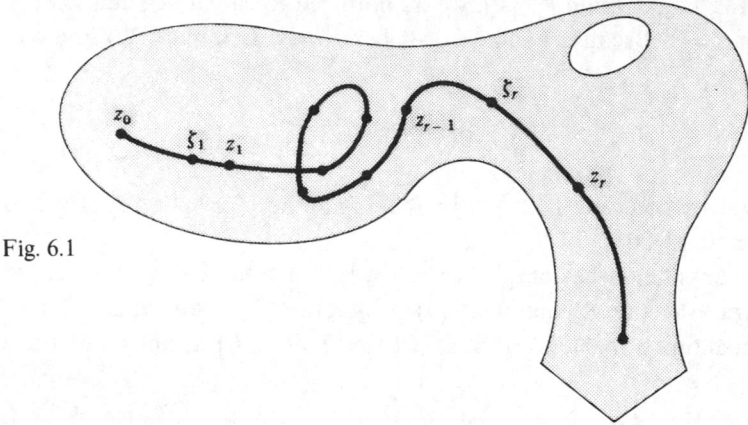

Fig. 6.1

$[a, b]$. (This means that if $\gamma(t) = x(t) + iy(t)$, then x' and y' are continuous on $[a, b]$, including the endpoints.)

A path satisfying condition (ii) is said to be *smooth*.

For any partition P given by $a = t_0 < t_1 < \cdots < t_n = b$ and $t_{r-1} \leqslant s_r \leqslant t_r$, form the sum

$$S(P, f, \gamma) = \sum_{r=1}^{n} f(\gamma(s_r))(\gamma(t_r) - \gamma(t_{r-1})).$$

By writing $z_r = \gamma(t_r)$, $\zeta_r = \gamma(s_r)$, this Riemann–Stieltjes sum becomes

$$S(P, f, \gamma) = \sum_{r=1}^{n} f(\zeta_r)(z_r - z_{r-1})$$

which exhibits the direct analogy with the real case.

We define the integral of f along γ to be the complex number K for which the following condition is satisfied:

Given $\varepsilon > 0$, there exists a partition P_ε such that for any finer partition P,

$$|S(P, f, \gamma) - K| < \varepsilon.$$

We define

$$\int_\gamma f(z)\,dz = K$$

and usually abbreviate to $\int_\gamma f$.

As in the real case, we rarely calculate the integral by this summation process. One method is to reduce it to two real integrals by writing the complex function of a real variable $\psi : [a, b] \to \mathbb{C}$ as

$$\psi(t) = U(t) + iV(t) \quad (a \leqslant t \leqslant b)$$

whence we may define

$$\int_a^b \psi(t)\,dt = \int_a^b U(t)\,dt + i \int_a^b V(t)\,dt.$$

With this convention we derive a complex version of Lemma 6.3:

THEOREM 6.4. For a continuous complex function f defined on a domain D and a smooth path $\gamma:[a, b] \to D$,

$$\int_\gamma f = \int_a^b f(\gamma(t))\gamma'(t)\,dt.$$

Proof. Let $\gamma(t) = x(t) + iy(t)$ $(a \leqslant t \leqslant b)$ and $f(z) = u(x, y) + iv(x, y)$ $(z = x + iy \in D)$. If we denote $u(x(t), y(t))$, $v(x(t), y(t))$, $x'(t)$, $y'(t)$ by u, v, x', y' for short, then our convention for complex integrals gives

$$\int_a^b f(\gamma(t))\gamma'(t)\,dt = \int_a^b (u + iv)(x' + iy')\,dt$$

$$= \int_a^b (ux' - vy' + i\,(ux' + vy'))\,dt$$

$$= \int_a^b ux'\,dt - \int_a^b vy'\,dt + i \int_a^b ux'\,dt + i \int_a^b vy'\,dt.$$

If we write $f(\gamma(s_r))$ as $u_r + iv_r$ and $\gamma(t_r)$ as $x_r + iy_r$, then

$$S(P, f, \gamma) = \sum_{r=1}^n (u_r + iv_r)[(x_r + iy_r) - (x_{r-1} + iy_{r-1})]$$

$$= \sum u_r(x_r - x_{r-1}) - \sum v_r(y_r - y_{r-1}) + i \sum u_r(x_r - x_{r-1})$$
$$+ i \sum v_r(y_r - y_{r-1}).$$

We now match the four integrals and four sums in pairs and use Lemma 5.3. For instance,

$$\sum u_r(x_r - x_{r-1}) = \sum_{r=1}^n u(x(s_r), y(s_r))(x(t_r) - x(t_{r-1}))$$

where both $\phi(t) = u(x(t), y(t))$ and $x'(t)$ are continuous on $[a, b]$. So given $\varepsilon > 0$, we can find a partition $P_1(\varepsilon)$ such that any partition P finer than $P_1(\varepsilon)$ implies

$$\left| \sum u_r(x_r - x_{r-1}) - \int_a^b ux'\,dt \right| < \varepsilon/4.$$

We then find partitions $P_2(\varepsilon)$, $P_3(\varepsilon)$, $P_4(\varepsilon)$ such that finer partitions give corresponding inequalities between corresponding pairs of integrals and sums. Taking P_ε to have as division points all those of $P_1(\varepsilon)$, $P_2(\varepsilon)$, $P_3(\varepsilon)$

and $P_4(\varepsilon)$, for P finer than P_ε, the four inequalities all hold and then

$$\left| S(P, f, \gamma) - \int_a^b f(\gamma(t))\gamma'(t) \, dt \right| < \tfrac{1}{4}\varepsilon + \tfrac{1}{4}\varepsilon + \tfrac{1}{4}\varepsilon + \tfrac{1}{4}\varepsilon = \varepsilon.$$

Hence

$$\int_\gamma f = \int_a^b f(\gamma(t))\gamma'(t) \, dt. \qquad \square$$

EXAMPLE. $f(z) = z^2$, $\gamma(t) = t^2 + it$ $(0 \leqslant t \leqslant 1)$. (Fig. 6.2)

$$\int_\gamma f = \int_0^1 f(\gamma(t))\gamma'(t) \, dt$$

$$= \int_0^1 (t^2 + it)^2(2t + i) \, dt$$

$$= \int_0^1 (t^4 + 2it^3 - t^2)(2t + i) \, dt$$

$$= \int_0^1 (2t^5 - 4t^3) \, dt + i \int_0^1 (5t^4 - t^2) \, dt$$

$$= [\tfrac{1}{3}t^6 - t^4]_0^1 + i[t^5 - \tfrac{1}{3}t^3]_0^1$$

$$= -\tfrac{2}{3} + \tfrac{2}{3}i.$$

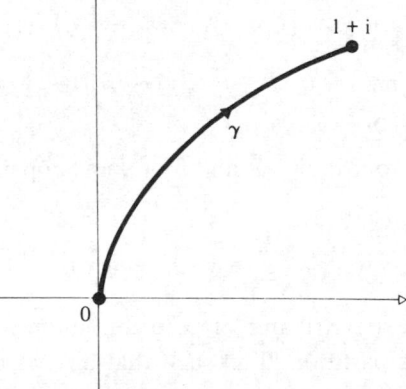

Fig. 6.2

3. The length of a smooth path

For an arbitrary path $\gamma:[a, b] \to \mathbb{C}$, we can take a subdivision P of $[a, b]$ given by $a = t_0 < t_1 < \cdots < t_n = b$ and calculate the length

$$L(\pi) = \sum_{r=1}^n |\gamma(t_r) - \gamma(t_{r-1})|$$

of the approximating polygonal curve π with vertices $\gamma(t_0), \gamma(t_1), \ldots, \gamma(t_n)$. (Fig. 6.3) The length $L(\gamma)$ of γ is defined to be the supremum (least upper bound) of the lengths $L(\pi)$ of all such approximating polygons. For an arbitrary path the length need not be finite.

Fig. 6.3

EXAMPLE 1. Suppose $\gamma(t)$ $(0 \leqslant t \leqslant 1)$ is defined as follows: let m_n be the midpoint of $[1/(n+1), \ 1/n]$ (so that $m_n = \frac{1}{2}(1/(n+1) + 1/n) = (2n+1)/(2n(n+1))$) and draw the graph of $y = \lambda(t)$ over $[1/(n+1), \ 1/n]$ as two straight line segments from $(1/(n+1), 0)$ up to $(m_n, 1/n)$ then down to $(1/n, 0)$. Define

$$\gamma(t) = \begin{cases} t + i\lambda(t) & (0 < t \leqslant 1) \\ 0 & (t = 0). \end{cases}$$

The graph of γ is drawn in Figure 6.4. The length of the graph from $t = 1/(n+1)$ to $t = 1/n$ exceeds $2/n$. Let P be the partition $0 < 1/(n+1) < m_n < 1/n < \cdots < \frac{1}{2} < m_1 < 1$, then the length of the polygonal path exceeds

$$(2/n) + (2/(n-1)) + \cdots + (2/1) = 2\left(1 + \frac{1}{2} + \cdots + \frac{1}{n}\right).$$

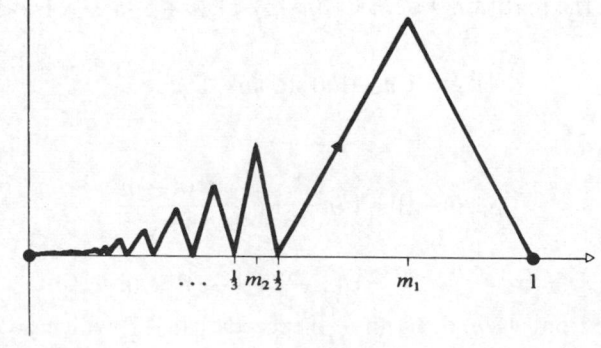

Fig. 6.4

Since the latter is twice the harmonic series, it increases without limit as n increases, so $L(\gamma)$ is infinite.

EXAMPLE 2. A 'nearly smooth path' is

$$\gamma(t) = \begin{cases} t + it \sin(\pi/t) & (0 < t \le 1) \\ 0 & (t = 0) \end{cases}$$

as in Figure 6.5. A calculation shows that although

Fig. 6.5

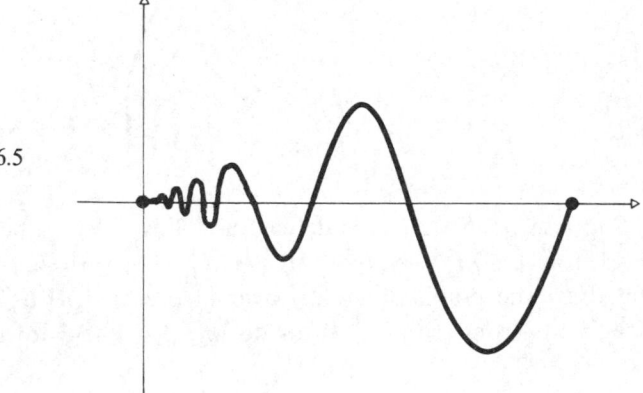

$$\gamma'(t) = 1 + i (\sin(1/t) + t \cos(\pi/t)(-\pi/t^2))$$

is continuous for $0 < t \le 1$, the limit of

$$\frac{\gamma(t) - \gamma(0)}{t} = 1 + i \sin(\pi/t)$$

as t tends down to 0 does not exist. Hence γ' is not continuous on the closed interval $[0, 1]$ and so is not smooth on $[0, 1]$ (though it *is* smooth on any subinterval $[k, 1]$ where $0 < k < 1$).

Let P_n be the partition $0 < 1/n < 1/(n - \frac{1}{2}) < 1/(n-1) < \cdots < 1/2 < 1/1\frac{1}{2} < 1$. Since

$$\gamma(1/n) = 1/n + (i/n) \sin \pi n = 1/n$$

and

$$\gamma(1/(n - \tfrac{1}{2})) = 1/(n - \tfrac{1}{2}) + i \frac{\sin(n - \frac{1}{2})\pi}{n - \frac{1}{2}}$$

$$= 1/(n - \tfrac{1}{2}) + i(-1)^{n-1}/(n - \tfrac{1}{2}),$$

the distance from $\gamma(1/n)$ to $\gamma(1/(n - \tfrac{1}{2}))$ exceeds $1/(n - \tfrac{1}{2})$ which exceeds $1/n$.

We need only compute the lengths of alternate segments of the polygonal approximation γ_n given by P_n to find

$$L(\gamma_n) > 1/n + 1/(n-1) + \cdots + 1$$

which again increases without limit.

If a path is *smooth*, then its length *is* finite and can be calculated by an integral:

PROPOSITION 6.5 The length of a smooth path $\gamma:[a, b] \to \mathbb{C}$ is

$$L(\gamma) = \int_a^b |\gamma'(t)|\, dt.$$

Before we prove this result, we first note that the integrand $|\gamma'(t)|$ is continuous on $[a, b]$, so the real integral

$$L = \int_a^b |\gamma'(t)|\, dt$$

certainly exists. We must show that L is the supremum of the lengths $L(\pi)$ of approximating polygons π.

Now L can be closely approximated by sums

$$S(P, \phi) = \sum_{r=1}^n |\gamma'(s_r)|(t_r - t_{r-1})$$

where P is the partition $a = t_0 < t_1 < \cdots < t_n = b$, $t_{r-1} \leqslant s_r \leqslant t_r$, and $\phi(t) = |\gamma'(t)|$.

The proof of Proposition 6.5 is greatly facilitated by

LEMMA 6.6. With the preceding notation, given any $\varepsilon > 0$, there exists a partition Q_ε such that for any partition P finer than Q_ε, the length $L(\pi)$ of the approximating polygon π corresponding to P satisfies

$$|S(P, \phi) - L(\pi)| < \varepsilon.$$

Proof. By definition,

$$S(P, \phi) = \sum_{r=1}^n |\gamma'(s_r)|(t_r - t_{r-1}) \quad \text{and}$$

$$L(\pi) = \sum_{r=1}^n |\gamma(t_r) - \gamma(t_{r-1})|.$$

Writing $\gamma(t) = x(t) + iy(t)$, then the mean value theorem of real analysis gives

$$x(t_r) - x(t_{r-1}) = x'(\sigma_r)(t_r - t_{r-1}) \quad \text{for some } \sigma_r \text{ in } t_{r-1} < \sigma_r < t_r,$$

$$y(t_r) - y(t_{r-1}) = y'(\tau_r)(t_r - t_{r-1}) \quad \text{for some } \tau_r \text{ in } t_{r-1} < \tau_r < t_r,$$

so
$$\gamma(t_r) - \gamma(t_{r-1}) = (x'(\sigma_r) + iy'(\tau_r))(t_r - t_{r-1}).$$

Now x', y' are continuous on $[a, b]$ and (from real analysis) they are *uniformly* continuous, which means that for any $\varepsilon > 0$, there exists $\delta > 0$ such that

$$s, t \in [a, b] \quad \text{and} \quad |s - t| < \varepsilon \quad \text{implies}$$

$$|x'(s) - x'(t)| < \varepsilon/2(b-a) \quad \text{and} \quad |y'(s) - y'(t)| < \varepsilon/2(b-a).$$

If we choose Q_ε to be *any* partition such that each subinterval $[t_{r-1}, t_r]$ has length less than δ, then any partition P finer than Q_ε has the same property. Because s_r, σ_r, $\tau_r \in [t_{r-1}, t_r]$, they are all within a distance δ of each other, so

$$\left| |\gamma'(s_r)|(t_r - t_{r-1}) - |\gamma(t_r) - \gamma(t_{r-1})| \right|$$

$$= \left| |\gamma'(s_r)| - |x'(\sigma_r) + iy'(\tau_r)| \right|(t_r - t_{r-1})$$

$$\leqslant |(x'(s_r) + iy'(s_r)) - (x'(\sigma_r) + iy'(\tau_r))|(t_r - t_{r-1})$$

(using $||w| - |z|| \leqslant |w - z|$)

$$\leqslant (|x'(s_r) - x'(\sigma_r)| + |y'(s_r) - y'(\tau_r)|)(t_r - t_{r-1})$$

$$< (\varepsilon/2(b-a) + \varepsilon/2(b-a))(t_r - t_{r-1})$$

$$= \varepsilon(t_r - t_{r-1})/(b-a).$$

Thus

$$|S(P, \phi) - L(\pi)| \leqslant \sum_{r=1}^{n} \left| |\gamma'(s_r)|(t_r - t_{r-1}) - |\gamma(t_r) - \gamma(t_{r-1})| \right|$$

$$< \sum_{r=1}^{n} \frac{\varepsilon}{(b-a)}(t_r - t_{r-1})$$

$$= \varepsilon,$$

as required.   □

We are now prepared for the

Proof of Proposition 6.5. By the definition of $L = \int_a^b \phi(t)\, dt$ where $\phi = |\gamma'|$, there exists a partition P_ε such that for any partition P finer than P_ε,

$$|S(P, \phi) - L| < \varepsilon. \tag{3}$$

Adding more division points to P_ε as necessary to obtain a finer partition Q_ε with each subinterval of length less than δ, then for P finer than Q_ε, Lemma 6.6 gives

$$|S(P, \phi) - L(\pi)| < \varepsilon. \tag{4}$$

Putting this with (1) gives

$$L-2\varepsilon<L(\pi)<L+2\varepsilon. \tag{5}$$

Hence for any positive k we can find an approximating polygon π where

$$L-k<L(\pi). \tag{6}$$

Given any partition Q of $[a,b]$ with approximating polygon κ, the addition of further vertices to κ can only increase its length; if P is finer than Q, then $L(\kappa)\leqslant L(\pi)$. We choose P finer than Q_ε so that (5) holds, then

$$L(\kappa)\leqslant L(\pi)<L+2\varepsilon,$$

and so

$$L(\kappa)<L+2\varepsilon.$$

But ε is any positive number, hence for *any* approximating polygon κ,

$$L(\kappa)\leqslant L. \tag{7}$$

The inequalities (6) and (7) exhibit L as the supremum of the lengths of all approximating polygons, completing the proof. □

EXAMPLE 1. The standard straight line path $\gamma=[z_1,z_2]$

$$\gamma(t)=z_1(1-t)+z_2t \quad (0\leqslant t\leqslant 1)$$

has length

$$L(\gamma)=\int_0^1 |\gamma'(t)|\,\mathrm{d}t=\int_0^1 |z_2-z_1|\,\mathrm{d}t=|z_2-z_1|,$$

as expected.

EXAMPLE 2. The circle $S(t)=z_0+r\mathrm{e}^{\mathrm{i}t}\,(0\leqslant t\leqslant 2\pi)$, centre z_0, radius $r>0$, has length

$$L(S)=\int_0^2 |r\mathrm{e}^{\mathrm{i}t}|\,\mathrm{d}t=\int_0^{2\pi} r\,\mathrm{d}t=2\pi r.$$

4. Contour integration

Some readers at this stage will have read the previous three sections and some will not. For the latter we define a path $\gamma:[a,b]\to\mathbb{C}$ to be *smooth* if γ' exists and is continuous on the closed interval $[a,b]$. The integral of a continuous function $f:D\to\mathbb{C}$ along a smooth path γ in a domain D is *defined* to be

$$\int_\gamma f=\int_a^b f(\gamma(t))\gamma'(t)\,\mathrm{d}t$$

and the length of γ is *defined* to be

$$L(\gamma) = \int_a^b |\gamma'(t)|\,dt.$$

(Note that the integrands $f(\gamma(t))\gamma'(t)$ and $|\gamma'(t)|$ are both continuous. The latter is real; the former may be written in real and imaginary parts $U(t) + iV(t)$ and the integral $\int_\gamma f$ calculated using two real integrals:

$$\int_\gamma f = \int_a^b U(t)\,dt + i\int_a^b V(t)\,dt.$$

Since all the integrals concerned are continuous, the real integrals all exist.)

For all readers we now generalize these integrals to paths made up of a finite number of smooth pieces. Using the notation of §2.4, we define a *contour* to be

$$\gamma = \gamma_1 + \cdots + \gamma_n$$

where $\gamma_1, \ldots, \gamma_n$ are smooth paths with the final point of γ_r coinciding with the initial point of γ_{r+1} for $r = 1, \ldots, n-1$.

EXAMPLE. If $\gamma_1(t) = t\ (\varepsilon \leqslant t \leqslant R)$, $\gamma_2(t) = \cos t + i\sin t\ (0 \leqslant t \leqslant \pi)$, $\gamma_3(t) = t$ $(-R \leqslant t \leqslant -\varepsilon)$, $\gamma_4(t) = -\cos t + i\sin t\ (0 \leqslant t \leqslant \pi)$, then $\gamma = \gamma_1 + \gamma_2 + \gamma_3 + \gamma_4$ is the (closed) contour in Figure 6.6.

Given a continuous function $f: D \to \mathbb{C}$ and a contour $\gamma = \gamma_1 + \cdots + \gamma_n$ in the domain D, we define

$$\int_\gamma f = \int_{\gamma_1} f + \cdots + \int_{\gamma_n} f,$$

and

$$L(\gamma) = L(\gamma_1) + \cdots + L(\gamma_n).$$

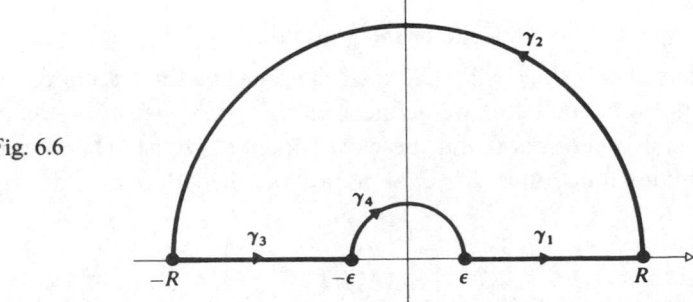

Fig. 6.6

It is trivially obvious that if a smooth path σ is subdivided $\sigma = \sigma_1 + \sigma_2$, then

$$\int_\sigma f = \int_{\sigma_1} f + \int_{\sigma_2} f,$$

so further subdivisions of the contours $\sigma_1, \ldots, \sigma_n$ in the above definitions will not affect the value of the integrals. The contour integrals are therefore well-defined.

The following standard properties hold, analogous to the real case:

$$\int_{\gamma_1 + \gamma_2} f = \int_{\gamma_1} f + \int_{\gamma_2} f \tag{8}$$

$$\int_\gamma (f_1 + f_2) = \int_\gamma f_1 + \int_\gamma f_2 \tag{9}$$

$$\int_\gamma cf = c \int_\gamma f \quad \text{for } c \in \mathbb{C}. \tag{10}$$

The first follows trivially from the definitions. The proofs of (9) and (10) depend on whether the reader has read §§1–3. From §2 we have

$$S(P, f_1 + f_2, \gamma) = S(P, f_1, \gamma) + S(P, f_2, \gamma)$$
$$S(P, cf, \gamma) = cS(P, f, \gamma)$$

for any partition P, from which (9) and (10) follow easily.

The reader who has omitted that section may base his verification on known properties of real integrals as follows. First note that it is sufficient to verify them for a smooth path. Formula (8) is given by

$$\int_\gamma (f_1 + f_2) = \int_a^b (f_1(\gamma(t)) + f_2(\gamma(t)))\gamma'(t) \, dt$$

$$= \int_a^b f_1(\gamma(t))\gamma'(t) \, dt + \int_a^b f_2(\gamma(t))\gamma'(t) \, dt$$

(using the additivity property for real integrals on the real and imaginary parts) so

$$\int_\gamma (f_1 + f_2) = \int_\gamma f_1 + \int_\gamma f_2.$$

Formula (9) is slightly longer (a minor penalty for skipping §§1–3). Let $c = \alpha + i\beta$, $f(\gamma(t))\gamma'(t) = U(t) + iV(t)$, then

$$\int_a^b (\alpha + i\beta)(U(t) + iV(t)) \, dt$$

$$= \int_a^b ([\alpha U(t) - \beta V(t)] + i[\alpha V(t) + \beta U(t)]) \, dt$$

$$= \int_a^b [\alpha U(t) - \beta V(t)] \, dt + i \int_a^b [\alpha V(t) + \beta U(t)] \, dt$$

$$= \alpha \int_a^b U(t) \, dt - \beta \int_a^b V(t) \, dt + i\alpha \int_a^b V(t) \, dt + i\beta \int_a^b U(t) \, dt$$

(using the corresponding property for real integrals)

$$= (\alpha + i\beta) \int_a^b (U(t) + iV(t)) \, dt,$$

as required.

Finally note that if $\gamma:[a, b] \to D$ is a contour, then the opposite path $-\gamma:[a, b] \to D$, defined by

$$(-\gamma)(t) = \gamma(a + b - t) \quad (a \leqslant t \leqslant b),$$

is also a contour. Computing the integral of f along $-\gamma$,

$$\int_{-\gamma} f = \int_a^b f(\gamma(a+b-t)) \frac{d}{dt} \gamma(a+b-t) \, dt$$

$$= -\int_a^b f(\gamma(a+b-t))\gamma'(a+b-t) \, dt$$

and substituting $s = a + b - t$, this becomes

$$-\int_b^a f(\gamma(s))\gamma'(s) \, (-ds)$$

$$= -\int_a^b f(\gamma(s))\gamma'(s) \, ds$$

$$= -\int_\gamma f.$$

Hence we have

$$\int_{-\gamma} f = -\int_\gamma f.$$

5. The Fundamental Theorem of Contour Integration

Integrals of complex functions are only occasionally computed by breaking them down into real and imaginary parts and calculating two real integrals as in the last section. This does prove to be necessary on occasion, but a far more efficient method is available for $\int_\gamma f$ if we can find an *antiderivative* F of $f:D \to \mathbb{C}$, by which we mean a function $F:D \to \mathbb{C}$ such that $F' = f$. An antiderivative, if it exists, is unique up to an added constant, for if $F' = G' = f$ in a domain D, then $(F - G)' = F' - G' = 0$ and $F - G$ is constant by Theorem 4.7. If we can find an antiderivative, then the integral can be computed immediately using

THEOREM 6.7 (The Fundamental Theorem of Contour Integration).
If $f:D \to \mathbb{C}$ is continuous, $F:D \to \mathbb{C}$ satisfies $F' = f$ and γ is a contour in D from z_0 to z_1, then

$$\int_\gamma f = F(z_1) - F(z_0).$$

Proof. If $w(t) = u(t) + iv(t)$, $W(t) = U(t) + iV(t)$ $(a \leqslant t \leqslant b)$ and $W' = w$, then $U' = u$, $V' = v$ and

$$\int_a^b w(t) \, dt = \int_a^b u(t) \, dt + i \int_a^b v(t) \, dt$$
$$= U(b) - U(a) + iV(b) - iV(a)$$
$$= W(b) - W(a).$$

Given $F' = f$, let $w(t) = f(\gamma(t))\gamma'(t)$, then

$$w(t) = F'(\gamma(t))\gamma'(t) = W'(t)$$

where $W(t) = F(\gamma(t))$. Hence

$$f = \int_a^b w(t) \, dt = W(b) - W(a)$$
$$= F(\gamma(b)) - F(\gamma(a))$$
$$= F(z_1) - F(z_0),$$

as required. □

EXAMPLE 1. If $f(z) = z^2$ and γ is any contour from $z_0 = 0$ to $z_1 = 1 + i$, then $F(z) = \frac{1}{3}z^3$ is an antiderivative of f and

$$\int_\gamma z^2 \, dz = \frac{1}{3}z_1^3 - \frac{1}{3}z_0^3$$
$$= \frac{1}{3}(1 + i)^3$$
$$= -\frac{2}{3} + \frac{2}{3}i.$$

This example is a visibly easier calculation than that performed at the end of §2 for the particular contour $\gamma(t) = t^2 + it$ $(0 \leqslant t \leqslant 1)$.

However, any euphoria we feel over this phenomenon must be tempered with the realization that, unlike the real case where a continuous function always has an antiderivative (Lemma 6.2 (ii)), in the complex case there are functions for which no antiderivatives exist whatever.

We shall prove later (Chapter 10) that a function differentiable once in a domain must be differentiable an infinite number of times. Granted this we see that an antiderivative F, being differentiable once, must be differentiable *twice*. So its derivative $F' = f$ must be differentiable. If we have a *non*-differentiable function f then it is futile to seek an antiderivative, for none can exist.

EXAMPLE 2. $f(z) = |z|^2$ is differentiable only at the origin, so it is useless to search for an antiderivative in computing $\int_\gamma |z|^2 \, dz$ along $\gamma(t) = t^2 + it$ $(0 < t < 1)$. In this case we return to the basic formula

$$\int_\gamma |z|^2 \, dz = \int_0^1 (t^4 + t^2)(2t + i) \, dt$$

$$= \int_0^1 (2t^5 + t^3) \, dt + i \int_0^1 (t^4 + t^2) \, dt$$

$$= [\tfrac{1}{3}t^6 + \tfrac{1}{2}t^4]_0^1 + i[\tfrac{1}{5}t^5 + \tfrac{1}{3}]_0^1$$

$$= \tfrac{5}{6} + \tfrac{8}{15}i.$$

Fortunately, many functions do have antiderivatives. For instance, a polynomial $p(z) = a_0 + a_1 z + \cdots + a_n z^n$ has an obvious antiderivative

$$P(z) = a_0 z + \tfrac{1}{2}a_1 z^2 + \cdots + a_n z^n/(n+1).$$

More generally, a power series has an antiderivative everywhere inside its disc of convergence.

THEOREM 6.8. If $f(z) = \Sigma_{n=0}^\infty a_n(z - z_0)^n$ converges for $|z - z_0| < R$, then

$$F(z) = \sum_{n=0}^\infty \frac{a_n}{n+1} (z - z_0)^{n+1}$$

also converges for $|z - z_0| < R$ and $F' = f$.

Proof. It is sufficient to show that $F(z)$ converges for $|z - z_0| < R$, for then we may differentiate it term by term (Theorem 4.12) to obtain $F' = f$.

The power series $\Sigma a_n(z - z_0)^n$ converges absolutely for $|z - z_0| < R$ (Lemma 3.7) and

$$|a_n(z - z_0)^{n+1}/(n+1)|/|a_n(z - z_0)^n| = |z - z_0|/(n+1)$$

tends to zero as $n \to \infty$ so, by the comparison test, $\Sigma a_n(z - z_0)^{n+1}/(n+1)$ converges.   □

If $f(z)=\sum a_n(z-z_0)^n$ has disc of convergence D, then for any contour γ in D from z_1 to z_2,

$$\int_\gamma f = \sum a_n(z_2-z_0)^{n+1}/(n+1) - \sum a_n(z_1-z_0)^{n+1}/(n+1).$$

In particular, for any contour γ from z_0 to z in D,

$$\int_\gamma f = \sum a_n(z-z_0)^{n+1}/(n+1).$$

6. The Estimation Lemma

Some results in mathematics are the handmaidens of the theory, of little intrinsic interest in themselves, perhaps even a little dull, yet quietly figuring in important decisions all the time. 'Always a bridesmaid, never a bride' here takes the form 'always a lemma, never a theorem'. One such result is the lemma we are about to prove. It is a simple idea, giving an upper bound for the size of an integral $|\int_\gamma f|$ in terms of an upper bound on $|f|$ and the length of γ, but it arises time and again in the theory, applying subtle pressure at critical points in the proofs of important theorems. Perhaps not a theorem of great stature, it is certainly worthy of a name. We call it

THE ESTIMATION LEMMA (Lemma 6.10). If $f:D\to\mathbb{C}$ is continuous, γ is a contour in D of length L and $|f(z)|\leqslant M$ for all z on γ, then

$$\left|\int_\gamma f\right|\leqslant ML.$$

Proof. It is poetic justice that we find it necessary to give two different versions, depending whether the reader has studied §§1–3 or not. Version A is the more natural, but it requires a knowledge of §§2–3. For readers who have taken the shortcut, version B is provided.

Version A. It is sufficient to prove the result for a smooth path $\gamma:[a, b]\to D$. Let P be the partition $a=t_0<t_1<\cdots<t_n=b$ with $t_{r-1}\leqslant s_r\leqslant t_r$. Then

$$|S(P, f, \gamma)| = \left|\sum_{r=1}^n f(\gamma(s_r))(\gamma(t_r)-\gamma(t_{r-1}))\right|$$

$$\leqslant \sum_{r=1}^n |f(\gamma(s_r))||\gamma(t_r)-\gamma(t_{r-1})|$$

$$\leqslant ML(\pi)$$

where π is the polygonal approximation to γ with vertices $\gamma(t_0), \gamma(t_1), \ldots,$

$\gamma(t_n)$. But $L(\pi) \leqslant L(\gamma)$, so

$$|S(P, f, \gamma)| \leqslant ML(\gamma) \tag{11}$$

for all partitions P.

Given any $\varepsilon > 0$, we can find a partition P_ε such that

$$\left| S(P_\varepsilon, f, \gamma) - \int_\gamma f \right| < \varepsilon$$

so

$$\left| \int_\gamma f \right| < |S(P_\varepsilon, f, \gamma)| + \varepsilon$$

and (11) gives

$$\left| \int_\gamma f \right| < ML(\gamma) + \varepsilon$$

for all positive ε. Hence

$$\left| \int_\gamma f \right| \leqslant ML(\gamma).$$

Version B. First we establish

$$\left| \int_a^b (u(t) + iv(t))\, dt \right| \leqslant \int_a^b |u(t) + iv(t)|\, dt \tag{12}$$

for continuous real functions u, v.

Let $\int_a^b u(t)\, dt = X$, $\int_a^b v(t)\, dt = Y$. Then

$$\int_a^b (u(t) + iv(t))\, dt = X + iY$$

and

$$X^2 + Y^2 = (X - iY)(X + iY) = \int_a^b (X - iY)(u(t) + iv(t))\, dt$$

$$= \int_a^b \left[Xu(t) + Yv(t) \right] dt + i \int_a^b \left[Xv(t) - Yu(t) \right] dt.$$

But $X^2 + Y^2 \in \mathbb{R}$, so

$$\int_a^b \left[Xv(t) - Yu(t) \right] dt = 0,$$

and

$$X^2 + Y^2 = \int_a^b \left[Xu(t) - Yv(t) \right] dt$$

where the integrand $Xu(t) - Yv(t)$ is the real part of $(X - iY)(u(t) + iv(t))$. So

$$Xu(t) - Yv(t) \leqslant |(X - iY)(u(t) + iv(t))|$$
$$= \sqrt{(X^2 + Y^2)}|u(t) + iv(t)|.$$

From real analysis,

$$\int_a^b [Xu(t) - Yv(t)]\,dt \leqslant \int_a^b \sqrt{(X^2 + Y^2)}|u(t) + iv(t)|\,dt$$

which gives

$$X^2 + Y^2 \leqslant \sqrt{(X^2 + Y^2)} \int_a^b |u(t) + iv(t)|\,dt$$

and this implies

$$\sqrt{(X^2 + Y^2)} \leqslant \int_a^b |u(t) + iv(t)|\,dt. \tag{13}$$

But

$$\sqrt{(X^2 + Y^2)} = |X + iY| = \left| \int_a^b u(t) + iv(t)\,dt \right|$$

and, with (13), this gives (12), as required.

We then have

$$\left| \int_\gamma f \right| = \left| \int_a^b f(\gamma(t))\gamma'(t)\,dt \right|$$

$$\int_a^b |f(\gamma(t))||\gamma'(t)|\,dt \quad \text{by (12)}$$

$$\leqslant \int_a^b M|\gamma'(t)|\,dt \quad \text{from real analysis}$$

$$= ML,$$

completing the proof. $\qquad\qquad\qquad\qquad\qquad\qquad\qquad\qquad\quad$ □

This result will be used in a variety of ways. For instance, suppose that γ is a fixed contour and f varies. If the maximum value of $|f|$ on γ tends to zero, then $|\int_\gamma f| \leqslant ML$ tends to zero. When f tends to zero we shall find $\int_\gamma f \to 0$. More generally, if $f \to f_0$, then $f - f_0 \to 0$, so $\int_\gamma (f - f_0) \to 0$ and $\int_\gamma f \to \int_\gamma f_0$.

On the other hand, if $|f|$ remains bounded and the length of γ tends to zero, then $\int_\gamma f \to 0$. In particular, if f is continuous at z_0, then $|f|$ is bounded in a neighbourhood of z_0. So, if the contour γ shrinks down to z_0 with the length of γ tending to zero, then $\int_\gamma f \to 0$.

7. Consequences of the Fundamental Theorem

If f is continuous and has an antiderivative F in a domain D, then we saw in §5 that, for any contour γ in D from z_0 to z_1,

$$\int_\gamma f = F(z_1) - F(z_0).$$

This has interesting consequences. For instance, if γ is a closed contour, then $z_0 = z_1$ and $\int_\gamma f = 0$. On the other hand, for general z_0 and z_1 in D, it does not matter which contour we use to go from z_0 to z_1; the integral will always be the same. These properties also characterize the existence of an antiderivative:

THEOREM 6.11. Let f be a continuous complex function defined on a domain D. Then the following conditions are equivalent:
  (i) f has an antiderivative F in D,
 (ii) $\int_\gamma f = 0$ for every closed contour γ in D,
(iii) $\int_\gamma f$ depends only on the endpoints of γ for any contour γ in D.

Proof. We have already established that (i) implies (ii) and (iii). To show (ii) implies (iii), suppose that γ_1, γ_2 are any two contours in D from z_0 to z_1. (Fig. 6.7)

Fig. 6.7

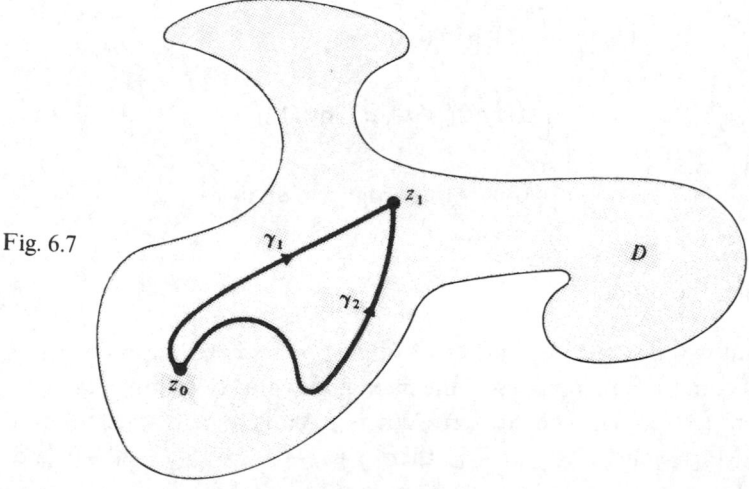

Then $\gamma_1 - \gamma_2$ is a closed contour, hence (ii) gives $\int_{\gamma_1 - \gamma_2} f = 0$. But

$$\int_{\gamma_1 - \gamma_2} f = \int_{\gamma_1} f - \int_{\gamma_2} f$$

and so

$$\int_{\gamma_1} f = \int_{\gamma_2} f$$

which is (ii).

Finally, to show that (iii) implies (i), we fix z_0 in D and for any z in D we choose a contour γ in D from z_0 to z_1 and define

$$F(z_1) = \int_{\gamma} f.$$

Because D is open, for some $\varepsilon_1 > 0$, if $|h| < \varepsilon_1$ then the line segment $\lambda(t) = z_1 + ht \ (0 \leqslant t \leqslant 1)$ lies in D and

$$F(z_1 + h) = \int_{\gamma} f + \int_{\lambda} f.$$

Thus

$$\frac{F(z_1 + h) - F(z_1)}{h} = \frac{1}{h} \int_{\lambda} f.$$

Now we know that, for any constant c and contour γ from z_1 to z_2,

$$\int_{\gamma} c \, dz = c(z_2 - z_1),$$

so

$$\int_{\lambda} (f(z_1)/h) \, dz = f(z_1).$$

We therefore have

$$\frac{F(z_1 + h) - F(z_1)}{h} - f(z_1) = \int_{\lambda} \frac{f(z) - f(z_1)}{h} \, dz.$$

Now we use the Estimation Lemma. We know from the continuity of f that, given $\varepsilon > 0$, there exists $\delta > 0$ (which we may take to be less than ε_1) such that

$$|z - z_1| < \delta \quad \text{implies} \quad |f(z) - f(z_1)| < \varepsilon,$$

so that, when $|h| < \delta$, for any z on the line segment λ the integrand satisfies

$$|(f(z) - f(z_1))/h| < \varepsilon/|h|.$$

The length of λ is $|h|$, so, whenever $|h| < \delta$,

$$\left| \int_{\lambda} \frac{f(z) - f(z_1)}{h} \, dz \right| \leqslant \frac{\varepsilon}{|h|} |h|,$$

which implies

$$\left|\frac{F(z_1+h)-F(z_1)}{h} -f(z_1)\right|\leqslant\varepsilon \quad (|h|<\delta).$$

But ε is arbitrary, so

$$\lim_{h\to0}\frac{F(z_1+h)-F(z_1)}{h}=f(z_1)$$

and $F'(z_1)=f(z_1)$ for z_1 in D, as required.   □

EXAMPLE. $f(z)=|z|$ is not differentiable, so $\int_\gamma f$ will depend on the choice of contour γ between points. For instance, consider γ, σ from 0 to i given by

$$\gamma(t)=\mathrm{i}t \ (0\leqslant t\leqslant1)$$

and

$$\sigma=\sigma_1+\sigma_2$$

where

$$\sigma_1(t)=t \ (0\leqslant t\leqslant1),$$
$$\sigma_2(t)=\mathrm{e}^{\mathrm{i}t} \ (0\leqslant t\leqslant\tfrac{1}{2}\pi).$$

(Fig. 6.8). Then

$$\int_\gamma |z|\,\mathrm{d}z=\int_0^1 |\mathrm{i}t|\cdot\mathrm{i}\,\mathrm{d}t$$
$$=[\mathrm{i}t^2/2]_0^1$$
$$=\tfrac{1}{2}\mathrm{i},$$

Fig. 6.8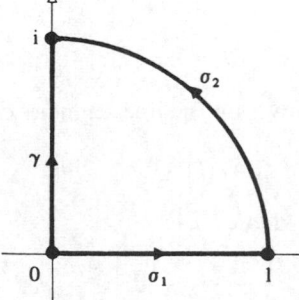

but

$$\int_\sigma |z|\,dz = \int_0^1 |t|\cdot 1\,dt + \int_0^{\pi/2} |e^{it}|\cdot(ie^{it})\,dt$$
$$= [t^2/2]_0^1 + [e^{it}]_0^{\pi/2}$$
$$= \tfrac{1}{2} + i - 1$$
$$= i - \tfrac{1}{2}.$$

While complex functions that are merely continuous may be integrated using the basic formula, they offer little of interest. From now on *differentiable* functions, and methods for integrating them will occupy centre stage. The resulting theory of integration has no natural real counterpart, because the real line lacks the flexible choice of path between points that occurs in the complex plane.

With minor exceptions, this chapter completes the natural analogies between the real and complex theories of differentiation and integration. From now on, new possibilities will unfold.

Exercises 6

1. Draw the contours $\gamma = [0,\,i]$, $\sigma = [0,\,1] + [1,\,i]$. Evaluate

$$\int_\gamma \mathrm{re}\,z\,dz, \quad \int_\sigma \mathrm{re}\,z\,dz.$$

2. Draw the contours $\gamma = [-i,\,i]$ and σ where

$$\sigma(t) = e^{it} \quad (-\pi/2 \leqslant t \leqslant \pi/2).$$

 Evaluate $\int_\gamma |z|\,dz$, $\int_\sigma |z|\,dz$.

3. Compute $\int_\gamma z^4\,dz$ for $\gamma(t) = (1+i)t$ $(0 \leqslant t \leqslant 1)$ using the formula

$$\int_\gamma f(z)\,dz = \int_0^1 f(\gamma(t))\gamma'(t)\,dt,$$

 multiplying out the right-hand side and integrating the real and imaginary parts. If $\gamma = [0,\,1] + [1,\,1+i]$, what is $\int_\gamma f(z)\,dz$?

4. Let n be a positive integer. If $\gamma(t) = z_0 + re^{it}$ $(0 \leqslant t \leqslant 2n\pi)$, describe the contours γ and $-\gamma$ geometrically. Compute

$$\int_\gamma 1/(z-z_0)\,dz \quad \text{and} \quad \int_{-\gamma} 1/(z-z_0)\,dz.$$

5. For $\gamma(t) = e^{it}$ $(0 \leqslant t \leqslant \pi)$, evaluate $\int_\gamma f$ for each of the following functions:
 (i) $1/z^2$ (ii) $1/z$ (iii) $\cos z$ (iv) $\sinh z$ (v) $\tan z$ (vi) $(\exp(z))^3$.

6. Given two smooth paths $\gamma:[a, b]\to\mathbb{C}$, $\sigma:[c, d]\to\mathbb{C}$, we say γ may be obtained from σ by a differentiable change in parameter if there exists a differentiable function $q:[a, b]\to[c, d]$ such that $\gamma=\sigma q$. For the purposes of this definition, q need not have an inverse. With this in mind, explain the geometric meaning of a differentiable change in parameter.

Show that if σ lies in the domain of a continuous function f then

$$\int_\gamma f=\int_\sigma f.$$

7. Verify that the length $L(\gamma)$ of a contour γ satisfies
   (i) $L(-\gamma)=L(\gamma)$ (ii) $L(\gamma+\sigma)=L(\gamma)+L(\sigma)$.

8. Let $\gamma(t)=z_0+re^{it}$ $(0\leqslant t\leqslant\theta)$ be the arc centre z_0, radius $r>0$ with parameter θ. Verify that

$$\theta=L(\gamma)/r,$$

   which is the usual definition of 'angle in radian measure'.

9. Show that the length of the parabolic arc $\gamma(t)=at^2+2ait$ $(0\leqslant t\leqslant1)$ is
$$a(\sqrt{2}+\log(1+\sqrt{2})).$$

10. Let $\sigma:[-2, 2]\to\mathbb{C}$ be any contour. Let $q:[-2, 2]\to[-2, 2]$ be given by $q(t)=t^3-3t$ and let $\gamma=\sigma q$ be the composite of q followed by σ. Show that $\int_\gamma f=\int_\sigma f$ for any continuous f with σ in its domain, but that $L(\gamma)=3L(\sigma)$. Explain the geometrical significance of this result.

11. If γ is obtained from σ by a differentiable change in parameter as defined in exercise 6, specify general conditions on q which imply $L(\gamma)=L(\sigma)$. Justify your answer.

12. If γ is a closed contour in $\mathbb{C}$, the *signed area* enclosed by γ is defined to be

$$S=\frac{1}{2i}\int_\gamma \bar{z}\,dz.$$

By writing out the integral explicitly in the form $\int_a^b (u-iv)(u'+iv')\,dt$, or otherwise, show that S is real. Show that its value for circular and triangular contours is $\pm$ the usual area and that both signs can occur. What is the geometric significance of the sign? What does the signed area represent for a contour such as Figure 6.9?

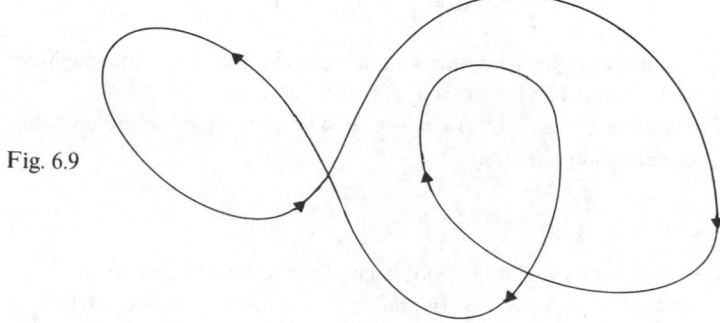

Fig. 6.9

13. Show that the signed area of exercise 12 may also be given by

$$\text{(i) } S = -\int_\gamma \operatorname{im} z \, dz \qquad \text{(ii) } S = \frac{1}{i}\int_\gamma \operatorname{re} z \, dz.$$

Write down the value of the integral $\int_\gamma \operatorname{im} z \, dz$ where γ is the square $\gamma = [0, 1] + [1, 1+i] + [1+i, i] + [i, 0]$.

14. *Integration by parts.* Suppose that f, g have continuous derivatives in the domain D and γ is a contour in D from z_1 to z_2. Prove that

$$\int_\gamma fg' = f(z_2)g(z_2) - f(z_1)g(z_1) - \int_\gamma f'g.$$

If $\gamma(t) = e^{it}$ $(0 \leqslant t \leqslant \pi/2)$, compute
   (i) $\int_\gamma z \sin z \, dz$ (ii) $\int_\gamma z \cos z \, dz$
   (iii) $\int_\gamma z \, e^{iz} \, dz$ (iv) $\int_\gamma z^2 \sin z \, dz.$

15. Let $C_r(t) = z_0 + re^{it}$ $(0 \leqslant t \leqslant 2\pi)$ be the circle centre z_0, radius $r > 0$. If f is continuous in the domain D and $z_0 \in D$, use the Estimation Lemma to prove
   (i) $\lim_{r \to 0} \int_{C_r} f(z) \, dz = 0$,   (ii) $\lim_{r \to 0} \int_{C_r} f(z)/(z - z_0) \, dz = 2\pi i f(z_0)$.

16. For each of the following functions $f : D \to \mathbb{C}$, either state an antiderivative $F : D \to \mathbb{C}$ or specify why an antiderivative cannot exist.
   (i) $f(z) = z^2$, $D = \mathbb{C}$,   (ii) $f(z) = 1/z^2$, $D = \mathbb{C} \setminus \{0\}$,
   (iii) $f(z) = 1/z$, $D = \mathbb{C} \setminus \{0\}$,   (iv) $f(z) = z \sin z$, $D = \mathbb{C}$,
   (v) $f(z) = |z|^2$, $D = \mathbb{C}$,   (vi) $f(z) = \bar{z}$, $D = \mathbb{C}$.

7

Angles, logarithms, and the winding number

If we attempt to define the logarithm of a complex number as some kind of 'inverse' to the exponential function, we have to face the fact that the latter is not a bijection, hence does not have an inverse in the technical sense. Unlike the real case, there is no very natural way to restrict its domain and codomain in such a way that it becomes a bijection – although a variety of more or less artificial such choices exist (such as the 'cut plane' $\mathbb{C}_\pi$ below) and are indeed useful.

In classical terms, the logarithm has to be 'multivalued'. The *way* in which its multiplicity of values fit together is closely analogous to the way that the measurement of an angle by radians gives not a single real number, but an infinite list differing only by multiples of 2π.

We shall discuss these ideas below, and apply them to a topological invariant known as the *winding number* of a curve relative to a point. In essence the total angle traversed by a point on the curve is measured as it moves continuously from one end to the other: if divided by 2π this gives the number of times that the curve winds around the point in question. This concept is extremely useful in the deeper parts of the subsequent theory.

1. Radian measure of angles

We first relate the 'power series' definition of sine to the usual geometric one, with the angle being measured in radians.

Let $0 \neq z \in \mathbb{C}$, and write z in polar coordinates, $z = re^{i\theta}$ where $r > 0$. Since $e^{i\theta} = \cos\theta + i\sin\theta$, the table in §5.5 implies that for fixed r, as θ increases from 0 to $\pi/2$, the real part of z decreases from r to 0 while the imaginary part increases from 0 to r. Since $r^2\cos^2\theta + r^2\sin^2\theta = r^2$, it follows that z traces out the first quadrant of a circle of radius r as θ moves from 0 to $\pi/2$. (Fig. 7.1)

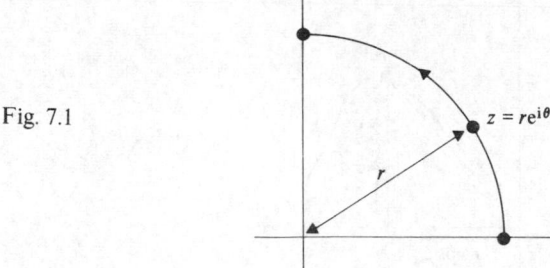

Fig. 7.1

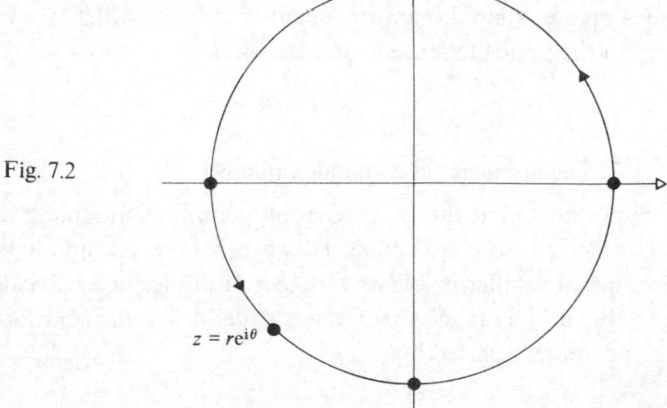

Fig. 7.2

Similarly, as θ moves from $\pi/2$ to π, the point z traces out the second quadrant of the circle; from π to $3\pi/2$ it traces out the third quadrant; from $3\pi/2$ to 2π the fourth: thereafter it continues to go round the circle. (Fig. 7.2)

We now compute the arc length from 1 to z along the circle. For $\theta \geqslant 0$, let $\gamma(t) = re^{it}$ $(0 \leqslant t \leqslant \theta)$. Then γ is a contour running along the relevant arc. The length of γ is given by

$$L(\gamma) = \int_0^\theta |\gamma'(t)| \, dt = \int_0^\theta |rie^{it}| \, dt = \int_0^\theta r \, dt = r\theta.$$

Thus $\theta = L(\gamma)/r$ which is the standard definition of 'angle measured in radians'. Figure 7.3 thus shows that the geometric definitions of $\sin\theta$ and

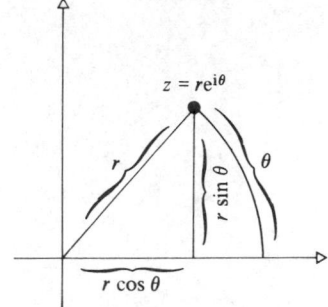

Fig. 7.3

$\cos \theta$ in terms of an angle of θ radians, agrees with our analytic definition via power series.

The established periodicity of sin and cos shows that this agreement extends to angles greater than 2π, and to negative angles (which are of course measured in the opposite sense round the circle).

2. The argument of a complex number

We now look in more detail at the expression of a complex number z in the form $re^{i\theta}$. Here $r = |z|$, so r is unique. However, there are infinitely many possible values of θ, differing only by integer multiples of 2π. Recall that the unique value in the range $-\pi < \theta \leqslant \pi$ is called the *principal value of the argument of z* and is denoted by

$$\arg z.$$

This defines a function

$$\arg : \mathbb{C} \setminus \{0\} \to \mathbb{R}.$$

This function is *not* continuous on the negative real axis. This is a result of the need to choose θ uniquely: just above the axis θ is close to π, just below it is close to $-\pi$.

We can sidestep this problem (inelegantly) by defining the *cut plane* (Fig. 7.4)

$$\mathbb{C}_\pi = \mathbb{C} \setminus \{x + iy \in \mathbb{C} \,|\, y = 0, \, x \leqslant 0\}.$$

Then arg is continuous in the cut plane. This is plausible geometrically, but we must provide a rigorous proof. There is one technical difficulty: the bad behaviour of inverse trigonometric functions. We circumvent it by using several overlapping domains, on each of which the behaviour is sufficiently good. The proof that follows is an inelegant 'bare hands'

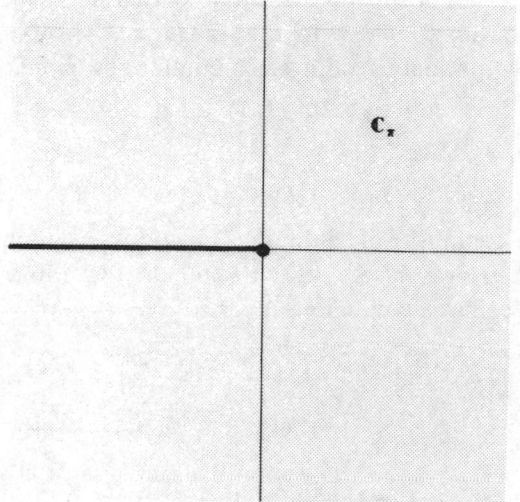

Fig. 7.4

reduction to properties of real functions: for a more elegant approach see §8.4.

Let

$$D_1 = \{x+iy \in \mathbb{C} \mid y > 0\}$$
$$D_2 = \{x+iy \in \mathbb{C} \mid x > 0\}$$
$$D_3 = \{x+iy \in \mathbb{C} \mid y < 0\}.$$

Then

$$\mathbb{C}_\pi = D_1 \cup D_2 \cup D_3$$

(draw a picture!), and in each of these three domains we have an easy way to find the desired value of arg z.

If

$$z = x+iy = re^{i\theta}$$

and we wish to solve for r, θ, we find that

$$r^2 = x^2 + y^2, \quad \text{so } r = \sqrt{(x^2+y^2)},$$

and

$$\cos\theta = \frac{x}{\sqrt{x^2+y^2}}$$

$$\sin\theta = \frac{y}{\sqrt{x^2+v^2}}$$

From the properties of sin and cos it follows that in D_1 there is a unique solution for θ with $0 < \theta < \pi$. In this range, cos is monotonic strictly decreasing and continuous; so it has a continuous inverse function

$$\cos^{-1} : (-1, 1) \to (0, \pi).$$

Now for $z \in D_1$ we have

$$\arg z = \cos^{-1} \operatorname{re}(z)/|z|)$$

which, being a composite of continuous functions, is continuous.

Similarly in D_2 we have $-\pi/2 < \theta < \pi/2$. In this range sin increases from -1 to 1, so has a continuous inverse

$$\sin^{-1} : (-1, 1) \to \left(-\frac{\pi}{2}, \frac{\pi}{2} \right).$$

Then

$$\arg z = \sin^{-1} (\operatorname{im}(z)/|z|)$$

is continuous on D_2.

Finally on D_3 we can use

$$\cos^{-1} : (-1, 1) \to (-\pi, 0)$$

(a *different* choice of an inverse cosine from that used in D_1, because we are inverting the restriction of cos to a different interval) and again arg is continuous.

Since arg is continuous on each of D_1, D_2, D_3 it is continuous on their union, which is $\mathbb{C}_\pi$.

It is sometimes convenient to proceed more generally. Let $\alpha \in R$, and let R_α be the ray

$$R_\alpha = \{ re^{i\alpha} \in \mathbb{C} \,|\, r \geqslant 0 \}.$$

Let

$$\mathbb{C}_\alpha = \mathbb{C} \setminus R_\alpha \quad \text{(Fig. 7.5)}$$

and choose

$$\theta = \arg_\alpha z \quad (z \in \mathbb{C})$$

by the rule

$$z = re^{i\theta}, \quad r > 0, \quad \alpha - 2\pi < \theta < \alpha.$$

Then, by a similar method, it may be shown that $\arg_\alpha$ is continuous in $\mathbb{C}_\alpha$.

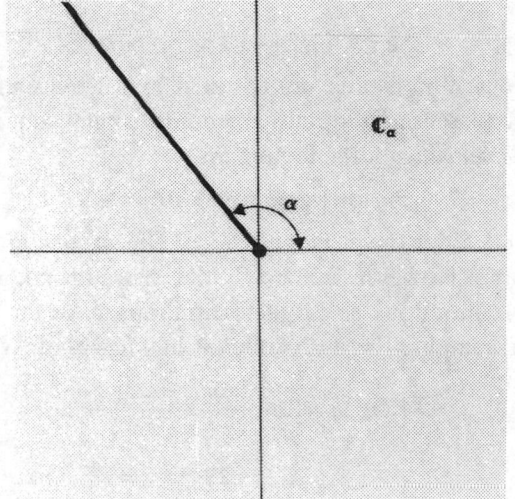

Fig. 7.5

3. The complex logarithm

This poses similar problems, because the exponential function is not one–one.

We wish to define $\log z$, for $0 \neq z \in \mathbb{C}$, by

$$w = \log z \quad \text{if } z = e^w.$$

Let $z = re^{i\theta}$, $w = u + iv$. (The mixture of polar and Cartesian coordinates is deliberate!) We assume $r > 0$, $-\pi < \theta \leqslant \pi$ so that $\theta = \arg z$. Then $z = e^w$ becomes

$$re^{i\theta} = e^{u+iv}. \tag{1}$$

Taking moduli,

$$r = e^u. \tag{2}$$

Since $r > 0$ and $r, u \in \mathbb{R}$ this has the unique solution

$$u = \log r$$

where log is the real natural logarithm. Then (1) and (2) imply that

$$e^{i\theta} = e^{iv}$$

so that

$$v = \theta + 2n\pi \quad (n \in \mathbb{Z}).$$

Hence

$$\log z = w = \log r + i(\theta + 2n\pi),$$

or

$$\log z = \log |z| + i(\arg z + 2n\pi). \tag{3}$$

The complex logarithm is thus 'multivalued' and therefore not a function in the set-theoretic sense. To obtain a genuine (single-valued) function we define the *principal value* of the logarithm by

$$\text{Log } z = \log |z| + i \arg z \quad (0 \neq z \in \mathbb{C}).$$

(Note the capital 'L' for the principal value.) This is *not* continuous on the negative real axis. However, for $z \in \mathbb{C}_\pi$ the real and imaginary parts of Log are clearly continuous, so Log is continuous in the cut plane.

Now we can compute the derivative of the logarithm. We have

$$\text{D Log } z_0 = \lim_{z \to z_0} \frac{\text{Log } z - \text{Log } z_0}{z - z_0}$$

$$= \lim_{w \to w_0} \frac{w - w_0}{e^w - e^{w_0}}$$

by setting $z = e^w$, $z_0 = e^{w_0}$ and using the continuity of exp,

$$= 1/e^{w_0}$$

$$= 1/z_0.$$

So in general

$$\text{D Log } z = 1/z \quad (z \in \mathbb{C}_\pi).$$

In the same way, in the cut plane $\mathbb{C}_\alpha$, we can define

$$\log_\alpha z = \text{Log } |z| + i \arg_\alpha z$$

and this is continuous, with derivative

$$\text{D} \log_\alpha z = 1/z.$$

Once the logarithm is defined, powers z^a where $z, a \in \mathbb{C}$ can also be defined, in a cut plane. The *principal value* of z^a, when $z \neq 0$, is

$$z^a = \exp(a \text{ Log } z).$$

Exercises 9, 10, 11, 12, 13, 14 below explore this function. For a more global view of z^a see Chapter 14.

4. The winding number

Suppose that $\gamma:[a, b] \to \mathbb{C} \setminus \{0\}$ is a closed path. Note that the choice of codomain here implies that the path does not pass through the origin. If we imagine t, the parameter, to be time increasing from a to b, and choose the argument $\theta(t)$ of $\gamma(t)$ to vary continuously with t, then as $\gamma(t)$

moves round the path, the total change in argument will be the number of times that γ winds around the origin, multiplied by 2π. We wish, of course, to make this idea precise.

To do so, define a *continuous choice of argument* along a path $\gamma:[a, b]\to$ $\mathbb{C}\setminus\{0\}$ (which for full generality, we no longer require to be a closed path) to be a *continuous* map $\theta:[a, b]\to\mathbb{R}$ such that

$$e^{i\theta(t)} = \gamma(t)/|\gamma(t)| \tag{4}$$

for all $t \in [a, b]$. The condition (4) says merely that $\theta(t)$ is *one* of the possible values for the argument of $\gamma(t)$.

EXAMPLE 1. Let $\gamma(t)=re^{it}$ $(0\leqslant t\leqslant\pi)$. (Fig. 7.6)

A continuous choice of argument is given by $\theta(t)=t$, because then $re^{it}/|re^{it}|=e^{it}=e^{i\theta(t)}$. This is not the only possible choice: $\theta(t)=t+2\pi$ would work equally well; or $\theta(t)=t+2n\pi$ for a fixed integer n.

What we *cannot* do is 'change horses in midstream', say by letting $\theta(t)$ be t for $0\leqslant t\leqslant\pi/2$, but $\theta(t)=t+2\pi$ for $\pi/2<t\leqslant\pi$. This θ makes condition (4) hold, but θ is of course not continuous.

The difficulty (though it does not take much getting used to) in handling continuous choices of argument is that it does *not* in general suffice to insist on some simple recipe for choosing a value of the argument, such as using only angles between 0 and 2π. Such a choice becomes *discontinuous* if we are dealing with a curve that has wound nearly once clockwise, so that the argument gets near to 2π, and then carries on clockwise across the real axis, forcing a jump in argument to something near 0. In such a case we would, to retain continuity, need to let the argument take a value a little larger than 2π.

What happens is that we can choose the argument at the starting point a to be whatever we wish from among the infinity of possible values; but

Fig. 7.6

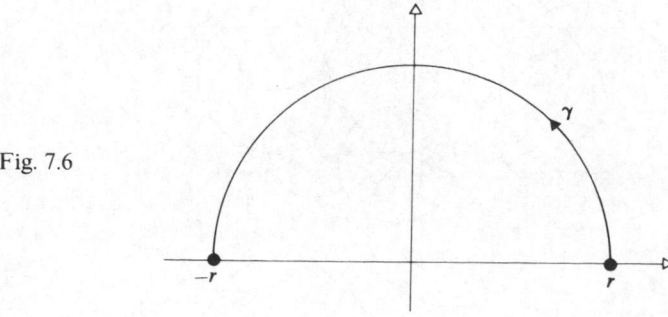

this choice determines all the rest uniquely – and they need not stay within any predetermined range. This is intuitively obvious, but its proof is a little tricky: it uses the Paving Lemma.

THEOREM 7.1. Let $\gamma:[a, b] \to \mathbb{C} \setminus \{0\}$ be a path not passing through the origin. Then there exists a continuous choice of argument for γ. Any other continuous choice of argument differs from this by a constant integer multiple of 2π.

Proof. By the Paving Lemma (Lemma 2.9) we can subdivide γ into finitely many subpaths $\gamma_r(r=1,\dots,n)$ such that each γ_r lies inside a disc D_r in $\mathbb{C} \setminus \{0\}$. If the centre of D_r is at $\rho_r e^{i\theta_r}$, then taking $\alpha_r = \theta_r + \pi$, we find that $D_r \subseteq C_{\alpha_r}$, so $\arg_{\alpha_r}$ is continuous in D_r. (Fig. 7.7)

This gives a continuous choice of argument across γ_r, for each r; but of course these choices may not fit together continuously. However, *any* choice of argument can be obtained from any other by adding a suitable integer multiple of 2π. We can adjust these $\arg_{\alpha_r}$s by adding suitable multiples $2n_r\pi$, as follows.

Let γ_r be defined on the parametric interval $[t_{r-1}, t_r]$. There is an integer n_2 such that

$$\arg_{\alpha_1}(t_1) = \arg_{\alpha_2}(t_1) + 2n_2\pi.$$

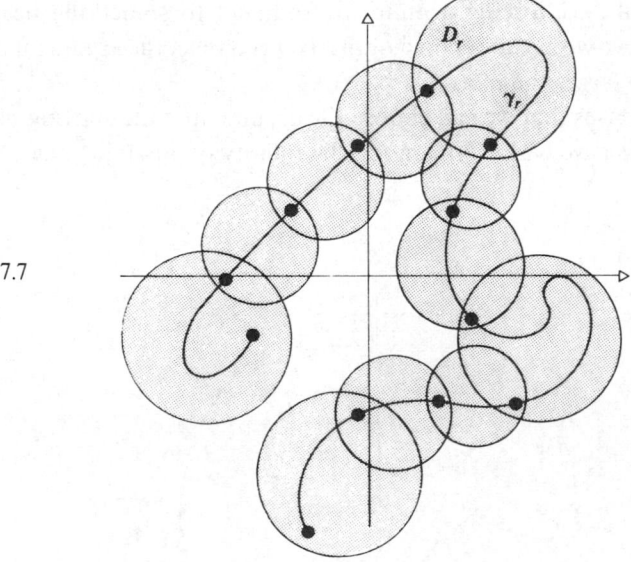

Fig. 7.7

Then there is an integer n_3 such that

$$\arg_{\alpha_2}(t_2) + 2n_2\pi = \arg_{\alpha_3}(t_2) + 2n_3\pi,$$

and so on inductively, choosing n_{r+1} so that

$$\arg_{\alpha_r}(t_r) + 2n_r\pi = \arg_{\alpha_{r+1}}(t_r) + 2n_{r+1}\pi.$$

Then we define

$$\theta(t) = \arg_{\alpha_r}(t) + 2n_r\pi \quad \text{if } t \in [t_{r-1}, t_r]$$

(with n_1 conventionally defined to be 0), and θ is continuous. This proves that a continuous choice of argument exists.

Next, suppose θ^* is another continuous choice of argument. Then we must have

$$\theta^*(t) = \theta(t) + 2n(t)\pi$$

for $n(t)$ an integer, possibly depending on t. But then $n(t)$ is a *continuous* function of t (since it equals $(\theta^*(t) - \theta(t))/2\pi$) taking only integer values; so $n(t)$ is constant. This completes the proof. $\qquad\square$

Note that a continuous choice of argument is defined on the parametric interval $[a, b]$, not on the image of γ. This means that, if the curve returns to the same point $\gamma(t_1) = \gamma(t_2)$, with $t_1 \neq t_2$, the arguments $\theta(t_1)$ and $\theta(t_2)$ may be different. (Intuitively it is clear that this will happen if the path winds round the origin between t_1 and t_2.)

EXAMPLE 2. $\gamma(t) = re^{4\pi i t}$ $(0 \leqslant t \leqslant 1)$.
A continuous choice of argument is given by $\theta(t) = 4\pi t + 2n\pi$, for an integer n, with $t \in [0, 1]$. Although $\gamma(t) = \gamma(t + \frac{1}{2})$ for $t \in [0, \frac{1}{2}]$, the choices of argument $\theta(t)$ and $\theta(t + \frac{1}{2})$ differ by 2π, because the path has travelled once round the origin (in an anticlockwise direction) between t and $t + \frac{1}{2}$. More significantly, $\gamma(0) = \gamma(1)$ but the difference in arguments is 4π, because the path has travelled *twice* round the origin in total.

We can take advantage of this idea by defining the *winding number* $w(\gamma, 0)$ of a path $\gamma : [a, b] \to \mathbb{C} \setminus \{0\}$ round the origin, to be

$$[\theta(b) - \theta(a)]/2\pi$$

where θ is a continuous choice of argument along γ.

By the second part of Theorem 7.1 the winding number is well defined. For arbitrary paths γ, $w(\gamma, 0)$ is a real number; for *closed* paths γ it is an *integer*, since $\theta(b) - \theta(a)$ is an integer multiple of 2π.

The winding number takes the *sense* of the path into account, in that anticlockwise turns are counted positive, clockwise turns negative.

EXAMPLE 3. $\gamma(t) = e^{-it}$ $(0 \leqslant t \leqslant 6\pi)$.

A continuous choice of argument is $\theta(t) = -t$. Then

$$w(\gamma, 0) = [\theta(6\pi) - \theta(0)]/2\pi = -3,$$

and γ winds round the origin three times in the clockwise direction.

The winding number is *additive* in the following sense:

THEOREM 7.2. Let γ_1 and γ_2 be two paths in $\mathbb{C} \setminus \{0\}$ such that the end point of γ_1 is the start of γ_2. Then

$$w(\gamma_1 + \gamma_2, 0) = w(\gamma_1, 0) + w(\gamma_2, 0).$$

Proof. We may assume that γ_1, γ_2, and $\gamma_1 + \gamma_2$ have the parametric intervals $[a, b]$, $[b, c]$, and $[a, c]$ respectively. Let θ be a continuous choice of argument on $\gamma_1 + \gamma_2$. Then

$$w(\gamma_1 + \gamma_2, 0) = [\theta(c) - \theta(a)]/2\pi,$$
$$w(\gamma_1, 0) = [\theta(b) - \theta(a)]/2\pi,$$
$$w(\gamma_2, 0) = [\theta(c) - \theta(b)]/2\pi.$$

The result follows. □

This is an extremely useful result, because it lets us compute the winding number of a complicated path by breaking it up into nicer pieces and adding the contributions. It extends easily to show that

$$w(\gamma_1 + \cdots + \gamma_r, 0) = w(\gamma_1, 0) + \cdots + w(\gamma_r, 0)$$

and that

$$w(-\gamma, 0) = -w(\gamma, 0).$$

5. The winding number as an integral

First let γ be a *closed* contour. Then the winding number $w(\gamma, 0)$ is given by the integral

$$w(\gamma, 0) = \frac{1}{2\pi i} \int_\gamma \frac{1}{z}\, dz. \tag{5}$$

To see this, subdivide γ into subpaths $\gamma_1, \ldots, \gamma_n$ as in Theorem 7.1, whose notation we now use. Each γ_r lies in a cut-plane $\mathbb{C}_{\alpha_r}$. If γ_r is defined on the interval $[t_{r-1}, t_r]$ then

$$\int_{\gamma_r} \frac{1}{z} dz = \log_{\alpha_r} \gamma(t_r) - \log_{\alpha_r} \gamma(t_{r-1})$$

$$= \log_{\alpha_r} |\gamma(t_r)| - \log_{\alpha_{r-1}} |\gamma(t_{r-1})|$$
$$+ i(\arg_{\alpha_r} \gamma(t_r) - \arg_{\alpha_{r-1}} \gamma(t_{r-1})).$$

As in Theorem 7.1 we make sure that $\arg_{\alpha_r}(t_r) = \arg_{\alpha_{r+1}}(t_r)$, that is, the choices of argument agree where the subpaths join together. Then summing the integrals for $r = 1, \ldots, n$, we find that the real parts cancel out (because γ is closed) and the imaginary parts add up to $2\pi w(\gamma, 0)$, as required.

If γ is not closed, a similar formula holds:

$$w(\gamma, 0) = \frac{1}{2\pi} \operatorname{im} \left[\int_\gamma \frac{1}{z} dz \right].$$

The proof is the same, except that the real parts do not cancel: they are removed by taking only the imaginary part of the integral.

6. The winding number round an arbitrary point

There is nothing very special about the origin. If $\gamma : [a, b] \to \mathbb{C}$ is a path, and if $z_0 \in \mathbb{C}$ does not lie on γ, we can define the winding number of γ around z_0. The easiest way to do this is to translate the origin, setting

$$\Gamma(t) = \gamma(t) - z_0 \quad (t \in [a, b]),$$

and defining

$$w(\gamma, z_0) = w(\Gamma, 0).$$

Exactly as above, if γ is closed then

$$w(\gamma, z_0) = \frac{1}{2\pi i} \int_\gamma \frac{1}{z - z_0} dz, \tag{6}$$

a formula proved most readily by substituting $z = \gamma(t)$ and using (5), as follows:

$$\frac{1}{2\pi i} \int_\gamma \frac{1}{z - z_0} dz = \frac{1}{2\pi i} \int_a^b \frac{\gamma'(t)}{\gamma(t) - z_0} dt$$

$$= \frac{1}{2\pi i} \int_a^b \frac{\Gamma'(t)}{\Gamma(t)} dt$$

$$= \frac{1}{2\pi i} \int_\pi \frac{1}{z} dz$$

$$= w(\Gamma, 0)$$

$$= w(\gamma, z_0).$$

7. Components of the complement of a path

We consider how the winding number $w(\gamma, z_0)$ of a given closed path γ can vary as z_0 varies.

By Proposition 2.11, the complement S of the image of γ is open, and also each connected component of S is open. For any $z_0 \in S$ we can define $w(\gamma, z_0)$, thereby obtaining an *integer*-valued function on S. It is geometrically obvious that this function is constant on each connected component of S: we prove this analytically by showing that $w(\gamma, z_0)$ is a *continuous* function of z_0. The desired result will then follow, since an integer-valued continuous function is constant on any connected set.

The proof that $w(\gamma, z_0)$ is continuous in z_0 is obtained by a direct estimate. Fix $z_0 \in S$. Since S is open there exists $k > 0$ such that $|z_1 - z_0| < k$ implies $z_1 \in S$. It follows from this that if z is on the image of γ then $|z - z_0| \geqslant k$; hence if $|z_1 - z_0| < k/2$, we have $|z - z_1| > k/2$. Now

$$|w(\gamma, z_0) - w(\gamma, z_1)| = \left| \int_\gamma \left[\frac{1}{z - z_0} - \frac{1}{z - z_1} \right] dz \right|$$

$$= \left| \int_\gamma \frac{z_1 - z_0}{(z - z_0)(z - z_1)} \, dz \right|$$

$$\leqslant \frac{|z_1 - z_0|}{\frac{1}{2} k^2} L(\gamma)$$

by the Estimation Lemma (6.10).

Given any $\varepsilon > 0$, take $\delta = \min (k/2, k^2 \varepsilon / 2L(\gamma))$. Then $|z_1 - z_0| < \delta$ implies $|w(\gamma, z_0) - w(\gamma, z_1)| < \varepsilon$. Hence $w(\gamma, z_0)$ is continuous in z_0.

For example, the path in Figure 7.8 has the winding numbers shown, around points z_0 in the components to which those numbers are assigned.

Fig. 7.8

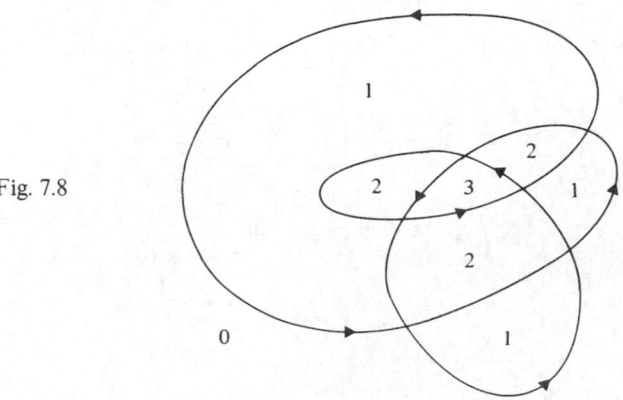

The complement S has only one unbounded component (Proposition 2.11), which we denote by $U(\gamma)$.

As Figure 7.8 shows, if $z_0 \in U(\gamma)$ then $w(\gamma, z_0)$ should be zero. This is easily proved from the integral formula (6), as follows. Let z_0 be 'far from the track of γ', that is, assume $|z - z_0| \geqslant K$ for all z on the track of γ. Then the Estimation Lemma (6.10) shows that

$$w(\gamma, z_0) \leqslant L(\gamma)/2\pi K$$

which tends to zero for large K. But the left-hand side is an integer; so it is *equal* to zero for large K.

8. Computing the winding number by eye

The somewhat complicated definition of the winding number may give the impression that it is complicated to calculate. This is not so, at least for the paths usually encountered. It is usually obvious on geometric grounds what the winding number is (run your finger round the path and count turns). The point of this section is that this process can easily be turned into a rigorous proof (and hence that in practice the proof is unnecessary: that which is 'obvious' is in this case also true!).

Consider, as a simple example, the case where the track of γ is a rectangle with vertices $\pm 2 \pm i$. (Fig. 7.9) If you want a formula, it's not hard to give one: for example let $t \in [0, 4]$ and set

$$\begin{aligned} \gamma(t) &= 2 - i + 2it & (0 \leqslant t \leqslant 1) \\ \gamma(t) &= 2 + i - 4(t - 1) & (1 \leqslant t \leqslant 2) \\ \gamma(t) &= -2 + i - 2i(t - 2) & (2 \leqslant t \leqslant 3) \\ \gamma(t) &= -2 - i + 4(t - 3) & (3 \leqslant t \leqslant 4). \end{aligned}$$

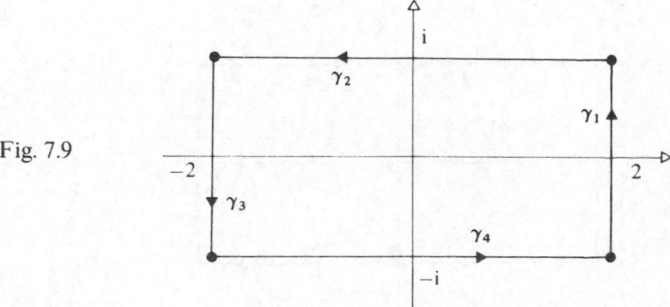

Fig. 7.9

First, here is how *not* to do the calculation.

Break γ up in the most natural way, into the subpaths $\gamma_1, \ldots, \gamma_4$ defined by $\gamma_r = \gamma|_{[r-1, r]}$. Then use the integral formula:

$$w(\gamma, 0) = \frac{1}{2\pi i} \int_\gamma \frac{1}{z} \, dz = \frac{1}{2\pi i} \sum_{r=1}^{4} \int_{\gamma_r} \frac{1}{z} \, dz.$$

Now (to take just one subpath)

$$\int_{\gamma_1} \frac{1}{z} \, dz = \int_0^1 \frac{\gamma'(t)}{\gamma(t)} \, dt = \int_0^1 \frac{2i}{2 - i + 2it} \, dt$$

$$= [\log (2 - i + 2it)]_0^1 = \mathrm{Log} \, (2 + i) - \mathrm{Log} \, (2 - i)$$

since γ_1 lies in $\mathbb{C}_\pi$, so the principal value Log is continuous on γ_1. Then

$$\mathrm{Log} \, (2 \pm i) = \log |2 \pm i| + i \arg (2 \pm i) \qquad (7)$$

$$= \log \sqrt{5} \pm i \sin^{-1} (1/\sqrt{5})$$

so

$$\int_{\gamma_1} \frac{1}{z} \, dz = 2i \sin^{-1} (1/\sqrt{5})$$

where the inverse sine is chosen between $-\pi/2$ and $\pi/2$.

You now have three similar integrals to evaluate. Add them, divide by $2\pi i \ldots$ and do some *very* careful book-keeping on the domains of the inverse trigonometric functions that occur. It *can* be done; in a sense it is not even difficult; but it is hardly to be recommended!

It is a little better (but not much) to work from the 'continuous choice of argument' definition: this starts you at stage (7) above for each subpath, with the same book-keeping problems at the end.

Here is a more civilized method. Divide γ into subpaths δ_1 and δ_2 as shown. (Fig. 7.10)

Now $w(\gamma, 0) = w(\delta_1, 0) + w(\delta_2, 0)$. Since δ_1 lies entirely within the cut plane $\mathbb{C}_\pi$, the principal value arg is continuous on δ_1. Hence, dotting all *is*

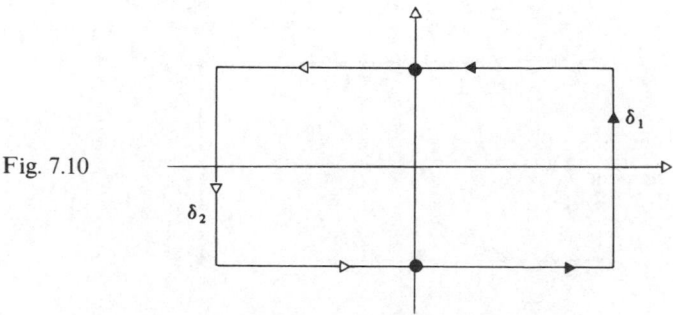

Fig. 7.10

and crossing all ts,

$$w(\delta_1, 0) = [\arg(i) - \arg(-i)]/2\pi$$
$$= [\pi/2 - (-\pi/2)]/2\pi$$
$$= 1/2.$$

Similarly δ_2 lies inside $\mathbb{C}_0$, so $\arg_0$ is continuous on δ_2; so

$$w(\delta_2, 0) = [\arg_0(-i) - \arg_0(i)]/2\pi$$
$$= [-\pi/2 - (-3\pi/2)]/2\pi$$
$$= 1/2.$$

Adding, we see that $w(\gamma, 0) = 1$.

Clearly this process can be telescoped: the arg calculations merely confirm, in a predictable way, the obvious.

We summarize this method as follows:

(1) Break γ into convenient pieces, *each lying in some cut plane.*

(2) For each piece, compute the contribution to the winding number as the difference between the arguments of the endpoints (using the relevant continuous arg) – or, geometrically, find the angle subtended at the origin by the subpath, with the appropriate sign.

(3) Add the contributions.

It helps if the dissection of γ into subpaths is performed by drawing a single line through the origin, because then the contributions are always 0 or $\pm 1/2$. Thus, for the path γ which the line L divides into four segments AB, BC, CD, DA, we have (Fig. 7.11)

$$w(\gamma, 0) = w(AB, 0) + w(BC, 0) + w(CD, 0) + w(DA, 0)$$
$$= \tfrac{1}{2} + \tfrac{1}{2} + \tfrac{1}{2} + \tfrac{1}{2}$$
$$= 2.$$

Similarly, for the path in Figure 7.12, we have

$$w(\gamma, 0) = w(AB, 0) + w(BC, 0) + w(CD, 0) + w(DE, 0) + w(EF, 0) + w(FA, 0)$$
$$= \tfrac{1}{2} + \tfrac{1}{2} + 0 + (-\tfrac{1}{2}) + 0 + \tfrac{1}{2}$$
$$= 1.$$

This method is essentially the mathematical equivalent of 'run your finger round the curve and count half-turns'. The ingredients needed to make it fully rigorous are *not* complicated evaluations of args: the real crunch comes in showing that the path really does cross the chosen line L at the points A, B, C, etc., that it passes through these points in the stated order, and that the sense (clockwise or anticlockwise) of each subpath is as in the diagram. But whenever (as in the sequel) γ is specified

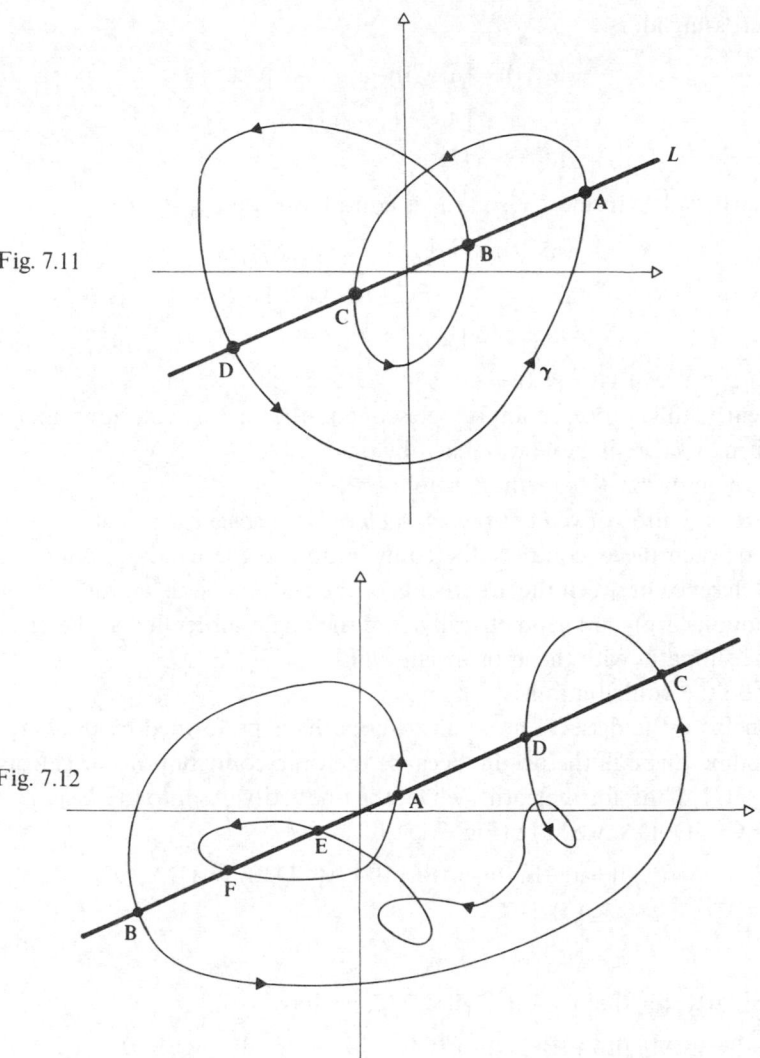

Fig. 7.11

Fig. 7.12

in a straightforward way (e.g. by a simple formula, or as a polygon, or a mixture of straight lines and circular arcs), these ingredients are easily supplied: we shall then make no further fuss when we assert that a certain winding number takes a certain value.

Comparison of the first 'bad' method shown with the final 'good' one gives a striking illustration of the dangers of blind 'formula-crunching' analysis. Complex analysis is a highly geometric subject, and the geometry should not be despised.

Exercises 7

1. Compute the principal values of the arguments of the following complex numbers: $1+i$, $(\sqrt{3/2})+i$, $(1+i)^3$, $((\sqrt{3/2})+i)^{243}$, $(1+i)^2((\sqrt{3/2}+i)^3$.

2. Let $\arg z$ denote the principal value of the argument of $z \neq 0$, i.e. $-\pi < \arg z \leqslant \pi$. For real x, y, where $x < 0$, show
   (i) $\lim_{y \to 0} \arg(x+i|y|) = \pi$ (ii) $\lim_{y \to 0} \arg(x-i|y|) = -\pi$.
   Compute the corresponding limits when $x = 0$ and when $x > 0$.

3. Compute the following principal logarithms:
   $\text{Log}(3i)$, $\text{Log}(-2i)$, $\text{Log}(1+i)$, $\text{Log}(-1)$, $\text{Log}(z^{10})$ where $z = 2e^{i\pi/3}$, $\text{Log}(x)$ for real $x \neq 0$.

4. For $z_1 \neq 0$, $z_2 \neq 0$, show that
$$\text{Log}(z_1 z_2) = \text{Log}(z_1) + \text{Log}(z_2) + 2n\pi i$$

   where n is an integer which need not be zero. Specify the values which n may take. Show that a logarithm of $z_1 z_2$ is of the form
$$\log(z_1) + \log(z_2)$$

   provided that appropriate values of the logarithms arc taken.

5. The 'Bernoulli Paradox' is as follows:
$$(-z)^2 = z^2,$$
   hence
$$2\log(-z) = 2\log(z)$$
   and so
$$\log(-z) = \log(z).$$

   What is the fallacy?

6. Let $f: \mathbb{C} \to \mathbb{C}$ be given by $f(z) = e^{i\theta} z$ for constant real θ. Show that f rotates the complex plane through an angle θ.
   Show that the transformation $f(z) = iz$ rotates the complex plane through a right angle and describe the transformations $f(z) = -z$, $f(z) = -iz$ as rotations.
   For any complex number $\lambda = re^{i\theta}$, describe the transformation $f(z) = \lambda z$ in geometrical terms.

7. In each of the following cases, for $z = re^{i\theta}$, draw z and $1/z$:
   (i) $3e^{i\pi/2}$ (ii) $2e^{i\pi/4}$ (iii) $\frac{1}{2}e^{i\pi/3}$ (iv) $3e^{-i\pi}$.
   Describe the transformation $f(z) = 1/z \ (z \neq 0)$ in geometrical terms.

8. Let n be a positive integer. A complex number ω is said to be an nth *root of unity* if $\omega^n = 1$.
   (i) Find all nth roots of unity in polar coordinates and draw a picture.
   (ii) For $n = 2, 3, 4$, express the nth roots of unity in the form $x + iy$.
   (iii) If ω_1, ω_2 are nth roots of unity, show that the following are also:
$$\omega_1^m, \quad \omega_1 \omega_2, \quad \omega_1/\omega_2.$$
   (iv) For given $r, \theta \in \mathbb{R}$, $r > 0$, find all $z \in \mathbb{C}$ such that
$$z^n = re^{i\theta}.$$
   (v) If $z_1^n = z_2^n$, show $z_1 = \omega z_2$ where ω is an nth root of unity.

9. For $z, \beta \in \mathbb{C}, z \neq 0$, define the principal value of z^β to be

$$z^\beta = \exp(\beta \operatorname{Log} z).$$

Compute the principal values of the following powers:

$$1^{\sqrt{2}}, (-2)^{\sqrt{2}}, i^i, 2^i, 1^{-i}, (3-4i)^{1+i}, (3+4i)^5.$$

10. Using the notation of question 9, for $z \in \mathbb{C}_\pi$, show $d(z^\beta)/dz = \beta z^{\beta-1}$. For fixed $a \in \mathbb{C}_\pi$, what is $d(a^z)/dz$?

11. Let $\alpha \in \mathbb{R}, \beta \in \mathbb{C}$. For $z \in \mathbb{C}_\alpha$, define

$$(z^\beta)_\alpha = \exp(\beta \log_\alpha z).$$

Compute all the possible values of $(i^i)_\alpha, (2^i)_\alpha, ((3-4i)^{1+i})_\alpha$ for various values of α. For fixed z, show that, as α varies, $(z^\beta)_\alpha$ takes on only a finite number of values when β is a rational number.

If $\beta = m/n$ where m, n are integers, $n > 0$, show that

$$((z^{m/n})_\alpha)^n = z^m.$$

What is $((z^n)^{m/n})_\alpha$?

12. Using the notation of question 11, for $z \in \mathbb{C}_\alpha$, compute $d((z^\beta)_\alpha)/dz$ and $d((a^z)_\alpha)/dz$. Describe a relation between $(z^{\beta+\gamma})_\alpha$, $(z^\beta)_\alpha$ and $(z^\gamma)_\alpha$.

13. Describe the image of the functions $f: \mathbb{C}_\pi \to \mathbb{C}$ geometrically where $f(z)$ is given by the principal values of the following:

(i) $z^{\frac{1}{2}}$ (ii) $z^{\frac{1}{3}}$ (iii) z^i.

14. Let $\alpha \in \mathbb{R}, \beta \in \mathbb{C}, z \in \mathbb{C}_\pi$. By writing $\beta = u + iv$, find

$$|(z^\beta)_\alpha|, \quad \arg((z^\beta)_\alpha).$$

Show that $|(z^\beta)_\alpha|$ is independent of α if and only if β is real. For positive integers m, n, let $f(z)$ be the principal value of $z^{m/n}$. Describe $f: \mathbb{C}_\pi \to \mathbb{C}$ geometrically.

15. Express $\operatorname{Log}(1+z)$ as a power series in z of the form

$$\operatorname{Log}(1+z) = \sum a_n z^n \quad |z| < R,$$

specifying the coefficients a_n and the radius of convergence R.

16. Show:

   (i) $\cos(-i \operatorname{Log}(z + \sqrt{(z^2-1)})) = z$,
   (ii) $\sin(-i \operatorname{Log}(i(z + \sqrt{(z^2-1)}))) = z$,
   (iii) $\tan((i/2) \operatorname{Log}((i+z)/(i-z))) = z$.

Use these properties to express $\cos^{-1} z, \sin^{-1} z, \tan^{-1} z$ in terms of the logarithm.

17. Show that all the values of $\cosh^{-1} z$ are given in the form

$$\cosh^{-1} z = \log(z + \sqrt{(z^2-1)})$$

where all possible values of the logarithm and square root are taken. In the same sense, show

$$\sinh^{-1} z = \log(z + \sqrt{(z^2+1)}),$$
$$\tanh^{-1} z = \tfrac{1}{2} \log((1+z)/(1-z)).$$

18. Let $D = \{z \in \mathbb{C} | (i+z)/(i-z) \in \mathbb{C}_\pi\}$. Describe D geometrically. Define the principal value of the inverse tangent $\tan^{-1}: D \to \mathbb{C}$ to be

$$\tan^{-1}(z) = \tfrac{1}{2}i \, \text{Log}\,((i+z)/(i-z))$$

(taking the principal value of Log). Show that $\tan^{-1}$ is differentiable in D with derivative $1/(1+z^2)$.

19. Draw the following paths and specify all the continuous choices of argument along them:
    (i) $\gamma(t) = 2e^{-it}$ $(0 \leqslant t \leqslant 4\pi)$,
    (ii) $\gamma(t) = t + i(1-t)$ $(0 \leqslant t \leqslant 1)$,
    (iii) $\gamma(t) = t - 1 + it^2$ $(-1 \leqslant t \leqslant 1)$,
    (iv) $\gamma(t) = \begin{cases} t + i(1-t) & (0 \leqslant t \leqslant 1) \\ 1 + i(t-1) & (1 \leqslant t \leqslant 2). \end{cases}$

    In each case compute to winding number of the path round the origin.

20. Find the winding number $w(\gamma, z_0)$ for each of the following choices of γ, z_0:
    (i) $\gamma(t) = 2e^{-it}$ $(0 \leqslant t \leqslant 2\pi)$, $\quad z_0 = 1,\, 3i$,
    (ii) $\gamma(t) = t + i(1-t)$ $(0 \leqslant t \leqslant 1)$, $\quad z_0 = 1+i,\, -i,\, 10i$.

21. In each of the following, draw a picture of the path and use a sensible method to compute $\int_\gamma 1/(z-z_0)\,dz$:
    (i) $\gamma(t) = te^{it}$ $(\pi \leqslant t \leqslant 5\pi)$, $\quad z_0 = 0$,
    (ii) $\gamma(t) = it$ $(0 \leqslant t \leqslant 1)$, $\quad z_0 = 1$,
    (iii) $\gamma(t) = it$ $(-1 \leqslant t \leqslant 1)$, $\quad z_0 = 1$,
    (iv) $\gamma = \sigma + [1, 2] + S + [-2, -1]$, $\quad z_0 = 0$,
    where $\sigma(t) = e^{i(\pi - t)}$ $(0 \leqslant t \leqslant \pi)$, and $S(t) = 2e^{it}$ $(0 \leqslant t \leqslant \pi)$.

22. Compute (by eye) the winding number of the given closed paths round a point in the domains A, B, C, ... (Fig. 7.13)

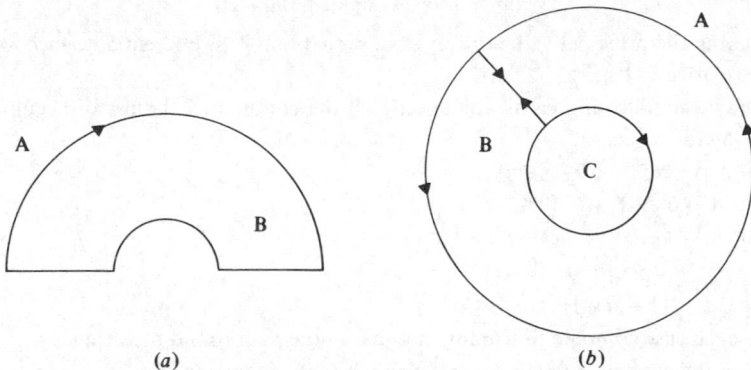

Fig. 7.13

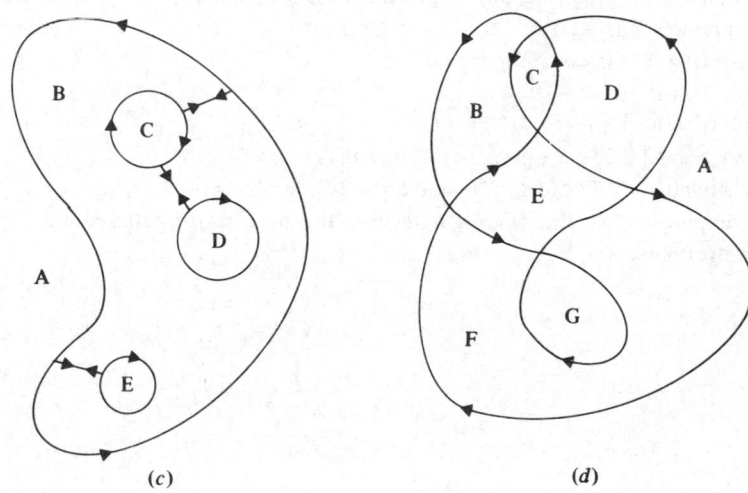

8

Cauchy's Theorem

The Fundamental Theorem of Contour Integration tells us that if f has an antiderivative in D, then we may evaluate the integral of f along a path in D from z_0 to z_1 by the formula

$$\int_\gamma f = F(z_1) - F(z_0).$$

In particular, if γ is closed, then

$$\int_\gamma f = 0.$$

Cauchy's Theorem puts forward conditions under which $\int_\gamma f = 0$ when there is no initial reason for f to have an antiderivative. There are many different versions of Cauchy's Theorem or, to be precise, many different theorems of this type due to Cauchy, who was first to publish such a result, announcing it in 1813 and getting it into print in 1825. Gauss was aware of the basic idea in 1811, but the accolade goes to Cauchy because he was the first to make it public.

Both Gauss and Cauchy realized the basic fact that if γ does not wind round points outside D, then $\int_\gamma f = 0$. For instance, $f(z) = 1/z$ has the single point 0 outside its domain D, so if the closed contour γ does not wind round the origin, then $\int_\gamma 1/z \, dz = 0$. (Fig. 8.1)

Our main aim in this chapter is to establish Cauchy's Theorem in the following form:

If f is differentiable in D and $w(\gamma, z) = 0$ for all z outside D, then

$$\int_\gamma f = 0.$$

We start the theory rolling by proving the special case where the contour is a triangle. Then we prove a theorem that requires a restriction on the domain D rather than the contour γ. We say that D is a *star-domain* if it contains a point z_* such that for every other point $z \in D$ the line segment

141

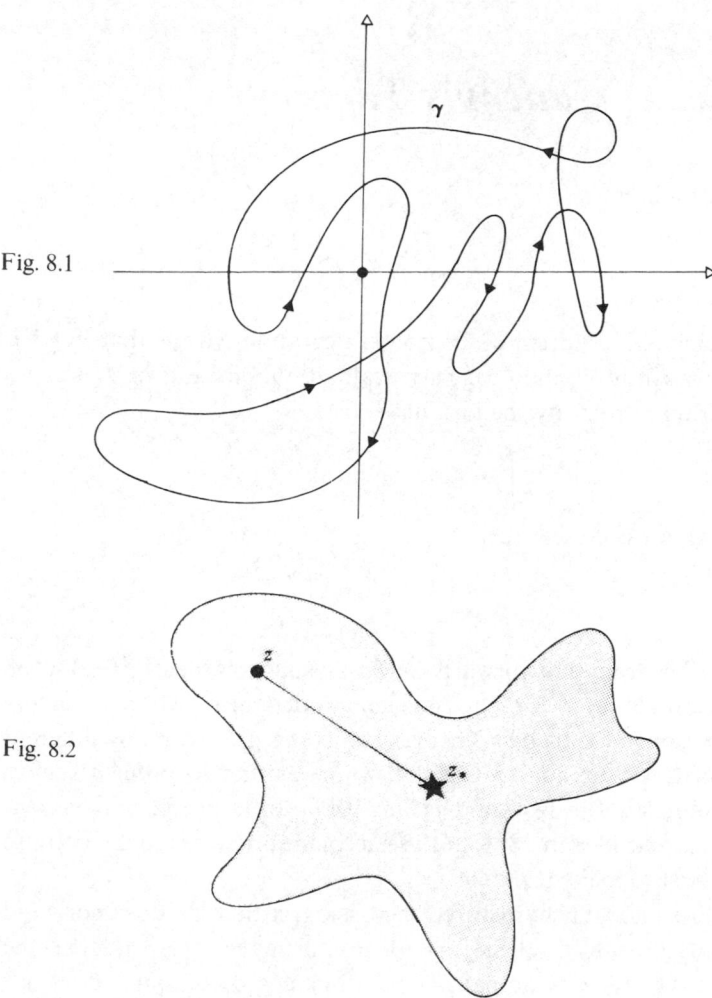

Fig. 8.1

Fig. 8.2

$[z_*, z]$ is in D. (Fig. 8.2) We then define $F(z) = \int_{[z_*, z]} f$ and use the triangle version of the theorem to show that F is an antiderivative of f. This means that $\int_\gamma f = 0$ for *any* closed contour in a star-domain. In particular a *disc* is a star-domain and this gives a very significant result. For a differentiable function f in a general domain D, we may not be able to find an anti-derivative $F:D \to \mathbb{C}$, but if we restrict our attention to any disc Δ in D, then there is an antiderivative $F:\Delta \to \mathbb{C}$. Thus an antiderivative may not exist *globally* throughout D but it does exist *locally* in any neighbourhood $N_r(z_0) \subseteq D$ for any $z_0 \in D$.

Using the Paving Lemma, any contour γ in an arbitrary domain D can be written as $\gamma = \gamma_1 + \cdots + \gamma_n$ where each subcontour γ_r lies in a disc $D_r \subseteq D$. In the (star-domain) D_r we can choose a step path σ_r with the same endpoints as γ_r and (by the existence of an antiderivative in D_r), $\int_{\sigma_r} f = \int_{\gamma_r} f$. If $\sigma = \sigma_1 + \cdots + \sigma_n$, then $\int_\sigma f = \int_\gamma f$. (Fig. 8.3)

This reduces the investigation of $\int_\gamma f$ to the case of an integral along a step path σ, which may be attacked by geometrically inspired methods.

Fig. 8.3

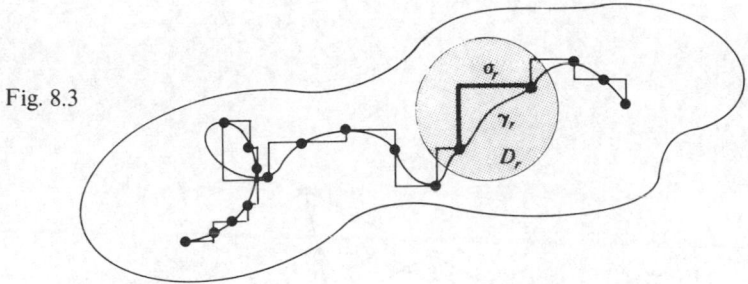

1. The Cauchy Theorem for a triangle

At the end of the nineteenth century, amongst many different versions of the Cauchy Theorem, a most ingenious proof for a triangular contour was conceived by E. H. Moore. Earlier proofs usually insisted that the function f had a *continuous* derivative f'. By restricting the contour to a triangle, Moore's proof requires only that f' *exists* throughout D. It will thus provide the basis for the theory for all differentiable functions.

For $z_1, z_2, z_3 \in \mathbb{C}$, let $T(z_1, z_2, z_3)$ be the set of points inside and on the triangle with vertices z_1, z_2, z_3. (Formally $T(z_1, z_2, z_3)$ is the set of points $\lambda_1 z_1 + \lambda_2 z_2 + \lambda_3 z_3$ for non-negative real numbers $\lambda_1, \lambda_2, \lambda_3$ satisfying $\lambda_1 + \lambda_2 + \lambda_3 = 1$). Let $\partial T(z_1, z_2, z_3)$ be the boundary contour composed of the three line segments $[z_1, z_2], [z_2, z_3], [z_3, z_1]$. Wherever there is no cause for confusion we shall denote the triangle by T and its boundary by ∂T.

THEOREM 8.1 (The Cauchy Theorem for a triangle).

Let f be a differentiable function in a domain D. If the triangle T lies in D, then $\int_{\partial T} f = 0$. (Fig. 8.4)

Proof. Let $|\int_{\partial T} f| = c \geq 0$.

We will show that $c = 0$ by an indirect argument. First we subdivide T into

Fig. 8.4

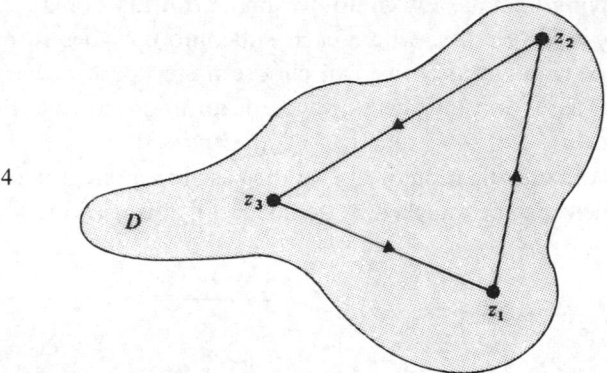

Fig. 8.5

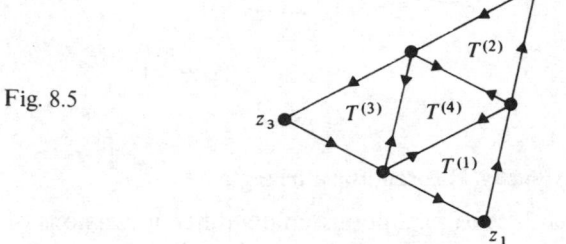

four triangles $T^{(1)}$, $T^{(2)}$, $T^{(3)}$, $T^{(4)}$ by joining the midpoints of the sides. (Fig. 8.5)

Then $\int_{\partial T} f = \Sigma_{r=1}^{4} \int_{\partial T^{(r)}} f$,

$$c = \left| \int_{\partial T} f \right| \leqslant \sum_{r=1}^{4} \left| \int_{\partial T^{(r)}} f \right|$$

and we must be able to choose r such that

$$\left| \int_{\partial T^{(r)}} f \right| \geqslant \tfrac{1}{4} c.$$

(If more than one r satisfies this inequality, choose the least.)

Define $T_1 = T^{(r)}$, then we have

$$\left| \int_{\partial T_1} f \right| \geqslant \tfrac{1}{4} c \quad \text{and} \quad L(\partial T_1) = \tfrac{1}{2} L(\partial T).$$

(where, as usual, $L(\gamma)$ denotes the length of γ).

We repeat the process of subdivision to give a sequence of triangles

$T \supseteq T_1 \supseteq T_2 \supseteq \cdots \supseteq T_n \supseteq \cdots$ satisfying

$$\left| \int_{\partial T_n} f \right| \geqslant (\tfrac{1}{4})^n c \quad \text{and} \quad L(\partial T_n) = (\tfrac{1}{2})^n L(\partial T). \tag{1}$$

Next we get another estimate for $\left| \int_{\partial T_n} f \right|$, using the fact that f is differentiable.

The nested sequence $T \supseteq T_1 \supseteq \cdots \supseteq T_n \supseteq \cdots$ contains a point z_0. Here f is differentiable, so, given $\varepsilon > 0$, there exists $\delta > 0$ such that

$$0 < |z - z_0| < \delta \quad \text{implies} \quad \left| \frac{f(z) - f(z_0)}{z - z_0} - f'(z_0) \right| < \varepsilon.$$

Hence

$$|z - z_0| < \delta \quad \text{implies} \quad |f(z) - f(z_0) - f'(z_0)(z - z_0)| \leqslant \varepsilon |z - z_0|.$$

For some integer N, every point in T_n is within ε of z_0 for $n \geqslant N$. Thus

$$|f(z) - f(z_0) - f'(z_0)(z - z_0)| \leqslant \varepsilon |z - z_0| \quad \text{for } z \in T_n, \quad n \geqslant N.$$

For $z \in T_n$ we trivially have $|z - z_0| \leqslant L(\partial T_n)$, so the Estimation Lemma gives

$$\left| \int_{\partial T_n} \{ f(z) - f(z_0) - f'(z_0)(z - z_0) \} \, dz \right| \leqslant \varepsilon L(\partial T_n) \cdot L(\partial T_n). \tag{2}$$

But $-f(z_0) - f'(z_0)(z - z_0)$ is of the form $a + bz$ where a and b are constants. This has an antiderivative $az + \tfrac{1}{2} bz^2$, so $\int_{\partial T_n} (a + bz) \, dz = 0$, and (2) reduces to

$$\left| \int_{\partial T_n} f \right| \leqslant \varepsilon L(\partial T_n)^2.$$

Comparing this with the earlier estimate in (1), we find

$$(\tfrac{1}{4})^n c \leqslant \left| \int_{\partial T_n} f \right| \leqslant \varepsilon L(\partial T_n)^2 = (\tfrac{1}{4})^n \varepsilon L(\partial T)^2$$

and this gives

$$c \leqslant \varepsilon L(\partial T)^2.$$

But ε is arbitrary and $c \geqslant 0$, so we must have $c = 0$, that is

$$\int_{\partial T} f = 0. \qquad \square$$

This proof deserves a commentary, because its analytic presentation obscures the fact that the basic idea is very simple and very geometric. It uses two facts. One is that the integral of $f(z)$ is *additive on contours* – that is, the contributions from the subdivided contours add up to that for the original one. The other is that a differentiable function is *approximately linear* (that is, of the form $a + bz$) near to any given point.

If it were possible for f to be *exactly* linear, locally, then we could take a fine subdivision making it linear on each subcontour; get zero for the integral on each subcontour by explicit computation using the anti-derivative $az + bz^2/2$; and add all these zeros to get zero for the original integral.

Unfortunately this can't happen, and we are faced with adding a larger and larger number of contributions, each getting closer and closer to zero. By *estimating* the rate of growth or shrinkage we show that the errors in assuming approximate linearity tend to zero fast enough to compensate for the increasing number of subcontours.

It is an interesting exercise to rewrite the proof in such a way that this informal description becomes a formal argument which keeps the geometry to the fore.

2. Existence of an antiderivative in a star-domain

Recall that D is a *star-domain* if there exists $z_* \in D$ (called a *star-centre*) such that the straight line $[z_*, z]$ lies in D for all $z \in D$. In a star-domain there is an obvious candidate for the antiderivative of a function f, namely the integral $F(z) = \int_{[z_*, z]} f$.

THEOREM 8.2. If f is differentiable in a star-domain D with star-centre z_*, then $F(z) = \int_{[z_*, z]} f$ is an antiderivative for f in D.

Proof. D is open, so for $z_1 \in D$ there exists $\varepsilon_1 > 0$ such that $N_{\varepsilon_1}(z_1) \subseteq D$. For $|h| < \varepsilon_1$, the triangle with vertices $z_0, z_1, z_1 + h$ lies entirely in D. (Fig. 8.6)

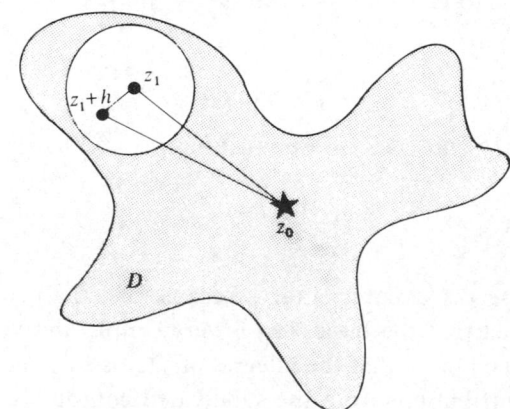

Fig. 8.6

Theorem 8. 1 gives

$$\int_{[z_\bullet,z_1]} f + \int_{[z_1,z_1+h]} f + \int_{[z_1+h,z_\bullet]} f = 0.$$

This can be written as

$$F(z_1) + \int_{[z_1,z_1+h]} f - F(z_1+h) = 0$$

or

$$\frac{F(z_1+h)-F(z_1)}{h} = \frac{1}{h} \int_{[z_1,z_1+h]} f.$$

The proof now proceeds in the same manner as Theorem 5.11.

For a constant $c \in \mathbb{C}$,

$$\int_{[z_1,z_1+h]} c \, dz = ch,$$

hence

$$\frac{F(z_1+h)-F(z_1)}{h} - f(z_1) = \int_{[z_1,z_1+h]} \frac{f(z)-f(z_1)}{h} \, dz \qquad (3)$$

From the continuity of f, given $\varepsilon > 0$ there exists $\delta > 0$ such that

$$|z - z_1| < \delta \quad \text{implies} \quad |f(z) - f(z_1)| < \varepsilon.$$

For z on the line segment $[z_1, z_1 + h]$,

$$|h| < \delta \quad \text{implies} \quad |z - z_1| < \delta \quad \text{and so} \quad |f(z) - f(z_1)| < \varepsilon.$$

The Estimation Lemma gives

$$\left| \int_{[z_1,z_1+h]} \frac{f(z)-f(z_1)}{h} \, dz \right| \leqslant \frac{\varepsilon}{|h|} |h|,$$

and from (3), if $|h| < \delta$, then

$$\left| \frac{F(z_1+h)-F(z_1)}{h} - f(z_1) \right| \leqslant \varepsilon.$$

Since ε is arbitrary, we have

$$\lim_{h \to 0} \frac{F(z_1+h)-F(z_1)}{h} = f(z_1)$$

so $F' = f$, as required. $\qquad\qquad\qquad\qquad\qquad\qquad\qquad\qquad\quad\square$

COROLLARY 8.3. If f is differentiable in a star-domain D, then $\int_\gamma f = 0$ for all closed contours γ in D, and the integral of f between any two points in D is independent of the choice of contour between the points.

Proof. This is an immediate deduction from Theorems 8.2 and 6.11.   □

3. An example – the logarithm

The function $f(z) = 1/z$ is differentiable in the domain $\mathbb{C} \setminus \{0\}$. The latter is not a star-domain, but if we restrict the function to a star-domain $D \subset \mathbb{C} \setminus \{0\}$, then the results of the last section apply in D.

The cut plane $\mathbb{C}_\pi = \mathbb{C} \setminus N$ where $N = \{x + iy \in \mathbb{C} \,|\, y = 0, \, x \leqslant 0\}$ is a star-domain with 1 as a star-centre. (Fig. 8.7)

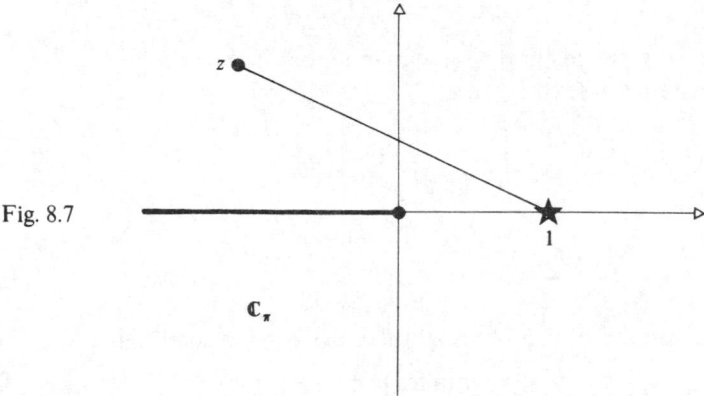

Fig. 8.7

Because $1/z$ is differentiable in $\mathbb{C}_\pi$, it has an antiderivative, which we denote by Log, given by

$$\text{Log } z_1 = \int_{[1,z_1]} 1/z \, dz \quad (z_1 \in \mathbb{C}_\pi).$$

We exploit the fact that the integral is independent of the path and integrate along a specially chosen contour. Let $z_1 = re^{i\theta}$ where $r > 0$ and $-\pi < \theta < \pi$, then define $\gamma = \gamma_1 + \gamma_2$ where γ_1 is the line segment $[1, r]$ and $\gamma_2(t) = re^{it} \, (0 < t < \theta)$. (For $r < 1$, then $[1, r]$ is the directed line segment from 1 back to r, and for $\theta < 0$, we take $\gamma_2(t) = re^{-it} \, (0 < t < \theta)$.) (Fig. 8.8)

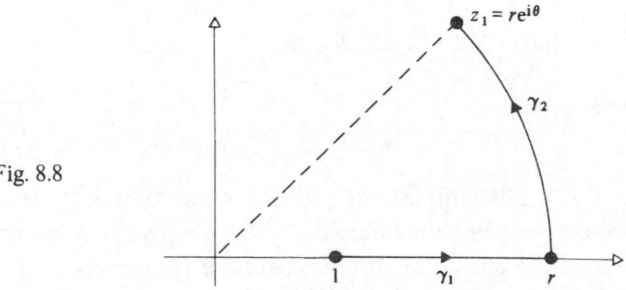

Fig. 8.8

Then

$$\text{Log } re^{i\theta} = \int_{\gamma_1} 1/z \, dz + \int_{\gamma_2} 1/z \, dz$$

$$= \int_1^r 1/t \, dt + \int_0^\theta \frac{1}{re^{it}} i r e^{it} \, dt$$

$$= \log r + i\theta.$$

This gives an alternative approach to the complex logarithm if we so desire. In particular it affords a much more satisfying proof of the continuity of the argument in the cut plane $\mathbb{C}_\pi$ than the prosaic version given in §7.2. Because the function Log is differentiable in $\mathbb{C}_\pi$, hence continuous, so its imaginary part, the argument of z, is also continuous there.

4. Local existence of an antiderivative

Let f be differentiable in an arbitrary domain D. Then f may not have an antiderivative which works throughout D. But D is open, so for any $z_0 \in D$ there is a disc $N_r(z_0) \subseteq D$. A function $F:N_r(z_0) \to \mathbb{C}$ such that $F'(z) = f(z)$ for all $z \in N_r(z_0)$ is called a *local antiderivative* of f. Because a disc is a star-domain, Theorem 8.2 tells us that there is a local antiderivative of f in every disc in D.

This immediately simplifies integration of a differentiable function along an arbitrary contour because we can integrate along a step path instead:

LEMMA 8.4. If γ is a contour in a domain D from z_0 to z_1, then there exists a step path σ in D from z_0 to z_1 such that $\int_\gamma f = \int_\sigma f$ for every function f differentiable in D.

Proof. By the Paving Lemma we have $\gamma = \gamma_1 + \cdots + \gamma_n$ where each γ_r lies in a disc $D_r \subseteq D$. Let σ_r be a step path in D_r from the initial point of γ_r to its final point, then in the star-domain D_r, $\int_{\gamma_r} f = \int_{\sigma_r} f$ (Corollary 8.3). If $\sigma = \sigma_1 + \cdots + \sigma_n$, then σ is a step path from z_0 to z_1 in D as in Figure 8.3 and

$$\int_\gamma f = \sum_{r=1}^n \int_{\gamma_r} f = \sum_{r=1}^n \int_{\sigma_r} f = \int_\sigma f.$$

5. Cauchy's Theorem

We build up to Cauchy's Theorem in stages. First we consider a rectangle

$$R = \{x + iy \in \mathbb{C} \mid a \leqslant x \leqslant b, c \leqslant y \leqslant d\}$$

with boundary contour

$$\partial R = [z_1, z_2] + [z_2, z_3] + [z_3, z_4] + [z_4, z_1]$$

where $z_1 = a + ic$, $z_2 = b + ic$, $z_3 = b + id$, $z_4 = a + id$. (Fig. 8.9)

Fig. 8.9

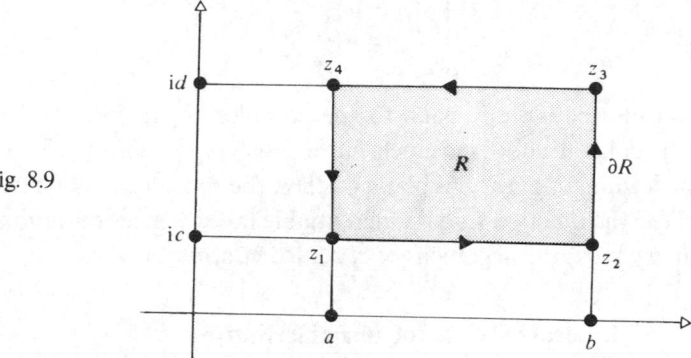

LEMMA 8.6. If $R \subseteq D$ and f is differentiable in D, then $\int_{\partial R} f = 0$.

Proof. Insert the opposite contours $[z_1, z_3]$, $[z_3, z_1]$ and use the Cauchy Theorem for a Triangle twice. (Fig. 8. 10)   □

Fig. 8.10

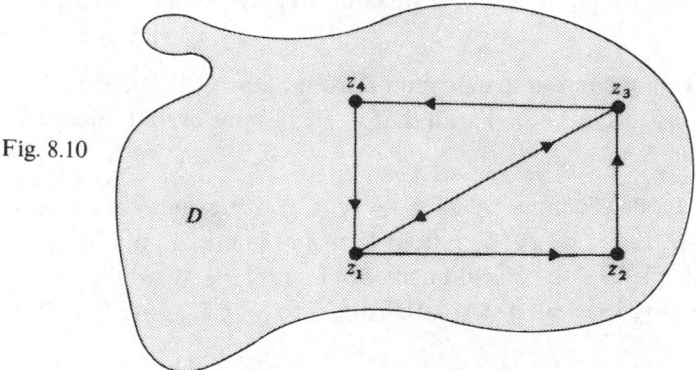

Now we take an arbitrary closed step path σ and insert extra line segments to make up a collection of rectangles. To do this we extend all the horizontal and vertical line segments of σ *ad infinitum*, breaking the plane up into a finite number of rectangles: some finite, $R_1, \ldots, R_k$ and some infinite, $R_{k+1}, \ldots, R_m$. (An example is drawn in Figure 8.11, where $k = 9$, $m = 25$.)

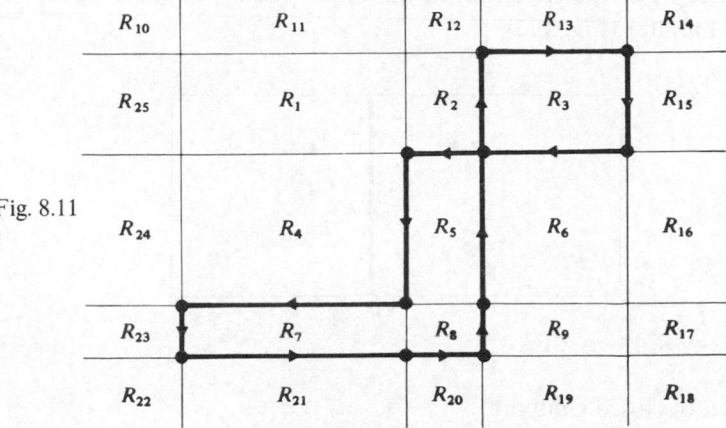

Fig. 8.11

In the interior of each R_n, we choose a point z_n and define

$$v_n = w(\sigma, z_n).$$

(This is independent of the particular choice of z_n inside R_n.)

We say that R_n is *relevant* if $v_n \neq 0$. Thus R_n is only relevant if σ winds round it. In particular the infinite rectangles $R_{k+1}, \ldots R_m$ are all irrelevant because they lie outside σ. (In Figure 8.11, the only relevant rectangles are R_3, R_5, R_7, R_8.)

We now demonstrate that we may express σ in terms of the boundaries of relevant rectangles by taking v_n copies of each boundary ∂R_n. (If v_n is negative, we take $-v_n$ copies of the opposite contour $-\partial R_n$.)

For instance, in Figure 8.11, we take $-R_3, R_5, R_7, R_8$; cancelling the opposite segments common to R_5, R_8 and to R_7, R_8, we finish up with the step contour σ.

To show that this works for an arbitrary closed step contour σ, it is convenient to use the notation $n\gamma$ to represent n copies of γ for $n \geqslant 0$ and $-n$ copies of $-\gamma$ for $n < 0$. The most straightforward process is then to start with the set of contours

$$A = \{v_1 \partial R_1, v_2 \partial R_2, \ldots, v_k \partial R_k, -\sigma\}$$

and show that the cancelling of opposite line segments L, $-L$, wherever they occur, removes them all.

Suppose that at the end there are q copies of some line segment L. Then L is a side of at least one finite rectangle R_s and, by allowing q to be negative if necessary, we may suppose that L is traced in the same direction as ∂R_s. Let R_r be the rectangle on the other side of L (where R_r may be finite or infinite). (Fig. 8.12)

Fig. 8.12

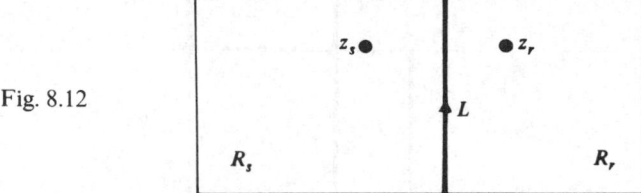

The set of closed contours

$$B = A \cup \{-q \partial R_s\}$$

then simplifies to have no copies of L. If we compute the winding numbers of the contours in B round z_s (inside R_s) and z_r (inside R_r), then the absence of L from the simplified set of contours tells us that the two winding numbers are the same. But the winding number around z_s is

$$v_1 w(\partial R_1, z_s) + \cdots + v_k w(\partial R_k, z_s) - w(\sigma, z_s) - q w(\partial R_s, z_s) = -q$$

and around z_r it is

$$v_1 w(\partial R_1, z_r) + \cdots + v_k w(\partial R_k, z_r) - w(\sigma, z_r) - q w(\partial R_s, z_s) = 0.$$

Hence $q = 0$, as required. This confirms that σ may be obtained by taking v_n copies of each relevant rectangular contour ∂R_n and deleting opposite line segments wherever they occur.   □

LEMMA 8.7. Let σ be a closed step path in a domain D such that $w(\sigma, z) = 0$ for all $z \notin D$. Then, for any function f differentiable in D, $\int_\sigma f = 0$.

Proof. We express σ in terms of relevant rectangles and show that every relevant rectangle R_n must lie entirely in D. Certainly, for z in the interior of R_n, $w(\sigma, z) = v_n \neq 0$ and so z must be in D. On the other hand, a point z on the boundary ∂R_n either lies on σ (and hence in D) or it is in the same component of the complement of σ as points in the interior of R_n, whence $w(\sigma, z) = v_n \neq 0$ and, again, $z \in D$.

By cancelling contributions along opposite contours,

$$\int_\sigma f = \sum_{n=1}^{k} v_n \int_{\partial R_n} f.$$

Integrals on the right need only be considered when $v_n \neq 0$; the relevant rectangle R_n then lies completely in D and (by Lemma 7.6),

$$\int_{\partial R_n} f = 0.$$

Hence $\int_\sigma f = 0$, as required. $\qquad\qquad\qquad\qquad\qquad\qquad\qquad\square$

We now reach the focal point of the chapter:

CAUCHY'S THEOREM (Theorem 8.8).
Let f be differentiable in a domain D and γ a closed contour in D which does not wind round any points outside D (meaning $w(\gamma, z) = 0$ for $z \notin D$). Then $\int_\gamma f = 0$.

Proof. There exists a step path σ in D such that $\int_\sigma \phi = \int_\gamma \phi$ for any function ϕ differentiable in D (using Lemma 8.4). In particular $\int_\sigma f = \int_\gamma f$.
For $z_0 \notin D$, the function $\phi(z) = 1/(z - z_0)$ is also differentiable in D, so

$$w(\sigma, z_0) = \frac{1}{2\pi i} \int_\sigma \phi = \frac{1}{2\pi i} \int_\gamma \phi = w(\gamma, z_0) = 0.$$

From Lemma 8.7 we have $\int_\sigma f = 0$. Hence

$$\int_\gamma f = \int_\sigma f = 0. \qquad\qquad\qquad\qquad\qquad\square$$

6. Applications of Cauchy's Theorem

The Cauchy Theorem we have just given has far wider applications than simply showing that the integral round certain closed contours is zero. It enables us to calculate non-zero integrals too. For example, suppose that γ_1 and γ_2 have the same winding number round all points outside D, (that is $w(\gamma_1, z) = w(\gamma_2, z)$ for all $z \notin D$). Let z_1 be the point where γ_1 begins and ends and z_2 the point where γ_2 begins and ends. Take any contour σ from z_1 to z_2. (Fig. 8.13)

Then $\gamma = \gamma_1 + \sigma - \gamma_2 - \sigma$ (round γ_1, along σ, round $-\gamma_2$ and back along $-\sigma$) is a closed contour in D and $w(\gamma, z) = 0$ for $z \notin D$. Using Cauchy's Theorem, $\int_\gamma f = 0$, hence

$$\int_{\gamma_1} f + \int_\sigma f - \int_{\gamma_2} f - \int_\sigma f = 0$$

and

$$\int_{\gamma_1} f = \int_{\gamma_2} f.$$

Fig. 8.13

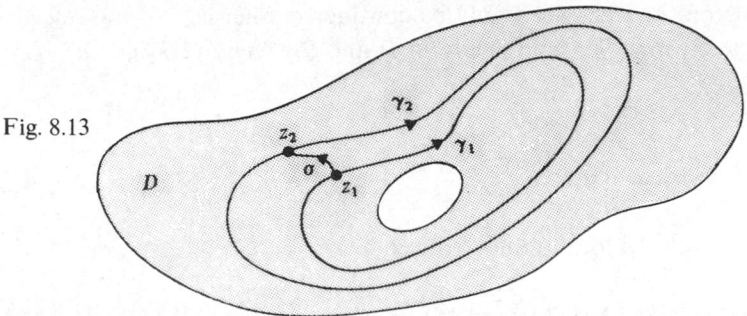

If we desire to compute $\int_{\gamma_1} f$, then we may be able to find another contour
γ_2, as above, where the integral $\int_{\gamma_2} f$ is easier to calculate. We shall exploit
this fact with telling effect later in the text.

The technique of introducing opposite contours σ, $-\sigma$ whose contributions eventually cancel is also a useful one. It enables us to prove a
much more powerful theorem.

THEOREM 8.9 (The generalized version of Cauchy's Theorem).

Suppose that $\gamma_1, \ldots, \gamma_n$ are closed contours in a domain D such that
$$w(\gamma_1, z) + \cdots + w(\gamma_n, z) = 0 \quad \text{for all } z \notin D.$$
Then, for f differentiable in D,
$$\int_{\gamma_1} f + \cdots + \int_{\gamma_n} f = 0.$$

Proof. Suppose that γ_r begins and ends at $z_r \, (1 \leqslant r \leqslant n)$. Choose any $z_0 \in D$
and contours $\sigma_1, \ldots, \sigma_n$ in D which join z_0 to $z_1, \ldots, z_n$ respectively.
(Fig. 8.14) Then
$$\gamma = \sigma_1 + \gamma_1 - \sigma_1 + \cdots + \sigma_n + \gamma_n - \sigma_n$$

Fig. 8.14

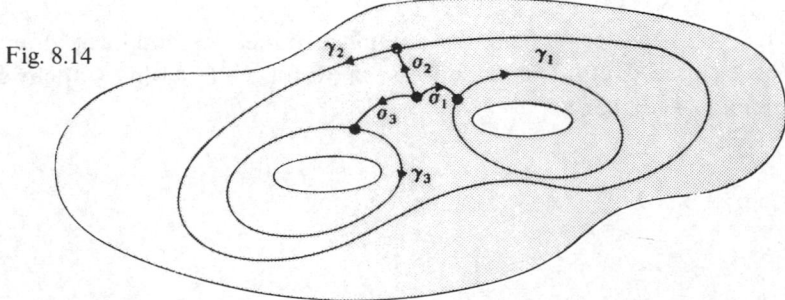

is a closed contour beginning and ending at z_0 and

$$w(\gamma, z) = 0 \quad \text{for all } z \notin D.$$

By Cauchy's Theorem, $\int_\gamma f = 0$, so

$$\sum_{r=1}^{n} \left(\int_{\sigma_r} f + \int_{\gamma_r} f + \int_{-\sigma_r} f \right) = 0,$$

that is

$$\sum_{r=1}^{n} \int_{\gamma_r} f = 0. \qquad\qquad \square$$

Given two closed contours γ_1, γ_2 in D and a contour σ in D from a point on γ_1 to a point on γ_2 (Fig. 8.13), then the pair of contours $\sigma, -\sigma$ is called a *cut* from γ_1 to γ_2. There is a historical reason for this. Earlier versions of Cauchy's Theorem were invariably proved for Jordan contours. A *closed Jordan contour* is a closed contour $\gamma:[a, b] \to \mathbb{C}$ which does not cross itself, that is,

$$a < t_1 < t_2 \leqslant b \quad \text{implies } \gamma(t_1) \neq \gamma(t_2).$$

It is intuitively obvious, but analytically difficult to prove, that every closed Jordan contour separates the plane into two components, the points $O(\gamma)$ outside γ and the points $I(\gamma)$ inside γ, and that $O(\gamma)$ and $I(\gamma)$ are both connected sets. (Fig. 8.15)

Earlier versions of Cauchy's Theorem stated that if γ and $I(\gamma)$ were in D, then $\int_\gamma f = 0$. In applications it was then necessary to introduce 'cuts' to manufacture Jordan contours. For instance, suppose that f is differentiable everywhere except at z_0 and two Jordan contours γ_1, γ_2 both wind once round z_0, as in Figure 8.16a. Two cuts are made in the picture to give

Fig. 8.15

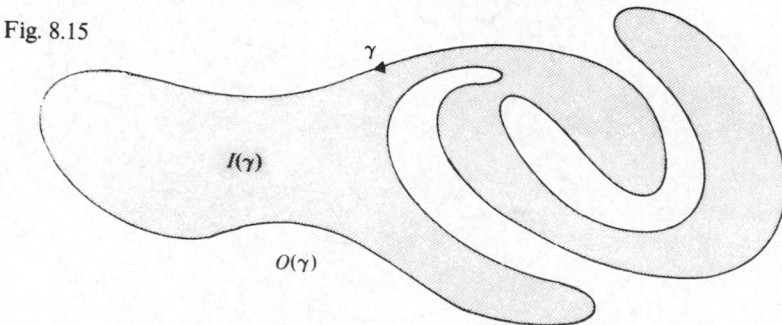

Fig. 8.16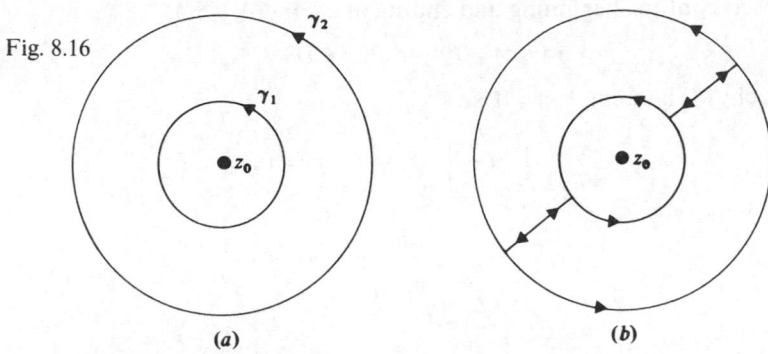

(a) (b)

two Jordan contours so that f is differentiable inside each of them (Fig. 8.16b). The integral round each Jordan contour is then known to be zero and on cancelling the contributions due to the cuts, the result $\int_{\gamma_1} f = \int_{\gamma_2} f$ is obtained. Such methods usually relied on geometrical intuition, sometimes unsupported by analytic proof. By introducing the winding number to link analysis and geometric intuition such pitfalls may be avoided, and with them the necessity to restrict the theory to Jordan contours. Instead we can define the *inside* of *any* closed contour γ to be

$$I(\gamma) = \{z \in \mathbb{C} \mid w(\gamma, z) \neq 0\}$$

and the *outside* to be

$$O(\gamma) = \{z \in \mathbb{C} \mid w(\gamma, z) = 0\}.$$

In general $I(\gamma)$ and $O(\gamma)$ need not be connected. (In Figure 8.17, $O(\gamma)$ has two components O_1 and O_2, whilst $I(\gamma)$ has three, I_1, I_2, I_3.)

Fig. 8.17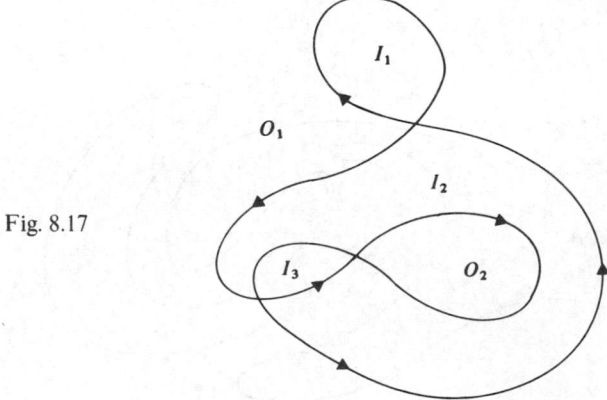

The *Jordan Contour Theorem* (which we do not prove), then says that the outside and inside of a closed Jordan contour *are* both connected.

For an arbitrary closed contour γ we can rephrase Cauchy's Theorem as given in Theorem 8.8 to get

THEOREM 8.10. Let f be differentiable in a domain D. If a closed contour γ and its inside $I(\gamma)$ lie in D then $\int_\gamma f = 0$. $\square$

7. Simply connected domains

We have known for some time (Theorem 6.11) that $\int_\gamma f = 0$ for all closed contours in a domain D precisely when f has an antiderivative. We can now state the precise conditions under which this happens for *all* functions differentiable in a given domain D. We say that a domain D is *simply connected* if $w(\gamma, z) = 0$ for every closed contour γ in D and $z \notin D$. Equivalently, if γ is a closed contour in D, then the inside $I(\gamma)$ lies in D. We then have

THEOREM 8.11. For a given domain D, $\int_\gamma f = 0$ for all closed contours γ in D and all functions f differentiable in D if and only if D is simply connected.

Proof. If D is simply connected, then Cauchy's Theorem implies $\int_\gamma f = 0$ for all γ in D and f differentiable in D. Conversely, if D is not simply connected, then there exists a closed contour γ_0 in D and $z_0 \notin D$ such that $w(\gamma_0, z_0) \neq 0$. Let $\phi(z) = 1/2\pi i(z - z_0)$, then ϕ is differentiable in D and $\int_{\gamma_0} \phi = w(\gamma_0, z_0) \neq 0$. $\square$

Exercises 8

1. State which of the following are star-domains, specifying a star-centre for those which are and justifying your response for those which are not:
   (i) $\{z \in \mathbb{C} \mid z \neq x + i0 \quad \text{where } x \leqslant -1 \quad \text{or } x \geqslant 1\}$,
   (ii) $\{z \in \mathbb{C} \mid |z| > 1\}$,
   (iii) $\{z \in \mathbb{C} \mid z \neq e^{it} \quad \text{for } 0 \leqslant t \leqslant \pi\}$,
   (iv) $\{z \in \mathbb{C} \mid |z| > 1 \quad \text{and either im } z > 0 \quad \text{or re } z > 0\}$.
2. Let $D = \mathbb{C} \setminus \{0\}$. For $z_0 \in D$, specify a local antiderivative in some neighbourhood of z_0 for each of the following functions:
   (i) $1/z$ (ii) $1/z^2$ (iii) $(z+1)/z^2$ (iv) $(\cos z)/z$ (v) $(\sin z)/z$,

3. Let

$$\gamma_1(t) = -1 + \tfrac{1}{2}e^{it} \quad (0 \leqslant t \leqslant 2\pi)$$
$$\gamma_2(t) = 1 + \tfrac{1}{2}e^{it} \quad (0 \leqslant t \leqslant 2\pi)$$
$$\gamma(t) = 2e^{-it} \quad (0 \leqslant t \leqslant 2\pi).$$

If $f(z) = 1/(z^2 - 1)$, use Theorem 8.9 to deduce

$$\int_\gamma f = \int_{\gamma_1} f + \int_{\gamma_2} f.$$

Interpret this statement in terms of winding numbers of γ, γ_1, γ_2 around 1, -1.

4. Show that $D = \{z \in \mathbb{C} \mid z \neq \pm 1\}$ is not simply connected. Let

$$L_1 = \{x + iy \in \mathbb{C} \mid y = 0, \ x \leqslant -1\},$$
$$L_2 = \{x + iy \in \mathbb{C} \mid y = 0, \ x \geqslant 1\},$$
$$D_0 = D \smallsetminus (L_1 \cup L_2).$$

Show that D_0 is simply connected. Is it a star-domain? Does $f(z) = 1/(z^2 - 1)$ have an antiderivative in D or D_0? In each case, justify your response.

5. Let $D = \{z \in \mathbb{C} \mid z \neq \pm i\}$ and let γ be a closed contour in D. Find all the possible values of $\int_\gamma 1/(z^2 + 1)\, dz$. If σ is a contour from 0 to 1, find all the possible values of $\int_\sigma 1/(z^2 + 1)\, dz$.

6. Let $\gamma_1 = S_1 + L - S_2 - L$, $\gamma_2 = S_1 + L + S_2 - L$ where

$$S_1(t) = e^{it} \quad (0 \leqslant t \leqslant 2\pi),$$
$$S_2(t) = 2e^{it} \quad (0 \leqslant t \leqslant 2\pi),$$
$$L = [1, 2].$$

Describe the inside and outside of γ_1 and γ_2.

Let $f(z) = (\cos z)/z$. By writing $\cos z$ as a power series and considering $f(z) = (1/z) + g(z)$, or otherwise, compute $\int_{\gamma_1} f$, $\int_{\gamma_2} f$. Compare these computations with Theorem 8.10.

7. Let $D = \{z \in \mathbb{C} \mid z \neq z_1, z \neq z_2, \ldots, z \neq z_k\}$ and suppose that f is differentiable in D. Show that for any closed contour γ in D

$$\int_\gamma f = \sum_{r=1}^{k} n_r \int_{S_r} f$$

where S_r is a circle centre z_r and n_r is an integer.

If $\lim_{z \to z_r} (z - z_r) f(z) = a_r \in \mathbb{C}$ for $r = 1, \ldots, k$, show that

$$\int_\gamma f = \sum_{r=1}^{k} 2\pi i n_r a_r.$$

9
Homotopy versions of Cauchy's Theorem

In this chapter we consider what happens to $\int_\gamma f$ when we allow γ to vary. This is not essential for any later work in the text, so the whole chapter may be omitted. However, it gives precise conditions under which we may vary γ continuously and yet leave $\int_\gamma f$ unchanged. There are two ways in which this will be done. One is to fix the endpoints z_1, z_2 and allow the contour from z_1 to z_2 to be continuously deformed. The other is to allow a continuous deformation of a closed contour. In both cases the deformations must occur in the domain D where f is differentiable. (Fig. 9.1)

Both of these are special cases of a single result (the Cauchy Theorem for a boundary) proved in §2. Before this, in §1, we show how the conditions on γ can be relaxed when f is differentiable to define $\int_\gamma f$ along an arbitrary path. This allows us to vary γ freely in $\int_\gamma f$ without the need to worry whether the intermediate variations are always contours.

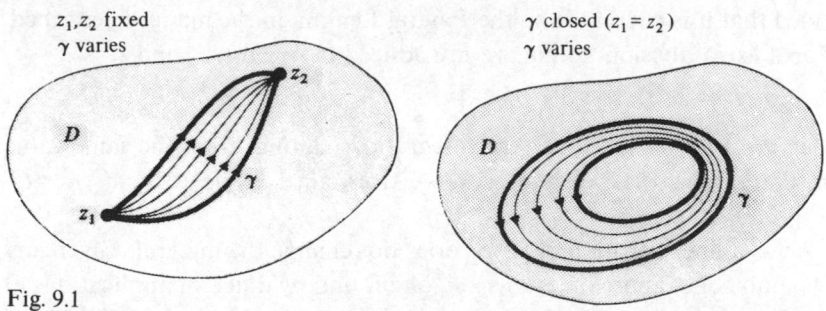

z_1, z_2 fixed
γ varies

γ closed $(z_1 = z_2)$
γ varies

Fig. 9.1

1. Integration along arbitrary paths

Suppose that f is differentiable in a domain D. Up to now we have always insisted that integration is performed along a *contour* in D. However, for a *differentiable* function, the precise path taken is not so important. We

159

saw this in Lemma 7.4 where we were able to replace a contour by a step path and get the same result. We can use the same technique to define the notion of an integral along an arbitrary path. Let $\pi:[a,\ b]\to D$ be an arbitrary path in D. Then the Paving Lemma gives a subdivision $a = t_0 < t_1 < \cdots < t_n = b$ such that each subpath π_r defined on $[t_{r-1},\ t_r]$ lies in a disc $D_r \subseteq D$. Let λ be the approximating polygon $\lambda = \lambda_1 + \cdots + \lambda_n$ where λ_r is the line segment from $\pi(t_{r-1})$ to $\pi(t_r)$. Then λ is a contour in D and we can define $\int_\pi f$ by

$$\int_\pi f = \int_\lambda f.$$

(See Fig. 9.2)

Fig. 9.2

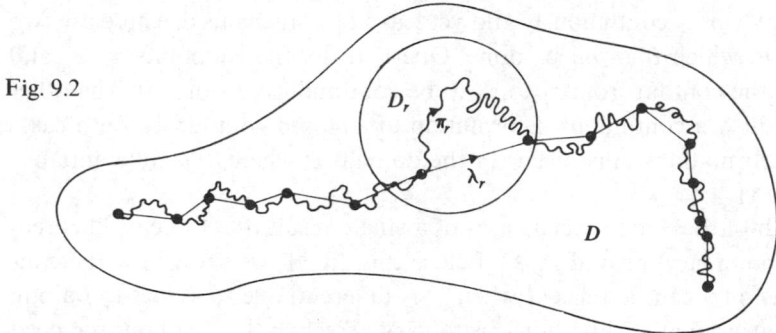

This definition may be seen to be independent of the choice of λ, provided that it is chosen using the Paving Lemma in the manner described. For if extra division points are introduced between t_{r-1} and t_r

$$t_{r-1} < s_1 < \cdots < s_k < t_r,$$

then $\pi(s_1), \ldots, \pi(s_k) \in D_r$, so the integral of f along λ_r has the same value as that along the polygon with vertices $\pi(t_{r-1})$, $\pi(s_1), \ldots, \pi(s_k)$, $\pi(t_r)$. (Fig. 9.3)

Thus a finer partition of $[a,\ b]$ does not change the integral. Given any two polygonal approximations λ, μ of π found by different applications of the Paving Lemma, let v be the polygon whose vertices are all of those of λ, μ taken together. Then $\int_\lambda f = \int_v f = \int_\mu f$ and the definition of $\int_\pi f$ is independent of the choice of polygonal approximation in the manner described.

An arbitrary polygonal approximation to π will not do. In Figure 9.4 we assume that $D = \mathbb{C} \setminus \{z_0\}$. The polygon λ with vertices $z_1, z_2, z_3, z_4,$

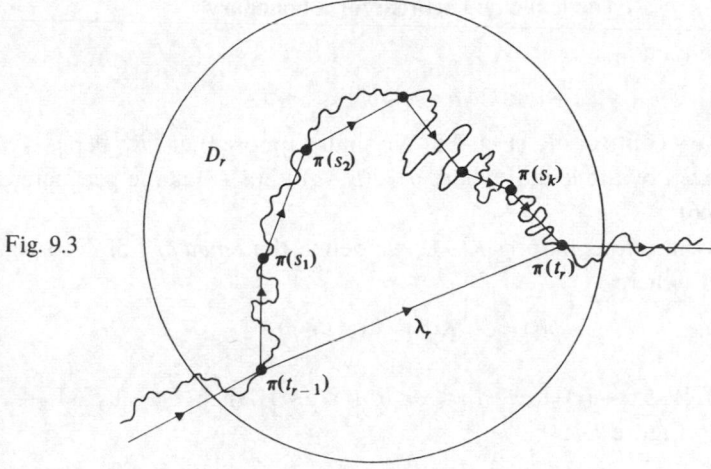

Fig. 9.3

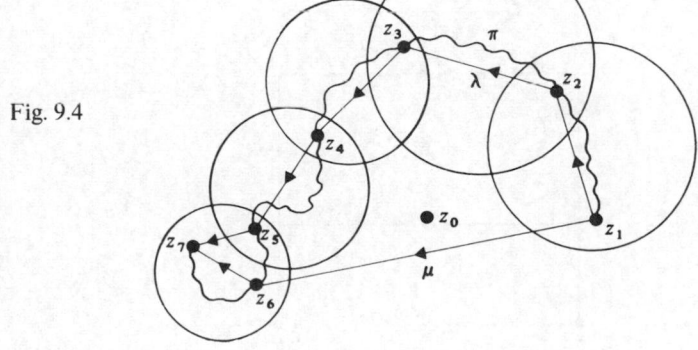

Fig. 9.4

z_5, z_7 is an approximation to π found using the Paving Lemma but the polygon μ with vertices z_1, z_6, z_7 is not. If $f = 1/(z - z_0)$, then $\int_\lambda f - \int_\mu f = 2\pi i$, so $\int_\lambda f \neq \int_\mu f$.

It is a straightforward matter to check that the theorems of Chapter 8 hold also for arbitrary paths, however wild (even 'space-filling' curves). For instance, if f is differentiable in D and π does not wind round points outside D, then $\int_\pi f = 0$. With such results at our disposal we can widen the scope of our ideas and introduce general notions from topology.

2. The Cauchy Theorem for a boundary

Let R be the rectangle

$$\{x+iy \in \mathbb{C} \mid a \leqslant x \leqslant b, \, c \leqslant y \leqslant d\}$$

with boundary contour ∂R. (Fig. 9.5) We shall suppose that $\partial R:[0, p] \to \mathbb{C}$ is parametrized by arc length where $p = 2(b-a) + 2(d-c)$ is the perimeter of R. (Fig. 9.6)

Given a continuous map $\phi:R \to \mathbb{C}$, we define the *boundary* of ϕ to be $\partial\phi:[0, p] \to \mathbb{C}$ where

$$\partial\phi(t) = \phi(\partial R(t)) \quad (0 \leqslant t \leqslant p).$$

EXAMPLE 1. If $\phi(x+iy) = xe^{i\pi y} \, (a \leqslant x \leqslant b, \, 0 \leqslant y \leqslant 1)$ then $\partial\phi = \Gamma_1 + \Gamma_2 + \Gamma_3 + \Gamma_4$ as in Figure 9.7.

EXAMPLE 2. Let $R = \{x+iy \mid 0 \leqslant x \leqslant 2, 0 \leqslant y \leqslant 2\}$. For $x+iy \in R$ and $x+y \geqslant 1$, define $\phi(x+iy) = x+iy$, and for $x+iy \in R$ and $x+y \leqslant 1$, define $\phi(x+iy)$ to be the reflection of $x+iy$ in the line $x+y = 1$. Then the effect

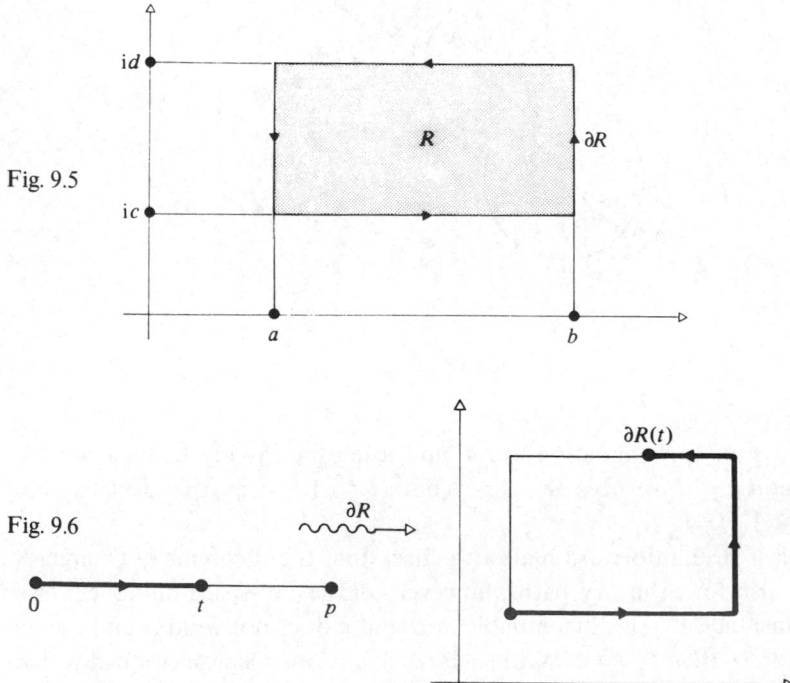

Fig. 9.5

Fig. 9.6

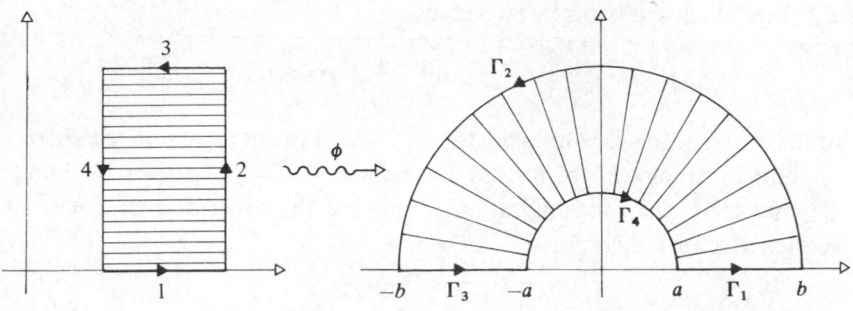

Fig. 9.7

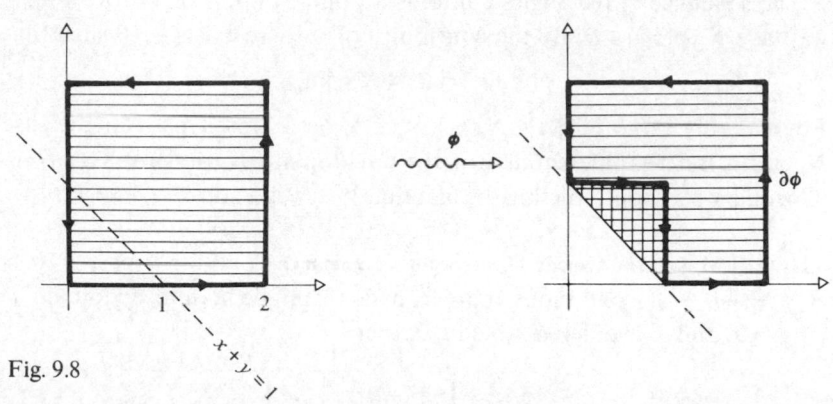

Fig. 9.8

of ϕ is to fold over the bottom left-hand corner of R. (Fig. 9.8). This shows that $\partial\phi$ need not be the boundary of the image $\phi(R)$.

However Figure 9.8 gives the clue that all points inside $\partial\phi$ lie in the image $\phi(R)$. We now prove this is true.

LEMMA 9.1. If $\phi:R\to\mathbb{C}$ is continuous, then $I(\partial\phi)\subseteq\phi(R)$ where $I(\partial\phi)=\{z\in\mathbb{C}\mid w(\partial\phi,z)\neq0\}$.

Proof. Suppose, to the contrary, that there exists $z_0\in I(\partial\phi)$, $z_0\notin\phi(R)$. Let $D=\mathbb{C}\setminus\{z_0\}$, then D is a domain and $\phi(R)\subseteq D$. If $f(z)=1/2\pi i(z-z_0)$ then f is differentiable in D and

$$\int_{\partial\phi}f=w(\partial\phi,z_0)$$

is a non-zero integer.

Subdivide the rectangle R into four equal rectangles $R^{(1)}$, $R^{(2)}$, $R^{(3)}$, $R^{(4)}$ and let $\phi^{(r)}$ be the restriction of ϕ to $R^{(r)}$. Then for $r=1,2,3,4$ the boundaries

$\partial \phi^{(r)}$ are all closed paths in D. Because

$$\int_{\partial \phi} f = \sum_{r=1}^{4} \int_{\partial \phi^{(r)}} f \quad \text{and} \quad \int_{\partial \phi^{(r)}} f = w(\partial \phi^{(r)}, z_0),$$

at least one of the four integrals is a non-zero integer. Denote the corresponding rectangle $R^{(r)}$ by R_1 and the restriction of ϕ to R_1 by ϕ_1. Dividing R_1 into four equal rectangles and repeating the process gives a nested sequence of rectangles

$$R \supseteq R_1 \supseteq \cdots \supseteq R_n \supseteq \cdots$$

where each $\int_{\partial \phi_n} f$ is a non-zero integer.

The sequence of rectangles contains a point $z_1 \in R$. For $\varepsilon = |\phi(z_1) - z_0|$, we have $N_\varepsilon(\phi(z_1)) \subseteq D$. By the continuity of ϕ there exists $\delta > 0$ such that

$$\phi(N_\delta(z_1) \cap R) \subseteq N_\varepsilon(\phi(z_1)).$$

For suitably large N, $R_N \subseteq N_\delta(z_1)$, so $\partial \phi_N$ is a closed path in the disc $N_\varepsilon(\phi(z_1))$. But f is differentiable in (the star-domain) $N_\varepsilon(\phi(z_1))$, so $\int_{\partial \phi_N} f = 0$. (Corollary 8.3), contradicting the fact that $\int_{\partial \phi_N} f$ is a non-zero integer.   □

THEOREM 9.2 (The Cauchy Theorem for a boundary).
If $\phi : R \to D$ is a continuous map from a rectangle R into a domain D (Fig. 9.9) and f is differentiable in D, then

$$\int_{\partial \phi} f = 0.$$

Proof. Cut R up into rectangles $\{R_{pq}\}$ such that $\phi(R_{pq})$ is contained in a disc $D_{pq} \subseteq D$. (Fig. 9.10) Let ϕ_{pq} be the restriction of ϕ to R_{pq}, then its

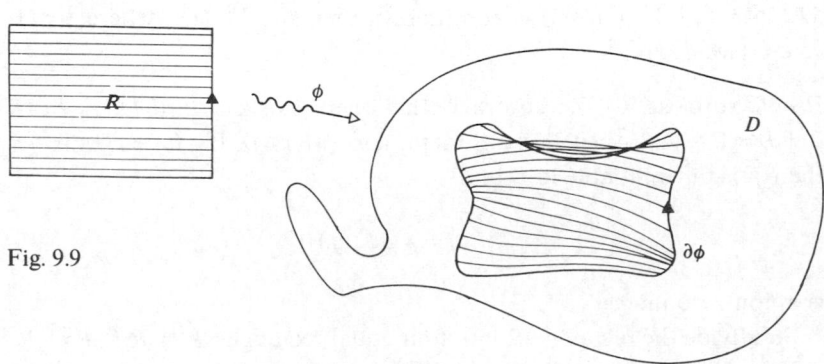

Fig. 9.9

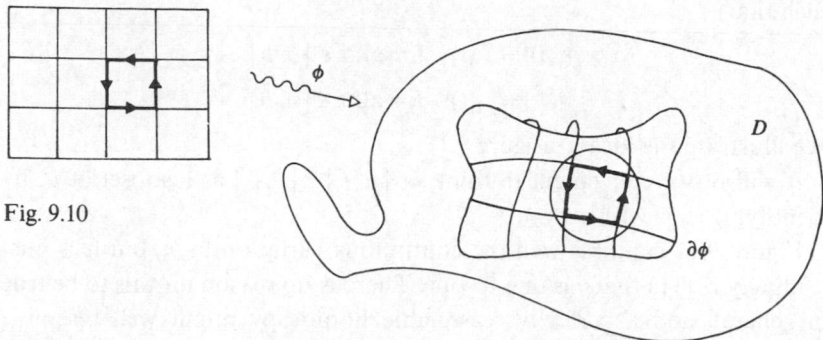

Fig. 9.10

boundary $\partial\phi_{pq}$ is a closed path in the disc D_{pq}, so

$$\int_{\partial\phi_{pq}} f = 0.$$

Adding up all the integrals for $1 \leqslant p \leqslant n$, $1 \leqslant q \leqslant m$, and cancelling integrals along opposite paths, we are left with

$$\int_{\partial\phi} f = 0. \qquad \square$$

3. Homotopy

A *homotopy* between two paths $\gamma_0:[a, b] \to D$ and $\gamma_1:[a, b] \to D$ is, roughly speaking, a *continuously varying family* of paths $\gamma_s:[a, b] \to D$, where s runs over the interval $[0, 1]$. At the start, $s = 0$ and $\gamma_s = \gamma_0$; at the end $s = 1$ and $\gamma_s = \gamma_1$.

How do we make this precise? Notice that the whole family depends on two variables: the parameter s, and the original variable t giving the position on the path. We can combine everything into a single function γ of (t, s) if we set

$$\gamma(t, s) = \gamma_s(t).$$

It is then natural to insist that γ, as a function of two real variables, be continuous. This will ensure that every intermediate γ_s defines a continuous path, and that these paths themselves vary continuously with s.

We are thus led to the following definition. A *homotopy* in D between γ_0 and γ_1, as above, is a continuous map

$$\gamma:[a, b] \times [0, 1] \to D$$

such that

$$\gamma(t, 0) = \gamma_0(t) \quad \text{for all } t \in [a, b]$$
$$\gamma(t, 1) = \gamma_1(t) \quad \text{for all } t \in [a, b].$$

We illustrate the idea in Figure 9.11.

It will prove convenient to think of $[a, b] \times [0, 1]$ as a subset of $\mathbb{C}$, by identifying (t, s) with $t + is$.

Figure 9.11 captures well the continuous variation in γ_s, but it is misleadingly nice in that γ is one-to-one. There is no reason for this to be true in general, and a perfectly reasonable homotopy might well resemble Figure 9.12.

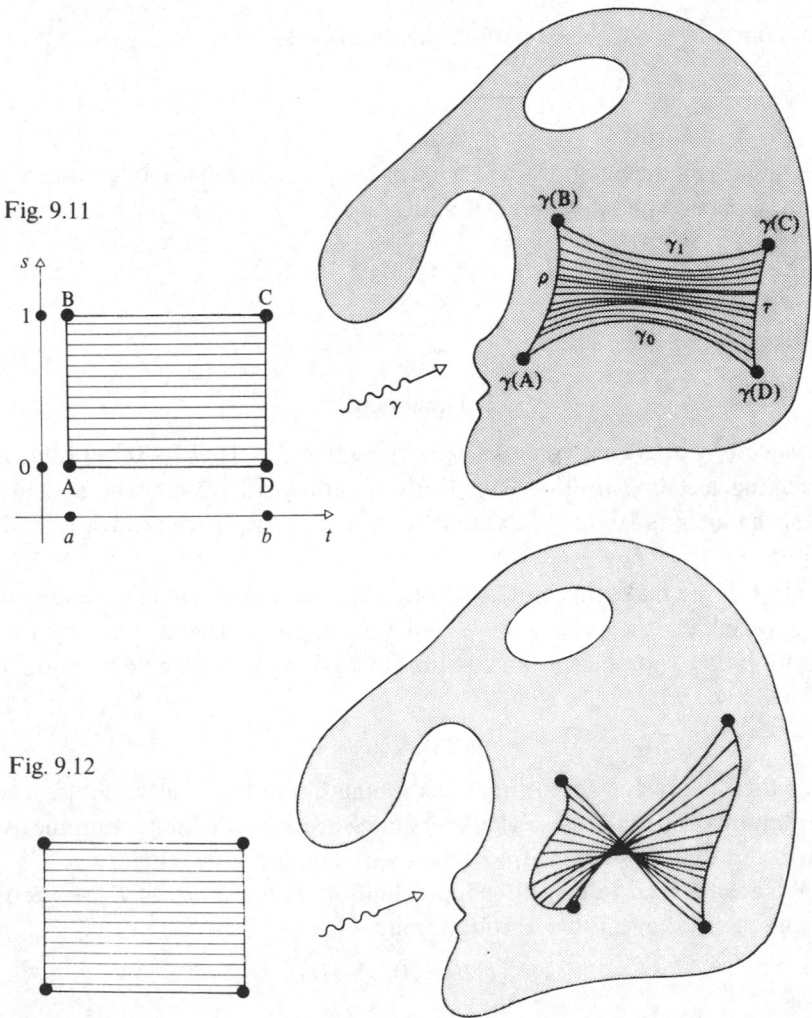

Fig. 9.11

Fig. 9.12

Geometrically, of course, $[a, b] \times [0, 1]$ is a rectangle, and its boundary is a closed path (with corners!). If we parametrize the four edges of the rectangle in some way, and then use γ to map the result into D, we obtain paths in D that join up to give a closed path, as for example in Figure 9.12.

Our aim is to apply Cauchy's Theorem for a boundary to this set of paths. Now, there is a problem: the two edges that give γ_0 and γ_1 are obviously useful to us, but the other two edges (marked ρ, τ in Figure 9.11) are going to be a nuisance. We therefore add conditions that eliminate them. There are two obvious ways to do this:

(*a*) Insist that each of ρ and τ goes to a single point in D.

(*b*) Insist that ρ and τ cancel each other out.

These two conditions yield two more restricted types of homotopy: fixed end point homotopy, and closed path homotopy. We describe these in detail in the next two sections.

4. Fixed end point homotopy

Let the rectangle R denote $\{t + is \in \mathbb{C} \,|\, a \leqslant t \leqslant b, \, 0 \leqslant s \leqslant 1\}$. Two paths $\gamma_0:[a, b] \to D$, $\gamma_1:[a, b] \to D$ are said to be *fixed end point homotopic* in D if there is a continuous map $\phi:R \to D$ such that (Fig. 9.13)

$$\phi(t, 0) = \gamma_0(t) \quad (a \leqslant t \leqslant b),$$

$$\phi(t, 1) = \gamma_1(t) \quad (a \leqslant t \leqslant b),$$

$$\phi(a, s) = z_0 \quad (0 \leqslant s \leqslant 1),$$

$$\phi(b, s) = z_1 \quad (0 \leqslant s \leqslant 1).$$

Fig. 9.13

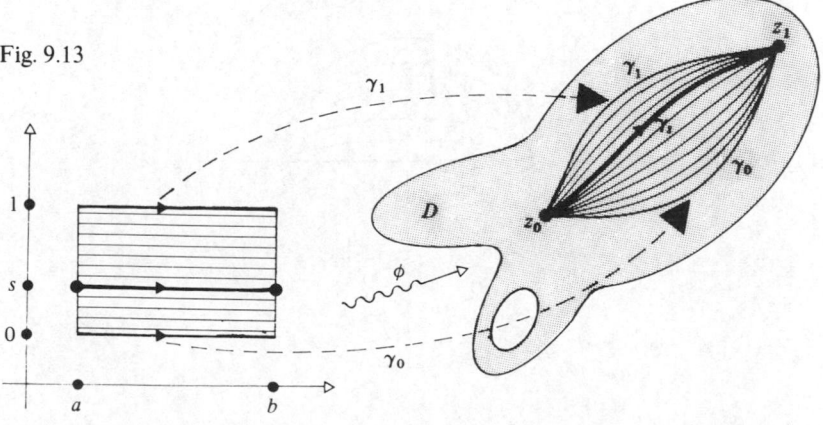

If we let $\gamma_s(t) = \phi(t, s)\,(a \leqslant t \leqslant b)$, then γ_s is a path in D from z_0 to z_1, and as s is increased from 0 to 1, γ_s is 'deformed continuously' in D from γ_0 to γ_1.

EXAMPLE. If $D = \{z \in \mathbb{C} \mid |z| < 2\}$, $\gamma_0(t) = t(-1 \leqslant t \leqslant 1)$, $\gamma_1(t) = e^{\frac{1}{2}\pi i(t-1)}$ $(-1 \leqslant t \leqslant 1)$, then γ_0 and γ_1 are fixed end point homotopic in D where $\phi(t) = (1-s)\gamma_0(t) + s\gamma_1(t)\,(-1 \leqslant t \leqslant 1, 0 \leqslant s \leqslant 1)$. (Fig. 9.14)

As a corollary of Theorem 9.2, we deduce:

THEOREM 9.3. If f is differentiable in a domain D and γ_0 is fixed end point homotopic in D to γ_1, then $\int_{\gamma_0} f = \int_{\gamma_1} f$.

Proof. We have a continuous map $\phi: \to D$ where

$$\phi(t, 0) = \gamma_0(t) \quad (a \leqslant t \leqslant b),$$
$$\phi(t, 1) = \gamma_1(t) \quad (a \leqslant t \leqslant b),$$
$$\phi(a, s) = z_0 \quad (0 \leqslant s \leqslant 1),$$
$$\phi(b, s) = z_1 \quad (0 \leqslant s \leqslant 1),$$

(as in Figure 9.13).

If $p_r:[0, 1] \to D$ is the point path $p_r(t) = z_r$ for $r = 0, 1$, then $\int_{p_r} f = 0$ and $\partial\phi = \gamma_0 + p_1 - \gamma_1 + p_0$. By Cauchy's Theorem for a Boundary,

$$\int_{\partial\phi} f = \int_{\gamma_0} f - \int_{\gamma_1} f = 0,$$

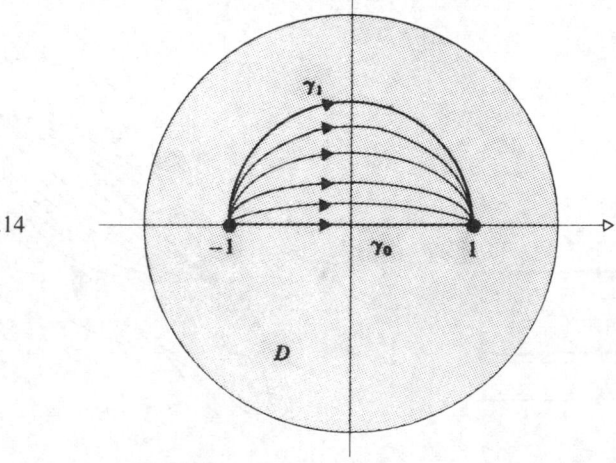

Fig. 9.14

so

$$\int_{\gamma_0} f = \int_{\gamma_1} f. \qquad \square$$

5. Closed path homotopy

Once more we take the rectangle R to be $\{t + is \in \mathbb{C} \mid a \leqslant t \leqslant b, 0 \leqslant s \leqslant 1\}$. Two paths $\gamma_0 : [a, b] \to D$, $\gamma_1 : [a, b] \to D$ are said to be *homotopic via closed paths* in D if there exists a continuous map $\phi : R \to D$ such that

$$\phi(t, 0) = \gamma_0(t) \qquad (a \leqslant t \leqslant b)$$
$$\phi(t, 1) = \gamma_1(t) \qquad (a \leqslant t \leqslant b)$$
$$\phi(a, s) = \phi(b, s) \quad (0 \leqslant s \leqslant 1).$$

(Fig. 9.15)

Again, if we define $\gamma_s(t) = \phi(t, s)$ $(a \leqslant t \leqslant b)$ then γ_s is a closed path in D and as s increases from 0 to 1, γ_s is 'deformed continuously' from γ_0 to γ_1.

EXAMPLE. For $|z_0| < K$, let $D = \{z \in \mathbb{C} \mid |z| < K, z \neq z_0\}$. For $|z_0| < \rho < K$, let $\gamma_0(t) = \rho e^{it}$ $(0 \leqslant t \leqslant 2\pi)$, for $0 < \varepsilon < K - |z_0|$, let $\gamma_1 = z_0 + \varepsilon e^{it}$ $(0 \leqslant t \leqslant 2\pi)$. (Fig. 9.16) Then γ_0 is homotopic to γ_1 through closed paths in D where

$$\phi(t, s) = (1 - s)\gamma_0(t) + s\gamma_1(t) \quad (0 \leqslant t \leqslant 2\pi, 0 \leqslant s \leqslant 1).$$

THEOREM 9.4. If f is differentiable in D and γ_0, γ_1 are closed paths which are homotopic via closed paths in D, then $\int_{\gamma_0} f = \int_{\gamma_1} f$.

Fig. 9.15

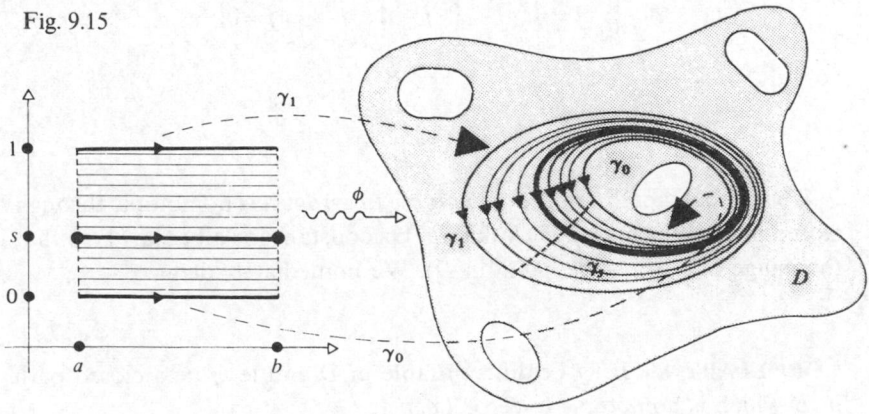

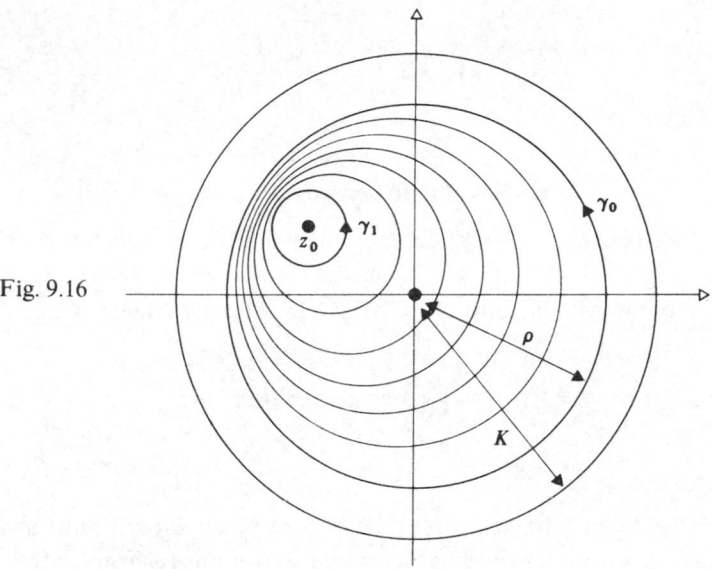

Fig. 9.16

Proof. We have a continuous map $\phi:R\to D$ such that

$$\phi(t, 0)=\gamma_0(t) \qquad (a\leqslant t\leqslant b)$$
$$\phi(t, 1)=\gamma_1(t) \qquad (a\leqslant t\leqslant b)$$
$$\phi(a, s)=\phi(b, s) \qquad (0\leqslant s\leqslant 1).$$

Let $\sigma(t)=\phi(a, t)\,(0\leqslant t\leqslant 1)$, then $\partial\phi=\gamma_0+\sigma-\gamma_1-\sigma$ (which means that σ is cut from γ_0 to γ_1 in the sense of §8.7). By Cauchy's Theorem for a boundary,

$$\int_{\partial\phi} f=\int_{\gamma_0} f+\int_{\sigma} f-\int_{\gamma_1} f-\int_{\sigma} f=0,$$

so

$$\int_{\gamma_0} f=\int_{\gamma_1} f. \qquad \square$$

We say that a path γ in D is *homotopic to zero* if it is homotopic through closed paths $\beta:[a, b]\to D$ such that $\beta(t)$ is constant for all $t\in[a, b]$ (so that the image of β is a single point in D). We immediately deduce:

COROLLARY 9.5. Let f be differentiable in D and let γ be a closed path in D which is homotopic to zero. Then $\int_{\gamma} f=0$. $\qquad \square$

The geometric significance of being homotopic to zero is that γ can be continuously deformed into a single point (or rather, to a path whose image is a single point $z_1 \in D$) as in Figure 9.17.

Starting with Cauchy's Theorem for a boundary we have deduced first the homotopy-invariance of the integral (for fixed end point or closed path homotopies) and hence that the integral is zero for a path homotopic to zero. However, we can also start with Corollary 9.5 and argue the other way if we wish, by virtue of

PROPOSITION 9.6. A closed contour γ in D is a boundary $\partial\phi$, up to parametrization, if and only if γ is homotopic to zero.

Proof. Note first that the possible need to reparametrize the interval on which γ is defined arises because of the way we have chosen a *specific* parametrization for a boundary $\partial\phi$. We can adjust the parameter provided the image of γ and the image of $\partial\phi$ coincide. This reduces the proof to a geometric argument. We now give the essence of the proof in a series of pictures, leaving to the reader the (routine) analytic definitions and verifications required to make it rigorous.

We define a map $H:R\to R$, where R is a rectangle, as shown in Figure 9.18. The definition proceeds in stages:
(1) Identify opposite vertical edges, to get a cylinder.
(2) Squash the top rim to a point, to get a cone.
(3) Open the cone out flat to get a disc.
(4) Stretch the disc to get a square.

Suppose that γ is a boundary, say $\gamma = \partial\phi$, where $\phi:R\to D$. Then $\phi \circ H:R\to D$ is a homotopy. The lower edge of R, marked by the heavy line in the first diagram, maps to (the image under ϕ of) $\partial\phi$. The top edge maps to a point. So γ is homotopic to zero.

Now suppose that γ is a closed curve, homotopic to zero. Then we can define a map of the cone into D such that the base circle goes to (the image

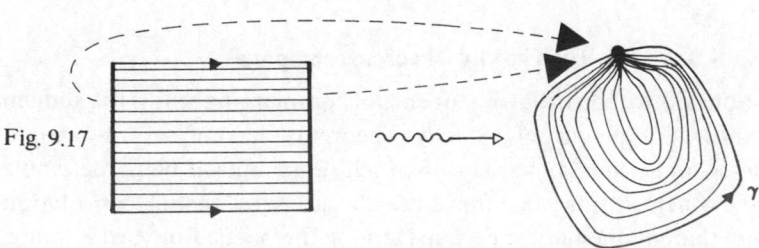

Fig. 9.17

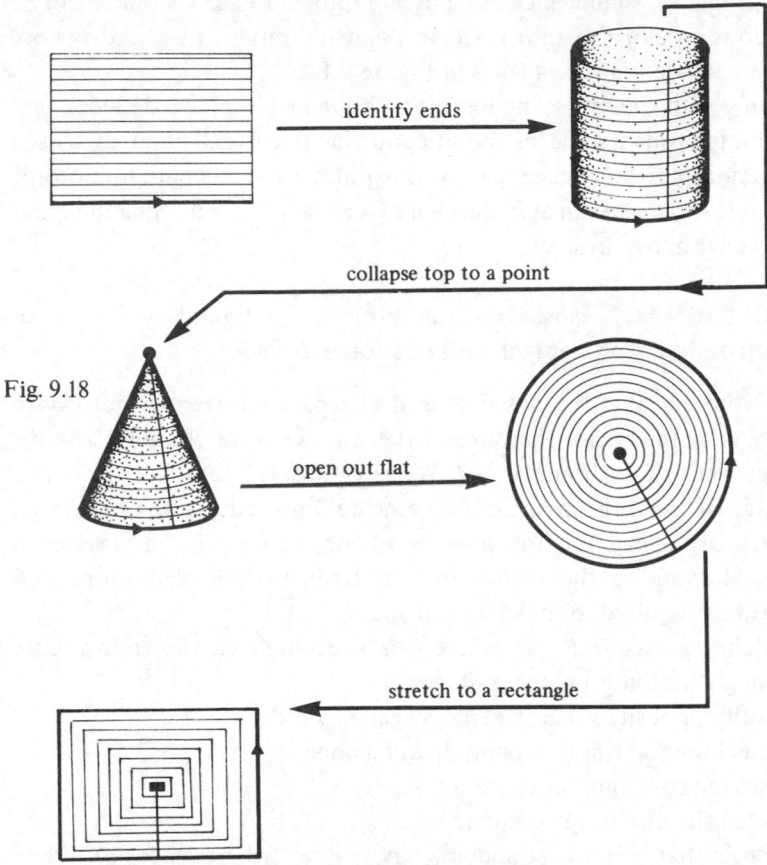

identify ends

collapse top to a point

Fig. 9.18

open out flat

stretch to a rectangle

of) γ. Therefore (reversing the last two steps in the definition of H) we can map R into D so that its perimeter goes to γ. Hence (up to parametrization) $\gamma = \partial \phi$.   □

6. The Cauchy Theorems compared

A common cause of distress for students of complex analysis is the sudden appearance of a plague of Cauchy Theorems, having several variant hypotheses and a similar variety of conclusions, but all being derivable from each other. At times like this it may be advisable to seek consolation elsewhere than mathematics: perhaps among the poets. Rudyard Kipling, in *In the Neolithic Age*, made the point admirably:

There are nine and sixty ways of constructing tribal lays,
And – every – single – one – of – them – is – right!

And it is much the same with the Cauchy Theorems: all of the different versions are essentially the same.

At the heart of all Cauchy-type theorems is the local existence of an antiderivative (§8.5). The theorems themselves supply, as hypotheses, conditions which permit this local result to be *globalized* in some way.

For example, in Chapter 8 the local result led to the central Cauchy Theorem, that if f is differentiable in D and a closed contour γ does not wind round points outside D, then $\int_\gamma f = 0$. This 'non-winding' condition in fact ensures that the 'local pieces' of antiderivative fit together well on the global level. The generalized version, with several contours $\gamma_1, \ldots, \gamma_n$, was a simple corollary found by making cuts between the contours.

In this chapter we studied what happens when a contour is deformed. Again the local existence of an antiderivative is involved: it lets us define the integral of f along an arbitrary path, and it gives the main result of that chapter, that the integral of f along a boundary is zero. This is also a globalization: associated with any boundary is a map of an entire rectangle, and the local antiderivatives all fit together properly across the image of this rectangle. Moreover, we saw that a path is a boundary if and only if it is homotopic to zero. Since homotopy leaves the integral round a closed contour invariant, this makes the reason why $\int_\gamma f = 0$ transparently clear.

We therefore have *two* main variants of the Cauchy Theorem: one for curves that don't wind round points outside D, and another for curves that are homotopic to zero. But, like the Colonel's Lady and Judy O'Grady, these are 'sisters under their skins'.

Obviously, if γ is homotopic to zero and $z_0 \notin D$, then

$$w(\gamma, z_0) = \frac{1}{2\pi i} \int_\gamma \frac{1}{z - z_0} \, dz = 0,$$

since $1/(z - z_0)$ is differentiable in D. So the 'non-winding' version of Cauchy's Theorem easily implies the 'homotopy' version.

It is in fact strictly stronger, in the following sense: the curve of Figure 9.19 does not wind round any $z_0 \notin D$, but it is manifestly *not* homotopic to zero. (Though this is harder to prove than it may appear.) So the 'non-winding' hypothesis is *weaker*, hence applies directly to more cases.

This surface difference vanishes when we look more deeply, however, because every curve that does not wind round any $z_0 \notin D$ can be transformed, by a series of cuts whose contributions to the integral cancel, into a closed path (or set of paths) homotopic to zero. For example, Figure 9.19 is so transformed in Figure 9.20. This fact is also not trivial to prove, but it shows that the *practical* consequences of the extra generality are largely spurious. (The theoretical consequences are more important: the 'non-winding' condition is part of 'homology theory' and what we have here is the relation between homotopy and homology. But that's another story.)

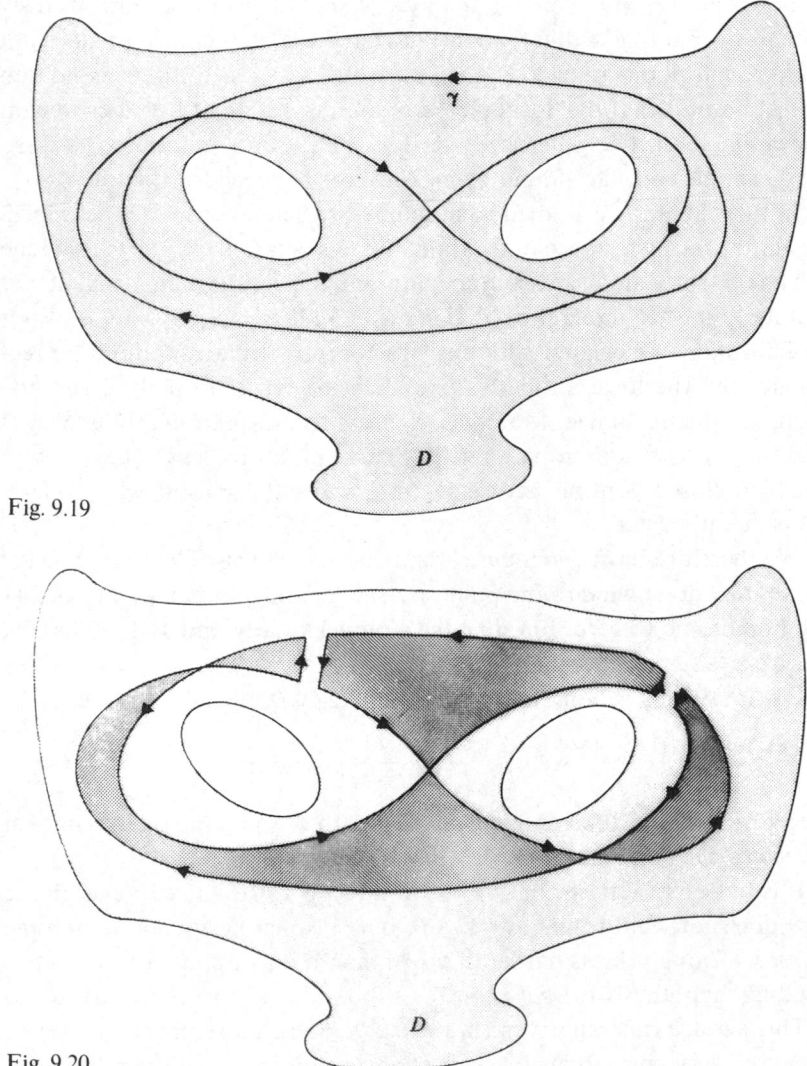

Fig. 9.19

Fig. 9.20

Exercises 9

1. Let $D = \{z \in \mathbb{C} \mid z \neq 0\}$ and $S_r(t) = re^{it}$ $(0 \leqslant t \leqslant 2\pi)$ for $r = 1, 2$. Define a continuous map $\phi : R \to D$ where R is a rectangle such that

$$\int_{\partial\phi} f = \int_{S_1} f - \int_{S_2} f$$

for any function f differentiable in D. Hence use Theorem 9.2 to deduce that

$$\int_{S_1} f = \int_{S_2} f.$$

Describe a homotopy via closed paths in D from S_1 to S_2.

2. Let γ_1, γ_2 be closed paths in D which are homotopic via closed paths in D. By making a suitable cut σ from γ_1 to γ_2, describe a fixed endpoint homotopy in D from γ_1 to $\sigma + \gamma_2 - \sigma$. Draw a picture to illustrate the continuous deformation.

3. Let the boundary $\partial\phi$ of a continuous map $\phi : R \to D$ be subdivided into two sub-paths $\partial = \gamma_1 + \gamma_2$. Describe a fixed endpoint homotopy from γ_1 to $-\gamma_2$ in D and draw a picture to illustrate the deformation from γ_1 to γ_2.

4. Draw the semicircle $\gamma(t) = e^{it}$ $(-\frac{1}{2}\pi \leqslant t \leqslant \frac{1}{2}\pi)$ in $D = \mathbb{C} \backslash \{0\}$. Define two explicit polygonal paths λ_1, λ_2 from $-i$ to i in D such that

$$\int_{\lambda} f = \int_{\gamma} f$$

is true for all f differentiable in D when $\lambda = \lambda_1$ but not when $\lambda = \lambda_2$.

5. Let $\gamma : [0, 1] \to \mathbb{C}$ be given by $\gamma(0) = \gamma(1) = 0$ and

$$\gamma(t) = \begin{cases} t + it \sin(\pi/t) & \text{for } 0 \leqslant t \leqslant \frac{1}{2} \\ (1-t) - i(1-t)\sin(\pi/(1-t)) & \text{for } \frac{1}{2} \leqslant t \leqslant 1. \end{cases}$$

Show that γ is a closed path but not a contour and draw a sketch.

Integrate the following functions around γ:

(i) $\cos^3(z^2)$ (ii) $\sum_{n=1}^{\infty} z^n/n$ (iii) $1/(z - \frac{1}{3}\sqrt{2})$.

7. Let f be a function differentiable in D. For a closed path γ in D, beginning and ending at z_0, the *integral value* I_γ is the complex number

$$I_\gamma = \int_\gamma f.$$

Show that the set of integral values forms a commutative group $I(f, D)$ under the operation

$$I_\gamma + I_\delta = I_{\gamma + \delta}.$$

Determine the group of integral values in the following cases:

(i) $f(z) = 1/z$, $D = \mathbb{C} \backslash \{0\}$,

(ii) $f(z) = \cos z$, $D = \mathbb{C} \backslash \{0\}$,

(iii) $f(z) = 2/(z-1) + 3/(z+1)$, $D = \mathbb{C} \backslash \{\pm 1\}$,

(iv) $f(z) = 1/(z-1) + 2/(z+1)$, $D = \mathbb{C} \backslash \{\pm 1\}$.

8. *The Fundamental Group.* Let D be a domain and $z_0 \in D$. For closed paths γ, δ in D which begin and end at z_0, define $\gamma \simeq \delta$ to mean that γ, δ are fixed endpoint homotopic in D. Show that $\simeq$ is an equivalence relation. Let $[\gamma]$ denote the equivalence class containing γ and let $\pi(D, z_0)$ be the set of equivalence classes. Define the operation * on $\pi(D, z_0)$ by

$$[\gamma]*[\delta] = [\gamma + \delta].$$

Check that * is well-defined and show that $\pi(D, z_0)$ is a group under *, specifying the identity element and the inverse of $[\gamma]$.

For any other point $z_1 \in \mathbb{C}$ and any path σ in D from z_0 to z_1, define $g : \pi(D, z_0) \to \pi(D, z_1)$ by

$$g([\gamma]) = [-\sigma + \gamma + \sigma].$$

Show that g is an isomorphism of groups and hence deduce that $\pi(D, z_0)$ is independent of the choice of $z_0 \in D$. (For this reason the group $\pi(D, z_0)$ is usually denoted by $\pi(D)$. It is called the *fundamental group* of D.)

Describe (without formal proof) the fundamental groups of the following domains:

(i) $\mathbb{C}$ (ii) $|z| < 1$ (iii) $1 < |z| < 2$ (iv) $\mathbb{C} \setminus \{0\}$ (v) $\mathbb{C} \setminus \{\pm 1\}$ (vi) $\mathbb{C} \setminus \mathbb{Z}$.

10. Let f be differentiable in the domain D, and let γ, δ be closed contours in D beginning and ending at the same point z_0. Show that if $\gamma \simeq \delta$ in the sense of question 9, then the integral values I_γ, I_δ are the same. Prove that the map $h : \pi(D) \to I(f, D)$ where $h([\gamma]) = I_\gamma$ is a well-defined group homomorphism. Describe the homomorphism h for each f, D given in question 7.

11. *Integrals along arbitrary paths.* For fixed z_1, $z_2 \in D$, suppose that γ_0 is a fixed path, and γ is a variable path, in D from z_1 to z_2. Show that

$$\int_\gamma f = \int_{\gamma_0} f + I_\sigma$$

for some $\sigma \in I(f, D)$ and that if γ is deformed continuously in a homotopy via closed paths in D, then I_σ remains constant. For $z_1 = -i$, $z_2 = i$, determine all the possible values of $\int_\gamma f$ for each f, D in question 7.

10
Taylor series

We now reach a stage of the theory where we can take a great leap forward and show, as promised repeatedly, that any differentiable complex function has a local power series expansion. On the slim assumption that the derivative of f exists throughout a domain D, we find that near any point z_0 in D we have a power series expansion

$$f(z_0+h) = \sum_{n=0}^{\infty} a_n h^n \quad \text{for } z_0+h \in N_r(z_0) \subseteq D,$$

valid in *any* disc $N_r(z_0)$ within D. (Fig. 10.1)

This releases a tidal wave of results. For instance, we know that a power series can be differentiated term by term as many times as we like and that

$$f(z_0+h) = \sum_{n=0}^{\infty} \frac{f^{(n)}(z_0)}{n!} h^n \quad \text{for } z_0+h \in N_r(z_0) \subseteq D.$$

Thus if we only insist that the first derivative f' exists in D, and no more, it follows that all the higher derivatives exist and the function is equal to its Taylor series inside $N_r(z_0)$. More subtle consequences follow in this and subsequent chapters. It is this sequence of results that gives complex analysis its own special flavour.

Fig. 10.1

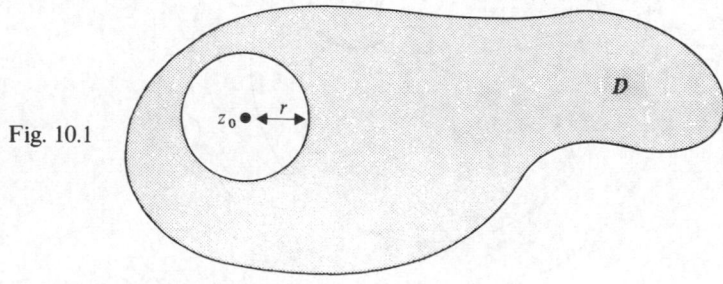

177

1. Cauchy's Integral Formula

The proof that any differentiable complex function can be expressed as a power series depends on a result of Cauchy, itself of intrinsic interest:

LEMMA 10.1 (Cauchy's Integral Formula for a circle).
Let f be differentiable in the disc $N_R(z_0) = \{z \in \mathbb{C} \mid |z - z_0| < R\}$. For $0 < r < R$, let C_r be the path

$$C_r(t) = z_0 + re^{it} \quad (0 \leqslant t \leqslant 2\pi),$$

then for $|w - z_0| < r$ we have

$$f(w) = \frac{1}{2\pi i} \int_{C_r} \frac{f(z)}{z - w} \, dz.$$

Proof. Fix w such that $|w - z_0| < r$. The function $F(z) = (f(z) - f(w))/(z - w)$ is differentiable in the domain

$$D = \{z \in \mathbb{C} \mid |z - z_0| < R, \ z \neq w\}$$

Let $0 < \varepsilon < r - |w - z_0|$. Then the circle S_ε, centre w, radius ε,

$$S_\varepsilon(t) = w + \varepsilon e^{it} \quad (0 \leqslant t \leqslant 2\pi)$$

lies in D, as do all the points inside C_r and outside S_ε. (Fig. 10.2)
  By the Generalized Cauchy Theorem (Theorem 8.9),

$$\int_{C_r} F(z) \, dz = \int_{S_\varepsilon} F(z) \, dz. \tag{1}$$

Fig. 10.2

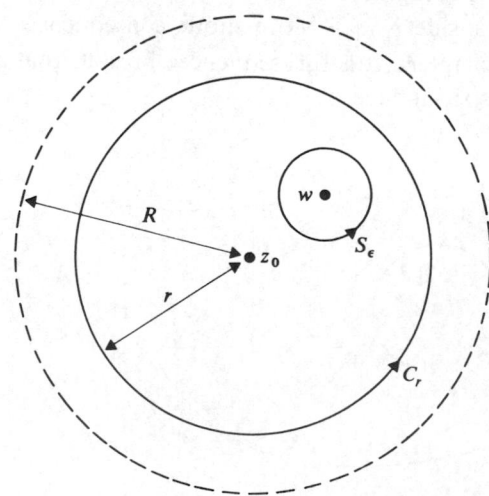

Now $\lim_{z \to w} F(z) = f'(w)$, so for some $\delta > 0$, $M \geqslant 0$, we have
$$0 < |z - w| < \delta \Rightarrow |F(z)| \leqslant M.$$
Thus for $\varepsilon < \delta$ we have
$$\left| \int_{S_\varepsilon} F(z)\, dz \right| \leqslant M \cdot 2\pi\varepsilon$$
by the Estimation Lemma (Lemma 6.10). From (1),
$$\left| \int_{C_r} F(z)\, dz \right| \leqslant 2M\pi\varepsilon$$
and, since ε is arbitrary, this implies
$$\int_{C_r} F(z)\, dz = 0$$
Therefore
$$\int_{C_r} \frac{f(z)}{z - w}\, dz = \int_{C_r} \frac{f(w)}{z - w}\, dz$$
$$= f(w) \int_{C_r} \frac{dz}{z - w}$$
$$= f(w) \cdot 2\pi i.$$
In other words,
$$f(w) = \frac{1}{2\pi i} \int_{C_r} \frac{f(z)}{z - w}\, dz. \qquad \square$$

2. Taylor series

Using the Cauchy Integral Formula we can now expand $f(z_0 + h)$ as a power series expansion (in powers of h) with coefficients expressed as integrals.

LEMMA 10.2. Let f be differentiable in $N_R(z_0)$. Then
$$f(z_0 + h) = \sum_{n=0}^{\infty} a_n h^n$$
where the series converges absolutely for $|h| < R$. Further, if $0 < r < R$,
$$C_r(t) = z_0 + re^{it} \quad (0 \leqslant t \leqslant 2\pi),$$

then

$$a_n = \frac{1}{2\pi i} \int_{C_r} \frac{f(z)}{(z-z_0)^{n+1}}\, dz.$$

Proof. Fix h such that $0 < |h| < R$ and initially suppose that r satisfies $|h| < r < R$. (Fig. 10.3)

Then Cauchy's Integral Formula gives

$$
\begin{aligned}
f(z_0 + h) &= \frac{1}{2\pi i} \int_{C_r} \frac{f(z)}{z-(z_0+h)}\, dz \\
&= \frac{1}{2\pi i} \int_{C_r} f(z) \left\{ \frac{1}{z-z_0} + \frac{h}{(z-z_0)^2} + \cdots + \frac{h^m}{(z-z_0)^{m+1}} \right. \\
&\quad \left. + \frac{h^{m+1}}{(z-z_0)^{m+1}(z-z_0-h)} \right\} dz \\
&= \sum_{n=0}^{m} a_n h^n + A_m
\end{aligned}
$$

where

$$a_n = \frac{1}{2\pi i} \int_{C_r} \frac{f(z)}{(z-z_0)^{n+1}}\, dz$$

and

$$A_m = \frac{1}{2\pi i} \int_{C_r} \frac{f(z)h^{m+1}}{(z-z_0)^{m+1}(z-z_0-h)}\, dz.$$

Fig. 10.3

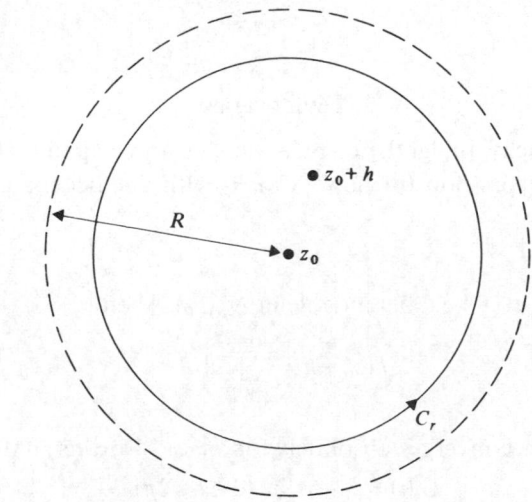

We demonstrate that $\lim_{m \to \infty} A_m = 0$. First we note that f is differentiable, hence continuous, so $\phi(t) = |f(C_r(t))|$ is a continuous real function on $[0, 2\pi]$. From real analysis, we know that ϕ is bounded, say $\phi(t) \le M$, so

$$|f(z)| \le M \quad \text{for } z \text{ on } C_r.$$

Now $|h| < r$, $|z - z_0| = r$ and

$$|z - z_0 - h| \ge |\,|z - z_0| - |h|\,| = r - |h|.$$

Therefore the Estimation Lemma gives

$$|A_m| \le \frac{1}{2\pi} \frac{M|h|^{m+1}}{r^{m+1}(r - |h|)} 2\pi r = \frac{M|h|}{r - |h|} \left(\frac{|h|}{r}\right)^n$$

and since we chose $|h| < r$, this tends to zero as m tends to infinity. Thus

$$\lim_{m \to \infty} \left(f(z_0 + h) - \sum_{n=0}^{m} a_n h^n \right) = 0,$$

which means that

$$f(z_0 + h) = \sum_{n=0}^{\infty} a_n h^n.$$

This expansion is valid for $|h| < R$, and

$$a_n = \frac{1}{2\pi i} \int_{C_r} \frac{f(z)}{(z - z_0)^{n+1}} \, dz$$

for $|h| < r$. The latter restriction can now be seen to be unnecessary, for the integrand is differentiable for $0 < |z - z_0| < R$, so the integral is unchanged if r is varied in the range $0 < r < R$. This completes the proof of the lemma. $\qquad \square$

Once we know that a power series expansion exists, we can then use our knowledge of power series to deduce:

THEOREM 10.3 (Taylor series).
If f is differentiable in a domain D, then all the higher derivatives of f exist throughout D and in any disc $N_R(z_0) \subseteq D$ the Taylor series expansion

$$f(z_0 + h) = \sum_{n=0}^{\infty} \frac{f^{(n)}(z_0)}{n!} h^n$$

is valid. Furthermore, if $0 < r < R$ and $C_r(t) = z_0 + re^{it}$ $(0 \le t \le 2\pi)$, then

$$f^{(n)}(z_0) = \frac{n!}{2\pi i} \int_{C_r} \frac{f(z)}{(z - z_0)^{n+1}} \, dz.$$

Proof. From Lemma 10.2,

$$f(z_0 + h) = \sum_{n=0}^{\infty} a_n h^n \quad \text{for } |h| < R.$$

In other words, putting $z = z_0 + h$,

$$f(z) = \sum_{n=0}^{\infty} a_n (z - z_0)^n \quad \text{for } |z - z_0| < R.$$

Now a power series may be differentiated as often as we please and, by Corollary 4.14, we have

$$f^{(n)}(z_0) = n! a_n = \frac{n!}{2\pi i} \int_{C_r} \frac{f(z)}{(z - z_0)^{n+1}} \, dz.$$

This gives the desired integral expression for $f^{(n)}(z_0)$ and substituting $a_n = f^{(n)}(z_0)/n!$ in the power series gives the Taylor expansion. $\qquad\square$

Theorem 10.3 was first proved by Cauchy in 1831 by the method given here. The series is named after Brooke Taylor who was the first to publish the idea that a function can be expanded as a power series of the form

$$f(x + h) = \sum_{n=0}^{\infty} \frac{f^{(n)}(x)}{n!} h^n$$

in 1715. The theory of Taylor was restricted to real functions and it will come as no surprise to the reader to know that the idea was known to others before Taylor's publication, specifically to Gregory who was aware of it some 45 years earlier, and to Isaac Newton in 1691. During the eighteenth century there were various attempts at basing the theory of real analysis on power series, of which the most famous was that of Lagrange in 1797. Cauchy used power series extensively in complex analysis. It is a curious quirk of fate that he quoted the counterexample $f(x) = e^{-1/x^2}$ to show that not all infinitely differentiable real functions are equal to their Taylor series in 1829, and just two years later went on to show that differentiable complex functions *all* have power series expansions.

A real function $f: D \to \mathbb{R}$ (where $D \subseteq \mathbb{R}$) or a complex function $f: D \to \mathbb{C}$ (where $D \subseteq \mathbb{C}$) is said to be *analytic* if for each $\alpha \in D$ it has a power series expansion

$$f(\alpha + h) = \sum_{n=0}^{\infty} a_n h^n$$

valid in some neighbourhood of α. Cauchy demonstrated that in the *real* case there are functions which are infinitely differentiable but not analytic. But in the *complex* case he showed that any function which is differentiable *once* in a domain is necessarily an analytic function. At a stroke he showed that complex analysis is simpler than real analysis by reducing the general study of differentiable complex functions to particular computations with power series, giving the sequence of theorems which unfold in the next few sections.

Note that a complex function f is differentiable *if and only if* it is analytic. The two words just emphasize different points of view, and may be used interchangeably.

3. Morera's Theorem

First we have a partial converse of Cauchy's Theorem, due to Morera (1889):

THEOREM 10.4. If f is continuous in a domain D and $\int_\gamma f = 0$ for all closed contours γ in D, then f is differentiable.

Proof. We already know (Theorem 6.11) that if $\int_\gamma f = 0$ for all closed contours in D, then there exists a differentiable function F in D whose derivative is f. But F is now known to be infinitely differentiable and $F'' = f'$, so f is differentiable. $\qquad\square$

This theorem explains why the reader was warned in Chapter 6 that it is futile to attempt to find an antiderivative F for a non-differentiable function f. It cannot have one. There is a class of functions, including $f(z) = |z|$, which are continuous but not differentiable. For such functions integration can be performed by using the formula

$$\int_\gamma f = \int_a^b f(\gamma(t))\gamma'(t)\,\mathrm{d}t$$

but the Fundamental Theorem of Contour Integration is no use whatsoever because f has no antiderivative (in stark contrast to the real case, where *all* continuous functions have an antiderivative.)

Summarizing our knowledge of the existence of derivatives and antiderivatives, we have the following:

If f is differentiable in a domain D, then all the higher derivatives of f exist. The function f can only have an antiderivative when f is itself differentiable and, even then, only *local* antiderivatives can be guaranteed. By the latter we mean that if $D_1 \subseteq D$ is simply connected, then f must have an antiderivative in D_1 (Theorem 8.11). In particular we can guarantee the existence of an antiderivative for a differentiable function in any disc in its domain D.

4. Cauchy's estimate

Included in Theorem 10.3 is a generalization of Cauchy's Integral Formula to the higher derivatives of a differentiable function:

$$f^{(n)}(z_0) = \frac{n!}{2\pi i} \int_{C_r} \frac{f(z)}{(z-z_0)^{n+1}} \, dz$$

where $N_R(z_0) \subseteq D$ and $0 < r < R$. With the standard conventions that $f^{(0)}(z) = f(z)$ and $0! = 1$, this formula is true for all integers $n \geqslant 0$. Using it we can give an upper bound for $|f^{(n)}(z_0)|$.

LEMMA 10.5 (Cauchy's Estimate).
If f is differentiable for $|z - z_0| < R$, $0 < r < R$ and $|f(z)| \leqslant M$ for $|z - z_0| = r$, then

$$|f^{(n)}(z_0)| \leqslant \frac{Mn!}{r^n}$$

for all integers $n \geqslant 0$.

Proof.
$$f^{(n)}(z_0) = \frac{n!}{2\pi i} \int_{C_r} \frac{f(z)}{(z-z_0)^{n+1}} \, dz$$

$$\leqslant \frac{n!}{2\pi} \frac{M}{r^{n+1}} \cdot 2\pi r$$

$$= \frac{Mn!}{r^n}. \qquad \qquad \square$$

Cauchy's Estimate yields an important theorem of Liouville, which has unexpected applications to a purely algebraic problem. First the theorem:

THEOREM 10.6 (Liouville's Theorem).
If f is differentiable and bounded in the whole complex plane, then f is a constant.

Proof. Suppose that $|f(z)| \leqslant M$. Then Cauchy's Estimate for the derivative gives

$$|f'(z)| \leqslant \frac{M}{r}.$$

Since f is differentiable in $\mathbb{C}$, we may let $r \to \infty$, then M/r may be made as small as we please and, because $f'(z)$ is independent of r, we obtain

$$|f'(z)| = 0.$$

Thus $f' = 0$ throughout $\mathbb{C}$ and f is constant. $\qquad\qquad\qquad\square$

Now the application:

THEOREM 10.7 (The 'Fundamental Theorem of Algebra').
Let $P(z) = z^n + a_1 z^{n-1} + \cdots + a_n$ be a polynomial where $n \geqslant 1$ and $a_1, \ldots, a_n \in \mathbb{C}$. Then there exists $w \in \mathbb{C}$ such that $P(w) = 0$.

Proof. If $P(z) \neq 0$ for all $z \in \mathbb{C}$, then $1/P(z)$ is differentiable throughout $\mathbb{C}$. For $z \neq 0$

$$P(z)/z^n = 1 + (a_1/z) + \cdots + (a_n/z^n) \to 1 \quad \text{as } |z| \to \infty$$

So there exists $k > 0$ such that

$$\left| \frac{P(z)}{z^n} \right| \geqslant \tfrac{1}{2} \quad \text{for } |z| > k.$$

Thus

$$\left| \frac{1}{P(z)} \right| \leqslant 2/|z^n| \leqslant 2/k^n \quad \text{for } |z| > k.$$

The same bound works for $1/P(z)$ throughout the complex plane. For $|z_0| \leqslant k$ we simply take a circle C_R, centre z_0, radius R, which is so large that $|z| > k$ for all z on C_R, then

$$|1/P(z)| \leqslant 2/k^n \quad \text{for all } z \text{ on } C_R,$$

and Cauchy's Estimate gives

$$|1/P(z_0)| \leqslant 2/k^n.$$

By Liouville's Theorem, $1/P(z)$ is constant; hence $P(z)$ is constant. But this contradicts $n \geqslant 1$. Hence $P(w) = 0$ for some $w \in \mathbb{C}$. $\qquad\square$

It follows in the usual way that any polynomial $P(z)$ with complex coefficients can be expressed as a product of terms of degree 1

$$P(z) = (z - \alpha_1)(z - \alpha_2) \ldots (z - \alpha_n).$$

5. Zeros

We now broaden our perspective from polynomials and look at the zeros of arbitrary differentiable functions. A zero of a differentiable function

$f : D \to \mathbb{C}$ is a point $z_0 \in D$ for which $f(z_0) = 0$.

Expanding f in a Taylor series about the zero z_0, we have

$$f(z) = \sum_{n=0}^{\infty} a_n (z - z_0)^n \quad \text{for } |z - z_0| < R$$

where $N_R(z_0) \subseteq D$. Then $a_0 = f(z_0)$ is zero and two distinctly different things can occur; either all the other a_i are zero, in which case $f(z) = 0$ in $N_R(z_0)$, or there exists m such that

$$a_0 = a_1 = \cdots = a_{m-1} = 0, \quad \text{but } a_m \neq 0.$$

In the latter case we say that z_0 is a *zero of (finite) order m*. It is easy to see that a zero of order m may also be characterized by the conditions

$$f(z_0) = f'(z_0) = \cdots = f^{(m-1)}(z_0) = 0, \quad f^{(n)}(z_0) \neq 0.$$

Another useful expression for such a zero is to write

$$f(z) = (z - z_0)^m g(z) \quad (|z - z_0| < R)$$

where $g(z) = \sum_{n=0}^{\infty} a_{m+n}(z - z_0)^n$ is differentiable for $|z - z_0| < R$ and $g(z_0) = a_m \neq 0$.

This leads to a fundamentally important idea. A zero of a differentiable function f is said to be *isolated* if some disc centred upon it contains no other zeros, that is there exists $\delta > 0$ such that

$$0 < |z - z_0| < \delta \quad \text{implies } f(z) \neq 0.$$

LEMMA 10.8. A zero of finite order is isolated.

Proof. Write $f(z) = g(z)(z - z_0)^m$ for $|z - z_0| < R$ where g is differentiable and $g(z_0) \neq 0$. Then g is certainly continuous at z_0, so, taking $\varepsilon = \frac{1}{2}|g(z_0)|$, there exists $\delta > 0$ such that

$$|z - z_0| < \delta \quad \text{implies } |g(z) - g(z_0)| < \varepsilon.$$

Therefore when $|z - z_0| < \delta$, we have

$$|g(z)| \geq | \, |g(z_0)| - |g(z_0) - g(z)| \, | > 2\varepsilon - \varepsilon = \varepsilon.$$

In particular, we have $g(z) \neq 0$. But if $0 < |z - z_0| < \delta$ then $|z - z_0|^m \neq 0$, so $f(z) = g(z)(z - z_0)^m \neq 0$, as required. $\qquad \square$

COROLLARY 10.9. Let S be a set of zeros of a differentiable function f in D, having a limit point $z_0 \in D$. Then in any disc $N_R(z_0) \subseteq D$, f is identically zero.

Proof. Because z_0 is a limit point of S, we can select a sequence $\{z_n\}_{n \geq 1}$ in S which tends to z_0. Then $f(z_0) = \lim_{n \to \infty} f(z_n) = 0$. So z_0 is a zero of f

which is not isolated; hence it does not have finite order and

$$f(z_0+h)=\sum_{n=0}^{\infty} a_n h^n$$

in any disc $N_R(z_0) \subseteq D$ with all the coefficients a_n are zero. ☐

From this we deduce:

PROPOSITION 10.10. If f is differentiable in a domain D and S is a set of zeros of f with a limit point z_0 in D, then f is identically zero in D.

Proof. Corollary 10.9 gives $f(z)=0$ for all z in any disc $N_R(z_0) \subseteq D$.

For any other $z \in D$, choose a path $\gamma:[a, b] \to D$ from z_0 to z. (Figure 10.4) We show $f(\gamma(t))=0$ for all $t \in [a, b]$. Certainly we can find $\delta>0$ such that

$$a \leqslant t < a+\delta \quad \text{implies } \gamma(t) \in N_R(z_0),$$

whence $f(\gamma(t))=0$ for $a \leqslant t < a+\delta$. Let s be the least upper bound of those $x \in [a, b]$ such that $f(\gamma(t))=0$ for $a \leqslant t < x$. Then $a+\delta \leqslant s \leqslant b$.

By continuity, $f(\gamma(s))=0$. If s were strictly less than b, then $\gamma(s)$ would be a non-isolated zero, so f would be identically zero in a neighbourhood of $\gamma(s)$ and we would find an interval $[s, s+\kappa]$ in which $f(\gamma(t))$ is zero, contradicting the definition of s. Hence $s=b$ and $f(z)=f(\gamma(b))=0$, as required. ☐

As an important consequence, we have

THEOREM 10.11 (The Identity Theorem).
If f and g are analytic in a domain D and $f(z)=g(z)$ for all $z \in S \subseteq D$ where S has a limit point in D, then $f=g$ throughout D. ☐

Fig. 10.4

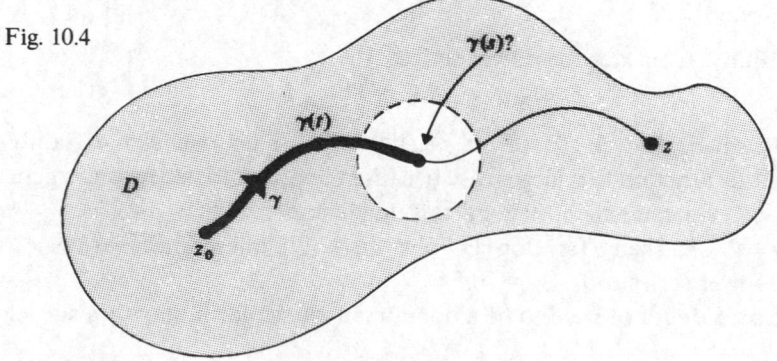

Proof. Apply Proposition 10.9 to $f-g$.   □

It is essential that the relevant limit point of S be in D: if not the theorem is false. For consider

$$f(z)=\sin(1/z)$$
$$g(z)=0$$

in $D=\mathbb{C}\setminus\{0\}$. Then $f(z)=g(z)$ for $z=\pm 1/(n\pi)$; $z_0=0$ is a limit point of $S=\{\pm 1/(n\pi)\}$, but $f\neq g$ in D.

6. Extension functions

A function $f:D\to\mathbb{C}$ is said to be an *extension function* of $h:S\to\mathbb{C}$ if $S\subseteq D$ and $f(z)=h(z)$ for all $z\in S$.

EXAMPLE 1. $f(z)=1/(1-z)$ on $D=\mathbb{C}\setminus\{1\}$ is an extension function of $h(z)=\Sigma_{n=0}^{\infty} z^n$ on $S=\{z\in\mathbb{C}\,|\,|z|<1\}$.

Suppose that D is a domain and S is a subset with a limit point in D. Then the Identity Theorem shows that if a function $f:S\to\mathbb{C}$ has an extension function $f:D\to\mathbb{C}$ which is differentiable then the extension is unique.

As an application, suppose that $f:D\to\mathbb{C}$ is differentiable and consider the Taylor expansion $f(z)=\Sigma\, a_n(z-z_0)^n$ of f in a disc $N_r(z_0)\subseteq D$. Then f is the only possible differentiable extension of the Taylor expansion to the whole of D. This means that the Taylor expansion of f about any point in its domain contains all the information required to determine f throughout the domain. (We will employ this fact to great advantage in Chapter 14.)

EXAMPLE 2. Consider the power series

$$z-\tfrac{1}{2}z^2+\cdots+(-1)^{n+1}z^n/n+\cdots$$

on the small disc $S=\{z\in\mathbb{C}\,|\,|z|<1/1000000\}$. Then we can certainly extend this function to a larger disc because the power series is convergent for $|z|<1$. We can extend beyond this, for instance, if $K=\{t\in\mathbb{R}\,|\,t\leqslant -1\}$ and $D=\mathbb{C}\setminus K$, then $f(z)=\log(1+z)$ is the (only) differentiable extension of the power series to D. (Fig. 10.5)

The set S does not need to be a domain, in particular it may be a subset of $\mathbb{R}$.

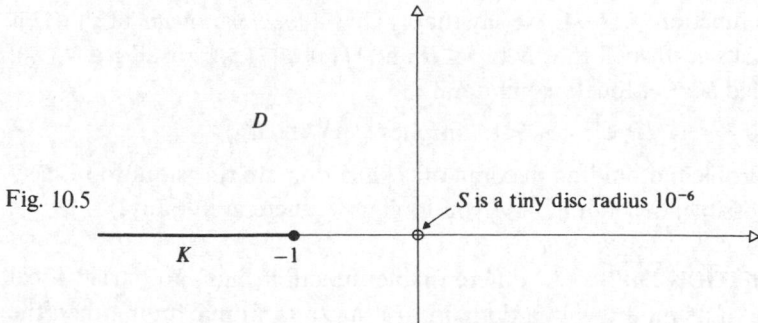

Fig. 10.5

S is a tiny disc radius 10^{-6}

K -1

EXAMPLE 3. $f(z) = \sin z$ for $z \in \mathbb{C}$ is the unique differentiable extension of $f(x) = \sin x$ for $x \in S = \mathbb{R}$.

Of course, if D is a domain, then it is an open set, so $D \cap \mathbb{R}$ is open in $\mathbb{R}$. A differentiable function $f : D \to \mathbb{C}$ has a power series expansion in a neighbourhood of any $x_0 \in D \cap \mathbb{R}$. So if a real function $h : S \to \mathbb{R}$ extends to a differentiable complex function on a domain, it must already have a power series expansion about any point in $S \subseteq D \cap \mathbb{R}$. Thus the only real functions which extend to differentiable complex functions are real analytic functions.

We may take the set S to be even more restricted still, provided that it has at least one limit point in D.

EXAMPLE 4. If $f(1/n) = 1/n^2$ for all positive integers n, then $f(z) = z^2$ is the unique analytic extension of f to the whole complex plane because

$$S = \{1/n \,|\, n \text{ is a positive integer}\}$$

has the limit point $0 \in \mathbb{C}$.

Because the Taylor expansion is valid on discs (Theorem 10.3), and may therefore be used to define extension functions, it is easy to see that the radius of convergence of the Taylor series expansion of a function f about a point z_0 is equal to the distance from z_0 to the *nearest* point z_1 at which no differentiable extension function of f may be defined. Such a point is called a *singularity* of f, and we discuss singularities further below. You should think of singularities as the obstacles which determine the size of the disc of convergence of Taylor series.

7. Local maxima and minima

Because the complex numbers are not ordered, we cannot speak about the maxima and minima of a complex function f. We can, however, consider the maximum and minimum values of the modulus $|f|$.

For a function $f: D \to \mathbb{C}$, we say that $|f|$ has a *local maximum* at $z_0 \in D$ if there exists $\varepsilon > 0$ such that $N_\varepsilon(z_0) \subseteq D$ and $|f(z)| \leqslant |f(z_0)|$ for all $z \in N_\varepsilon(z_0)$. It is called a *strict* local maximum if

$$0 < |z - z_0| < \varepsilon \quad \text{implies} \quad |f(z)| < |f(z_0)|.$$

The problem of finding maxima of $|f|$ in a domain turns out to be easy (or impossible, depending how you look at it: there aren't any!)

PROPOSITION 10.12.   A differentiable function has no strict local maxima of its modulus in a domain. If it has a local maximum, then the function is constant.

Proof. Suppose that $N_\varepsilon(z_0) \subseteq D$ and $|f(z)| \leqslant |f(z_0)|$ for all $z \in N_\varepsilon(z_0)$. For $0 < r < \varepsilon$, the circle $C_r(t) = z_0 + re^{it}$ $(0 \leqslant t \leqslant 2\pi)$ lies inside $N_\varepsilon(z_0)$, so $|f(z_0 + re^{it})| \leqslant |f(z_0)|$ for all $t \in [0, 2\pi]$. The Cauchy Integral Formula gives

$$
\begin{aligned}
f(z_0) &= \frac{1}{2\pi i} \int_{C_r} \frac{f(z)}{z - z_0}\, dz \\
&= \frac{1}{2\pi i} \int_0^{2\pi} \frac{f(z_0 + re^{it})}{re^{it}}\, ire^{it}\, dt \\
&= \frac{1}{2\pi} \int_0^{2\pi} f(z_0 + re^{it})\, dt
\end{aligned}
$$

and so

$$|f(z_0)| \leqslant \frac{1}{2\pi} \int_0^{2\pi} |f(z_0 + re^{it})|\, dt \leqslant \frac{1}{2\pi} \int_0^{2} |f(z_0)|\, dt \leqslant |f(z_0)|.$$

Hence

$$|f(z_0)| = \frac{1}{2\pi} \int_0^{2\pi} |f(z_0 + re^{it})|\, dt. \tag{2}$$

If the *strict* inequality $|f(z_0 + re^{it})| < |f(z_0)|$ were to hold for any $t \in [0, 2\pi]$, by continuity it would hold in a small interval and give the strict inequality

$$\frac{1}{2\pi} \int_0^{2\pi} |f(z_0 + re^{it})|\, dt < |f(z_0)|$$

which would contradict (2). Hence $|f(z_0 + re^{it})| = |f(z_0)|$ for all $t \in [0, 2\pi]$. This holds for any r smaller than ε, so $|f|$ is constant in $N_\varepsilon(z_0)$. By Proposition 4.8, f is constant in $N_\varepsilon(z_0)$ and the Identity Theorem shows f to be constant throughout D. $\square$

If we consider the complementary notion of local minima of $|f|$, then it is clear that where a non-constant function has a zero then the zero is isolated and $|f|$ has a strict local minimum there. However, if f is non-zero, then applying Proposition 10.12 to the function $1/f$, we get

PROPOSITION 10.13. A non-zero differentiable function has no strict local minima of its modulus in a domain. If it has a local minimum then the function is constant. ☐

8. The Maximum Modulus Theorem

The question of maxima or minima of $|f|$ on an arbitrary subset in its domain is somewhat different. We say that $|f|$ has a local maximum on $S \subseteq D$ at the point $z_0 \in S$ if
(i) z_0 is a limit point of S
(ii) for some $\varepsilon > 0$, $|f(z)| \leqslant |f(z_0)|$ whenever $z \in N(z_0) \cap S$.
Condition (i) is essential, for otherwise some neighbourhood $N_\varepsilon(z_0)$ contains no points of S other than z_0 and then condition (ii) would be vacuously true.

EXAMPLE 1. If $f(z) = e^z$ and $S = \{z \in \mathbb{C} \mid |z| \leqslant 1\}$, then $|f(x+iy)| = e^x$ and $|f|$ has a local maximum on S at the point $z_0 = 1$. ☐

Using Proposition 10.12, we see that if $N_\varepsilon(z_0) \subseteq S$, then $|f|$ cannot have a maximum at z_0. We define z_0 to be a *boundary point* of S if every neighbourhood of z_0 contains points in S and points not in S, other than z_0 itself. In other words, a boundary point is a limit point both of S and its complement $\mathbb{C} \setminus S$. We define the *boundary* of S to be its set of boundary points.

EXAMPLE 2. The boundary of $S = \{z \in \mathbb{C} \mid |z| \leqslant 1\}$ and of its complement $\{z \in \mathbb{C} \mid |z| > 1\}$ is the circle $\{z \in \mathbb{C} \mid |z| = 1\}$.

We can now rephrase Proposition 10.12 to give:

THEOREM 10.14 (The Maximum Modulus Theorem).
If a differentiable function is not constant, then the maximum value of its modulus on an arbitrary set occurs on the boundary of that set. ☐

We also have from Proposition 10.13

THEOREM 10.15 (The Minimum Modulus Theorem).
If a differentiable function is not constant, then the minimum value of its modulus on an arbitrary set occur either where the function is zero or on the boundary of the set.   □

EXAMPLE 3. If $f(z)=z^2$ on the set $S=\{z \in \mathbb{C} \mid |z| \leqslant 1\}$, then the maximum value of $|f(x+iy)|=x^2+y^2$ occur all round the boundary of S whilst the minimum value of $|f|$ occurs at the origin.

Exercises 10

1. Find the Taylor series at 0 of $f(z)=\text{Log}\,(1+z)$, where Log is the principal value. What is the disc of convergence? Answer the same questions for $g(z)=\exp\,(\alpha \cdot \text{Log}\,(1+z))$, where $\alpha \in \mathbb{C}$.

2. Find the first three terms and radius of convergence for the Taylor series at 0 of $f(z)=[1+\text{Log}\,(1-z)]^{-1}$.

3. Taylor expand the following around 0, and find the radius of convergence:

   (i) $\sin^2 (z)$
   (ii) $z^2(z+2)^{-2}$
   (iii) $(az+b)^{-1}$ $(a, b \in \mathbb{C}, b \neq 0)$
   (iv) $\int_0^z e^{w^2}\, dw$
   (v) $(\sin z)/z$ $(z \neq 0), 1\ (z=0)$
   (vi) $\int_0^z (\sin w)/w\, dw$.

4. Define the numbers c_n by the Taylor series

$$\sec (z)= \sum_{n=0}^{\infty} (-1)^n \frac{c_{2n}}{(2n)!} z^{2n}.$$

   Prove that

$$c_0=1$$

$$c_0+c_2\binom{2n}{2}+c_4\binom{2n}{4}+ \cdots +c_{2n}\binom{2n}{2n}=0.$$

   Show that c_n is always rational, and calculate it for $n \leqslant 8$.

5. Let

$$(1-z-z^2)^{-1}=\sum F_n z^n.$$

   Prove that

$$F_0=F_1=1; \quad F_n=F_{n-1}+F_{n-2}.$$

   (This is the recursive definition of the *Fibonacci numbers* 1, 1, 2, 3, 5, 8, ...) By expanding $(1-z-z^2)^{-1}$ in partial fractions, show that

$$F_n=\frac{1}{\sqrt{5}}\left[\left(\frac{1+\sqrt{5}}{2}\right)^{n+1} -\left(\frac{1-\sqrt{5}}{2}\right)^{n+1}\right].$$

6. Investigate analogous results to question 5 on the expansion of $(1-az-bz^2)^{-1}$, for $a, b \in \mathbb{C}$.

7. If

$$\exp\left(\frac{z}{1-z}\right)=\sum a_n z^n$$

prove that

$$a_0=1, \quad a_n=\sum_{s=1}^{n}\frac{1}{s!}\binom{n-1}{s-1}.$$

8. Let $f(z)$ have Taylor series $\sum a_n z^n$ for $|z| < R$. Let $\omega=e^{2\pi i/3}$ and define

$$g(z)=\tfrac{1}{3}(f(z)+f(\omega z)+f(\omega^2 z)).$$

Show that

$$g(z)=\sum a_{3n}z^{3n}$$

for $|z| < R$. Find similar expressions for $\sum a_{3n+1}z^{3n+1}$ and $\sum a_{3n+2}z^{3n+2}$.
[Hint: $1+\omega+\omega^2=0$].

9. Define three functions by

$$\alpha(z)=\sum\frac{z^{3n}}{(3n)!}, \quad \beta(z)=\sum\frac{z^{3n+2}}{(3n+2)!}, \quad \gamma(z)=\sum\frac{z^{3n+1}}{(3n+1)!}.$$

Prove the series converge for all z. Using question 8, prove the following:
   (i) $\alpha'(z)=\beta(z)$; $\beta'(z)=\gamma(z)$; $\gamma'(z)=\alpha(z)$.
   (ii) $\alpha(z)=\tfrac{1}{3}[e^z+2e^{-z/2}\cos(z\sqrt{3}/2)]$ (find similar expressions for $\beta(z)$, $\gamma(z)$)
   (iii) $\alpha(z+w)=\alpha(z)\alpha(w)+\beta(z)\gamma(w)+\gamma(z)\beta(w)$
   (iv) $\alpha^3(z)+\beta^3(z)+\gamma^3(z)-3\alpha(z)\beta(z)\gamma(z)=1$
   (v) $\alpha(z)=\beta(z)\Leftrightarrow z=(3n-1)\cdot 2\pi/3\sqrt{3}$ $(n\in\mathbb{Z})$.

10. Generalize question 8 to give an expression for $\sum a_{pn}z^{pn}$ $(p=2,3,\ldots)$ and (harder) for $\sum a_{pn+k}z^{pn+k}$ $(p=2,3,\ldots; k=0,1,\ldots,p-1)$.

11. Show that the Cauchy estimate is an equality if and only if $f(z)=Kz^n$ for some $K\in\mathbb{C}, n=1,2,3,\ldots$.

12. Let D be a disc centre z_0, and let f be analytic in a domain containing D. Prove that the 'mean value' of $f(z)$ as z runs over ∂D (defined by a suitable integral) is equal to $f(z_0)$.

13. Let f be analytic throughout $\mathbb{C}$, and suppose that $|f(z)|\leqslant K|z|^c$ for a real constant K and positive integer c. Prove that f is a polynomial function of degree $\leqslant c$. What if c is not an integer?

14. Let f and g be analytic on $\{z\in\mathbb{C}|-2<\text{im } z<2\}$. Suppose that $f(z)=g(z)$ for all z such that $|z|<0.01$. By considering Taylor expansions first about 0, then 1, and so on by induction, prove that $f(z)=g(z)$ in the whole strip.

15. Suppose that $f(z)=\sum a_n(z-z_0)^n$ in a disc D centre z_0, radius R. If $0\leqslant r<R$, show that

$$\frac{1}{2\pi}\int_0^{2\pi}|f(z_0+re^{i\theta})|^2\,d\theta=\sum|a_n|^2r^{2n}$$

(*Parseval's Equality*). Hence show that

$$\sum |a_n|^2 r^{2n} \leqslant M(r) = \sup_\theta |f(z_0 + re^{i\theta})|.$$

From this, give an alternative proof of the Maximum Modulus Theorem.

16. If $p(z)$ is a polynomial of degree n, show that for each $R > 0$ the 'level curve' of $|p(z)|$, defined to be $\{z \in \mathbb{C} \mid |p(z)| = R\}$, has at most n connected components.

17. Suppose that $x^2 + y^2 \leqslant a^2$. Prove that $(1+x)^2 + y^2$ attains its maximum value when $x = a$, $y = 0$. [Hint: apply the Maximum Modulus Theorem to $1 + z$ on the disc $|z| \leqslant a$.]

18. Suppose that $x^2 + y^2 \leqslant 1$. Prove that $(x^2 - y^2 - 1)^2 + 4x^2 y^2$ attains its maximum value when $x = 0$, $y = \pm 1$.

19. If f is differentiable in a domain D, prove that the zeros of f are either all of finite order and isolated, or f is identically zero in D.

20. Let

$$f(x) = \int_x^\infty t^{-1} e^{x-t} \, dt$$

where x is real and positive. By repeated integration by parts, show that if

$$h_n(x) = (-1)^n n! x^{-(n+1)}$$

then

$$f(x) = h_0(x) + h_1(x) + \cdots + h_{n-1}(x) + (-1)^n n! \int_x^\infty e^{x-t} t^{-(n+1)} \, dt.$$

Show that the series

$$\sum_{n=0}^\infty h_n(x) \tag{3}$$

diverges for all x. Show also that

$$\left| f(x) - \sum_{n=0}^N h_n(x) \right| < N! x^{-(N+1)},$$

so that *for large enough x* the series (3) provides a good approximation to $f(x)$, even though it diverges. (The catch is that, the better the approximation required, the larger x has to be: *no* particular choice of x gives an arbitrarily good approximation if we take N large enough. A series with this property is called an *asymptotic series*.)

11
Laurent series

The Taylor series expansion is too limited for many applications. A useful generalization was given by Laurent, who considered 'power series' involving negative powers as well as positive. The benefits which accrue are hinted at by the following example. The function $f(z) = e^{-1/z^2}$ is very badly behaved as regards Taylor series expansion. We have seen that, restricted to the real line, its Taylor series about the origin, is $0 + 0x + 0x^2 + \ldots$ which does not converge to $f(x)$; and on the whole complex plane it is, if such a statement makes sense, even less capable of being represented by a Taylor series! The natural series representation is obtained by starting with the series for e^z and replacing z by $-1/z^2$, which gives

$$f(z) = 1 - z^{-2} + \frac{1}{2!} z^{-4} - \frac{1}{3!} z^{-6} + \cdots$$

and this is a series of the 'negative powers' type. It converges for all z for which $-1/z^2$ is defined, namely $z \neq 0$.

1. Series involving negative powers

The general series of this type can be written in the form

$$\sum_{n=-\infty}^{\infty} a_n(z - z_0)^n,$$

where this is to be thought of as a compact notation for

$$\left(\sum_{n=0}^{\infty} a_n(z - z_0)^n \right) + \left(\sum_{n=1}^{\infty} a_{-n}(z - z_0)^{-n} \right)$$

and hence converges if and only if the two bracketed series converge. We know that power series converge inside a disc. Consequently power series with negative powers alone should converge *outside* a disc (for instance, that for e^{-1/z^2} converges outside the disc $|z| = 0$) and those with

195

both positive and negative powers should converge in the region *between* two concentric circles. Such a region is called an *annulus*: more precisely, if R_1 and R_2 are real numbers or ∞, with $0 \leqslant R_1 < R_2 \leqslant \infty$, and if $z_0 \in \mathbb{C}$, then

$$\{z \in \mathbb{C} : R_1 \leqslant |z - z_0| \leqslant R_2\}$$

is an annulus.

We begin with an existence theorem for series expansions of the above kind.

THEOREM 11.1 (Laurent's Theorem).
If f is differentiable in the annulus $R_1 \leqslant |z - z_0| \leqslant R_2$ where $0 \leqslant R_1 < R_2 \leqslant \infty$ then

$$f(z_0 + h) = \sum_{n=0}^{\infty} a_n h^n + \sum_{n=1}^{\infty} b_n h^{-n}$$

where $\Sigma \, a_n h^n$ converges for $|h| < R_2$, $\Sigma \, b_n h^{-n}$ converges for $|h| > R_1$, and in particular both series converge on the interior of the given annulus.

Further, if $C_r(t) = z_0 + re^{it}$ where $R_1 < r < R_2$, $0 \leqslant t \leqslant 2\pi$, then

$$a_n = \frac{1}{2\pi i} \int_{C_r} \frac{f(z)}{(z - z_0)^{n+1}} \, dz,$$

$$b_n = \frac{1}{2\pi i} \int_{C_r} f(z)(z - z_0)^{n-1} \, dz.$$

{*Remark:* in the more compact notation, we set $c_n = a_n \, (n \geqslant 0)$ and $c_{-n} = b_n, \, n \geqslant 1)$ and then

$$f(z_0 + h) = \sum_{n=-\infty}^{\infty} c_n h^n$$

which converges on the interior of the annulus where

$$c_n = \frac{1}{2\pi i} \int_{C_r} \frac{f(z)}{(z - z_0)^{n+1}} \, dz$$

for all $n \in \mathbb{Z}$.}

Proof. If $R_1 < |h| < R_2$ we choose r_1, r_2 such that

$$R_1 < r_1 < |h| < r_2 < R_2$$

and let (Fig. 11.1)

$$C_{r_1}(t) = z_0 + r_1 e^{it}$$
$$C_{r_2}(t) = z_0 + r_2 e^{it}.$$ $(0 \leqslant t \leqslant 2\pi)$

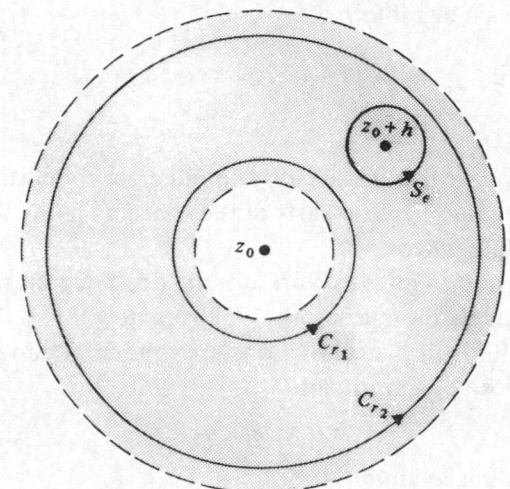

Fig. 11.1

We show first that

$$f(z_0+h) = \frac{1}{2\pi i} \int_{C_{r_2}} \frac{f(z)}{z-(z_0+h)}\,dz - \frac{1}{2\pi i} \int_{C_{r_1}} \frac{f(z)}{z-(z_0+h)}\,dz. \qquad (1)$$

To do this, enclose z_0+h in a small circle

$$S_\varepsilon(t) = (z_0+h) + \varepsilon e^{it} \quad (0 \leqslant t \leqslant 2\pi).$$

Then the function

$$F(z) = \frac{f(z)}{z-(z_0+h)}$$

is differentiable in

$$S = \{z \in \mathbb{C} : R_1 < |z-z_0| < R_2, \ z \neq z_0+h\}$$

and the contours $-C_{r_2}$, C_{r_1}, S_ε satisfy the hypotheses of Theorem 8.9. Therefore

$$\int_{-C_{r_2}} F(z)\,dz + \int_{C_{r_1}} F(z)\,dz + \int_{S_\varepsilon} F(z)\,dz = 0.$$

Hence

$$\int_{S_\varepsilon} F(z)\,dz = \int_{C_{r_2}} F(z)\,dz - \int_{C_{r_1}} F(z)\,dz.$$

But by Cauchy's integral formula

$$\int_{S_\varepsilon} F(z)\,dz = 2\pi i \cdot f(z_0+h)$$

and so we get (1).

(Alternatively, we can make cuts and integrate round the two halves, as shown in Figure 11.2. The parts of the contours along the cuts cancel in pairs when integrated.)

All we now need to do is to work out the two integrals in (1) as power series, and calculate the coefficients. Unfortunately this is the longer part of the proof, although the calculations are routine in nature.

First we choose ρ_1 and ρ_2 with

$$r_1 < \rho_1 < |h| < \rho_2 < r_2$$

which enforces the conditions

(i) $|z-(z_0+h)| > r_2-\rho_2$ for z on C_{r_2},
(ii) $|z-(z_0+h)| > \rho_1-r_1$ for z on C_{r_1}.

As in the proof of the Taylor series expansion (Lemma 10.2) we get

$$\frac{1}{2\pi i}\int_{C_{r_2}} \frac{f(z)}{z-(z_0+h)}\,dz = \sum_{n=0}^{\infty} a_n h^n$$

for $|h| < \rho_2$, where

$$a_n = \frac{1}{2\pi i}\int_{C_{r_2}} \frac{f(z)}{(z-z_0)^{n+1}}\,dz.$$

The treatment of the second integral is similar, but will be given in full. Since

$$\frac{1}{h} + \frac{z-z_0}{h^2} + \cdots + \frac{(z-z_0)^{n-1}}{h^n} - \frac{(z-z_0)^n}{h^n(z-z_0-h)}$$

$$= \frac{-1}{z-z_0-h}$$

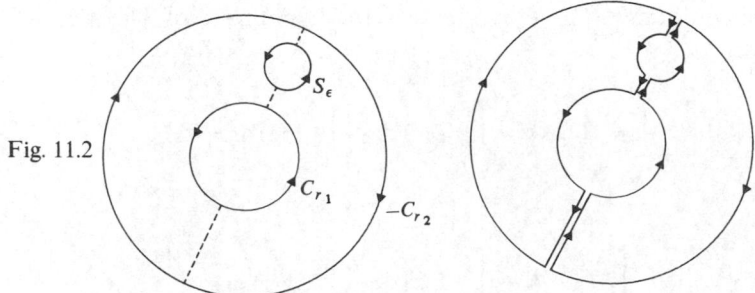

Fig. 11.2

(summing a geometric progression) we have

$$-\frac{1}{2\pi i}\int_{C_{r_1}}\frac{f(z)}{z-z_0-h}\,dz$$

$$=\frac{1}{2\pi i}\int_{C_{r_1}} f(z)\left\{\frac{1}{h}+\cdots+\frac{(z-z_0)^{n-1}}{h^n}-\frac{(z-z_0)^n}{h^n(z-z_0-h)}\right\}dz$$

$$=\sum_{m=1}^{n} b_m h^{-m}-B_n$$

where

$$b_m=\frac{1}{2\pi i}\int_{C_{r_1}} f(z)(z-z_0)^{m-1}\,dz,$$

$$B_n=\frac{1}{2\pi i}\int_{C_{r_1}}\frac{f(z)(z-z_0)^n}{h^n(z-z_0-h)}\,dz.$$

Finally we estimate the size of B_n. There exists $M>0$ such that $|f(z)|\leq M$ on C_{r_1} by compactness, and by (ii) above we have $|z-z_0-h|>\rho_1-r_1$. Also $|h|>\rho_1$, $|z-z_0|\leq r_1$. Hence

$$|B_n|\leq\frac{1}{2\pi}\frac{Mr_1^n}{\rho_1^n(\rho_1-r_1)}\cdot 2\pi r_1$$

$$=\frac{Mr_1}{(\rho_1-r_1)}\left(\frac{r_1}{\rho_1}\right)^n$$

which $\to 0$ as $n\to\infty$ since $r_1/\rho_1<1$. It follows that

$$-\frac{1}{2\pi i}\int_{C_{r_1}}\frac{f(z)}{z-(z_0+h)}\,dz=\sum_{m=1}^{\infty} b_m h^{-m}.$$

To finish the proof we have to replace C_{r_1} and C_{r_2} by C_r in the expressions for a_n and b_n. Since all three paths are homotopic inside the annulus, or by using cuts, this is immediate; and the theorem is proved.   □

Note that we can no longer assert that $a_n=f^{(n)}(z_0)/n!$ since $f(z)$ need not be differentiable for $|z-z_0|<R_1$ under our hypotheses.

The *Laurent series of* $f(z)$ *about* z_0 is the series

$$\sum_{n=-\infty}^{\infty} c_n h^n$$

where $h=z-z_0$ and c_n is as defined above. We also refer to it as the Laurent *expansion* of $f(z)$. The Laurent expansion is unique, for suppose we have any series

$$f(z) = \sum_{n=-\infty}^{\infty} d_n(z-z_0)^n.$$

Then

$$(z-z_0)^{-m-1}f(z) = \sum_{n=-\infty}^{\infty} d_n(z-z_0)^{n-m-1}$$

$$= f_1(z) + \frac{d_m}{z-z_0} + f_2(z)$$

where

$$f_1(z) = \sum_{n=-\infty}^{m-1} d_n(z-z_0)^{n-m-1}$$

$$f_2(z) = \sum_{n=m+1}^{\infty} d_n(z-z_0)^{n-m-1}.$$

But each of f_1 and f_2 has an antiderivative in $R_1 < |z-z_0| < R_2$. To see this, put

$$F_1(z) = \sum_{n=-\infty}^{m-1} \frac{1}{n-m} d_n(z-z_0)^{n-m}$$

$$F_2(z) = \sum_{n=m+1}^{\infty} \frac{1}{n-m} d_n(z-z_0)^{n-m}.$$

Then it is easy to check that the series converge absolutely for $R_1 < |z-z_0| < R_2$ and that $F'_1(z) = f_1(z)$, $F'_2(z) = f_2(z)$. Hence

$$\int_{C_r} f(z)(z-z_0)^{-m-1}\,dz = \int_{C_r} \frac{d_m}{z-z_0}\,dz = 2\pi i d_m$$

and so $d_m = c_m$ as defined above.

EXAMPLE 1. Let $f(z) = e^z + e^{1/z}$. We have

$$e^z = \sum_{n=0}^{\infty} \frac{1}{n!} z^n \quad \text{for all } z$$

$$e^{1/z} = \sum_{n=0}^{\infty} \frac{1}{n!} z^{-n} \quad \text{for } z \neq 0$$

and so

$$f(z) = \sum_{m=-\infty}^{\infty} c_m z^m$$

where

$$c_m = 1/m! \qquad (m \geqslant 1)$$
$$c_0 = 2$$
$$c_m = 1/(-m)! \quad (m \leqslant -1)$$

and the expansion is valid for $z \neq 0$.

EXAMPLE 2. $f(z) = \dfrac{1}{z} + \dfrac{1}{1-z}$. In a similar way, we have $f(z) = \Sigma\, c_m z^m$ where

$$c_m = \begin{cases} 0 & (m < -1) \\ 1 & (m \geqslant -1) \end{cases}$$

and the series converges absolutely for $0 < |z| < 1$.

EXAMPLE 3. $f(z) = \dfrac{1}{z-1} - \dfrac{1}{z-2}$. Writing this as

$$\left[1/z \left(1 - \frac{1}{z} \right) \right] + \left[1/2 \left(1 - \frac{z}{2} \right) \right]$$

we obtain an expansion $\Sigma\, c_m z^m$ where

$$c_m = \begin{cases} 1 & (m \leqslant -1) \\ 2^{-(m+1)} & (m \geqslant 0) \end{cases}$$

valid in the (genuinely annular !) annulus $1 < |z| < 2$.

2. Isolated singularities

If f is differentiable in a *punctured disc*

$$0 < |z - z_0| < R$$

we say that z_0 is an *isolated singularity* of f. We can use the Laurent expansion to study such singularities. There is a Laurent series

$$f(z) = \sum_{n=0}^{\infty} a_n (z - z_0)^n + \sum_{n=1}^{\infty} b_n (z - z_0)^{-n}$$

valid for $0 < |z - z_0| < R$; and this series can behave in three radically different ways.

(1) *All $b_n = 0$.* By defining $f(z_0) = a_0$ we obtain a function which is differentiable on the whole disc $|z - z_0| < R$, with Taylor series

$$\sum_{n=0}^{\infty} a_n (z - z_0)^n.$$

In this case z_0 is said to be a *removable singularity*. It arises more from our domain of definition of f than from any intrinsic feature of f.

For example, consider

$$f(z) = \frac{\sin(z)}{z} \quad (z \neq 0).$$

Around $z_0 = 0$ we have

$$f(z - z_0) = 1 - \frac{z^2}{3!} + \frac{z^4}{5!} - \cdots$$

and so by defining $f(0) = 1$ we get a function differentiable for all $z \in \mathbb{C}$.

(2) *Only finitely many b_n non-zero.* Then

$$f(z) = \frac{b_m}{(z-z_0)^m} + \cdots + \frac{b_1}{z-z_0} + \sum_{n=0}^{\infty} a_n(z-z_0)^n$$

where $b_m \neq 0$. In this case we say that f has a *pole* of *order m*. For example

$$f(z) = z^{-4} \cdot \sin(z) \quad (z \neq 0)$$

$$= \frac{1}{z^3} - \frac{1}{3!z} + \sum (-1)^n \frac{z^{2n+1}}{(2n+5)!}$$

has a pole of order 3 at $z_0 = 0$.

Poles of orders 1, 2, 3, ... are often called *simple, double, triple,* ... poles.

(3) *Infinitely many b_n non-zero.* Then we say that z_0 is an *isolated essential singularity.*

For example, $f(z) = \sin(1/z) \ (z \neq 0)$

$$= \frac{1}{z} - \frac{1}{3!z^3} + \frac{1}{5!z^5} - \cdots$$

has an isolated essential singularity at $z_0 = 0$.

We shall investigate the behaviour of an analytic function near an isolated singularity, according to the three possibilities.

3. Behaviour near an isolated singularity

Removable singularities are trivial and uninteresting – which makes it all the more important to be able to recognize them. The next lemma is usually sufficient for this purpose.

LEMMA 11.2. The following are equivalent for a function f differentiable in $0 < |z-z_0| < R$:

(i) z_0 is a removable singularity,

(ii) $\lim_{z \to z_0} f(z)$ exists and is finite,

(iii) There exist $M > 0$, $\delta > 0$ such that $|f(z)| < M$ for $0 < |z-z_0| < \delta$.

Proof. Trivially (i)$\Rightarrow$(ii)$\Rightarrow$(iii), so we need prove only that (iii)$\Rightarrow$(i). Suppose (iii) holds, and take a Laurent expansion

$$f(z) = \sum_{n=0}^{\infty} a_n(z-z_0)^n + \sum_{n=1}^{\infty} b_n(z-z_0)^{-n}.$$

Now

$$b_n = \frac{1}{2\pi i} \int_{C_r} f(z)(z-z_0)^{n-1} \, dz$$

where $C_r(t) = z_0 + re^{it}$ $(0 \leqslant t \leqslant 2\pi)$ for $0 < r < R$. So

$$|b_n| \leqslant \frac{1}{2\pi} Mr^{n-1} \cdot 2\pi r = Mr^n.$$

If we let $r \to 0$ it follows that $|b_n| = 0$ for all $n \geqslant 1$. Thus z_0 is a removable singularity and (i) holds. $\qquad \square$

We immediately deduce the useful

COROLLARY 11.3. If any coefficient $b_n \neq 0$ $(n \geqslant 1)$ then f is unbounded on every open disc with centre z_0. $\qquad \square$

For example, if

$$f(z) = z^2/((e^z - 1) \sin z) \quad (z \neq 0)$$

then

$$\lim_{z \to 0} f(z) = \lim_{z \to 0} \left(\frac{z}{e^z - 1} \right) \left(\frac{z}{\sin z} \right)$$

$$= \lim_{z \to 0} \left(1 + \frac{z}{2!} + \cdots \right)^{-1} \left(1 - \frac{z^2}{3!} + \cdots \right)^{-1}$$

$$= 1,$$

so that $z_0 = 0$ is a removable singularity.

There is a similar, slightly more complicated, criterion for poles.

PROPOSITION 11.4. If f is differentiable in $0 < |z - z_0| < R$, then f has a pole of order m at z_0 if and only if

$$\lim_{z \to z_0} (z - z_0)^m f(z) = l \neq 0.$$

Proof. If f has a pole of order m then

$$(z - z_0)^m f(z) = b_m + \cdots + b_1 (z - z_0)^{m-1} + \sum_{n=0}^{\infty} a_n (z - z_0)^{m+n}$$

so that

$$\lim_{z \to z_0} (z - z_0)^m f(z) = b_m \neq 0.$$

Conversely if the limit is $l \neq 0$, then $g(z) = (z - z_0)^m f(z)$ has a removable singularity at z_0 by Lemma 10.2, and hence there is a series

$$g(z) = \sum_{n=0}^{\infty} a_n (z - z_0)^n$$

valid for $|z-z_0| < R$, and $a_0 = l \neq 0$. But now,

$$f(z) = \frac{a_0}{(z-z_0)^m} + \cdots + \frac{a_{m-1}}{z-z_0} + \sum_{n=0}^{\infty} a_{n+m}(z-z_0)^n$$

where $a_0 \neq 0$; so f has a pole of order m at z_0.   □

For example, consider

$$f(z) = \frac{5z+3}{(1-z)^3 \sin^2 z} \quad (0 < |z| < 1).$$

Then $\lim_{z \to 0} z^2 f(z) = 3 \neq 0$, so there is a double pole at the origin. Also, $\lim_{z \to 1} (z-1)^3 f(z) = -8/(\sin^2 (1)) \neq 0$, so there is a triple pole at $z_0 = 1$.

COROLLARY 11.5. The function $f(z)$ has a pole of order m at z_0 if and only if $1/f(z)$ has a zero of order m at z_0 (or more accurately if $1/f(z)$ has a removable singularity at z_0 which, when removed, gives rise to a zero of order m).

Proof. If $f(z)$ has a pole of order m then $g(z) = (z-z_0)^m f(z)$ has a removable singularity at z_0. Further there exists $\delta > 0$ such that $g(z) \neq 0$ for $|z-z_0| < \delta$. So

$$1/f(z) = (z-z_0)^m/g(z)$$

and $1/g(z)$ is differentiable for $|z-z_0) < \delta$. Therefore $1/f(z)$ has a zero of order m at z_0. The converse implication is proved by an almost identical argument.   □

COROLLARY 11.6. If $f(z)$ has a pole of order m at z_0 then

$$\lim_{z \to z_0} |f(z)| = +\infty.$$

Proof. $\lim_{z \to z_0} |f(z)| = \lim_{z \to z_0} |g(z)/(z-z_0)^m|$

$$= |l| \cdot \lim_{z \to z_0} |1/(z-z_0)^m|$$

$$= +\infty.$$ □

Thus near a pole the behaviour of f is really quite good. However, near an isolated essential singularity the behaviour is much wilder. The following result is classical:

THEOREM 11.7 (Weierstrass—Casorati).
In every neighbourhood of an isolated essential singularity z_0 a differentiable function f takes values arbitrarily close to any assigned complex

number. Specifically, given $r>0$, $\varepsilon>0$, and $w \in \mathbb{C}$, then there exists z_1 such that $0<|z_1-z_0|<r$ and $|f(z_1)-w|<\varepsilon$.

Proof. We have $f(z)=\Sigma_{n=0}^{\infty} a_n(z-z_0)^n + \Sigma_{n=1}^{\infty} b_n(z-z_0)^{-n}$ where infinitely many $b_n \neq 0$. If we define $\phi(z)=f(z)-w$ then the Laurent series of $\phi(z)$ differs from that of $f(z)$ only in the coefficient a_0, which is decreased by an amount w. Hence $\phi(z)$ also has an isolated essential singularity at z_0, and we need prove only that $\phi(z_1)$ can be made arbitrarily small in any neighbourhood of z_0, i.e. that there exists z_1 with

$$0<|z_1-z_0|<r, \quad |\phi(z_1)|<\varepsilon.$$

If z_0 is a limit of zeros of ϕ this is trivial. Failing this, there exists $\rho>0$ such that $\phi(z) \neq 0$ for $0<|z-z_0|<\rho$. Either there exists z_1 with $0<|z_0-z_1|<r$ and $|1/\phi(z_1)|>1/\varepsilon$, or else $1/\phi(z)$ is bounded for $|z-z_0|<r$. If the former, then the theorem is proved. If the latter, then $1/\phi(z)$ has a removable singularity at z_0 by Theorem 11.2, and so by Corollary 11.5 $\phi(z)$ has at worst a pole at z_0, which contradicts z_0 being essential. Thus the latter case cannot occur, and this completes the proof. $\qquad\square$

In point of fact a stronger and more satisfying result is true. Picard (1856–1941) proved that in every neighbourhood of an isolated essential singularity a differentiable function takes every value, *with at most one exception.* For instance, $\sin(1/z)$ takes every value in $0<|z|<r$ for any $r>0$, as can easily be verified. The exception can occur: $e^{1/z}$ misses the value 0 but attains all others in $0<|z|<r$ for any $r>0$. But the proof of Picard's theorem requires machinery (elliptic modular functions) considerably beyond the reach of this text.

4. The extended complex plane, or Riemann sphere

We now consider a way of describing the behaviour of a complex function 'at infinity' by adjoining to the complex plane $\mathbb{C}$ an extra point '∞', much as we extended the real line $\mathbb{R}$ by adjoining $+\infty$ and $-\infty$. That a single point is required is due to the geometry of the plane: the distinction between the two 'ends' of the line becomes blurred when we deal with the whole plane. The idea is due to Riemann and has a good geometric realization.

We think of $\mathbb{C}$ as being embedded as the (x, y)-plane in $\mathbb{R}^3$, so that a point $x+iy \in \mathbb{C}$ is identified with $(x, y, 0) \in \mathbb{R}^3$. Let

$$S^2 = \{(\xi, \eta, \zeta) \in \mathbb{R}^3 : \xi^2 + \eta^2 + \zeta^2 = 1\}$$

be the 'unit sphere'. A line joining the 'North pole' $(0, 0, 1)$ to $(x, y, 0)$ cuts S^2 in a unique point (ξ, η, ζ), so we have a one-to-one correspondence between $\mathbb{C}$ and the points of S^2 other than the North pole. (Fig. 11.3) As the reader may verify,

$$(\xi, \eta, \zeta) \quad \text{corresponds to} \quad \left(\frac{\xi}{1-\zeta}\right) + i\left(\frac{\zeta}{1-\zeta}\right).$$

As (ξ, η, ζ) gets near to $(0, 0, 1)$ it follows (as is obvious geometrically) that $|x+iy|$ becomes very large: thus it is reasonable to introduce the symbol

$$\infty$$

to correspond to $(0, 0, 1) \in S^2$. Then we have a one-to-one correspondence between S^2 and $\mathbb{C} \cup \{\infty\}$, which latter we call the *extended complex plane*. It may be identified with S^2 which is then known as the *Riemann Sphere*. We can now think of $\mathbb{C} \cup \{\infty\}$ either as a plane plus an extra point, or as a sphere: correspondingly we think of $\mathbb{C}$ as a plane, or as a sphere without a North pole. Both viewpoints are valuable, depending on the problem in hand.

Since $\{x+iy:|x+iy|>R\}$ corresponds to a 'cap' between a line of latitude and the North pole (Fig. 11.4) it makes sense to think of $\{z:|z|>R\}$ as a 'neighbourhood of ∞'. Such neighbourhoods get smaller as R gets larger. Doing this leads to a concept of continuity on S^2 agreeing with one's geometrical intuition. (Those familiar with topology can render this in more precise terms: we obtain a topology on the Riemann sphere identical to its usual topology as a subspace of $\mathbb{R}^3$.)

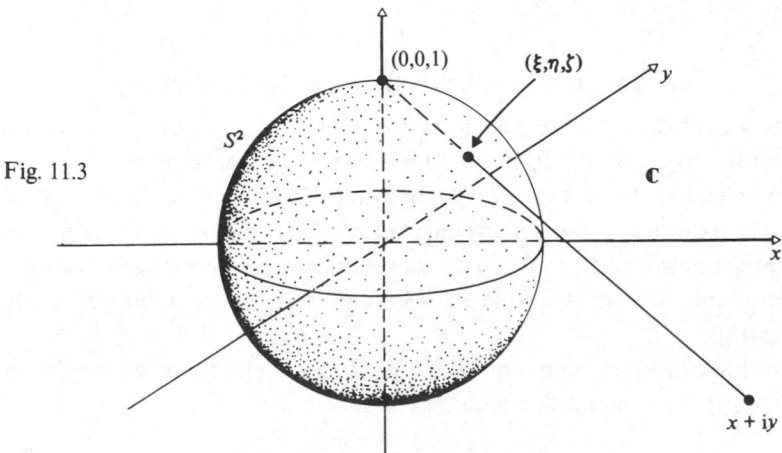

Fig. 11.3

Fig. 11.4

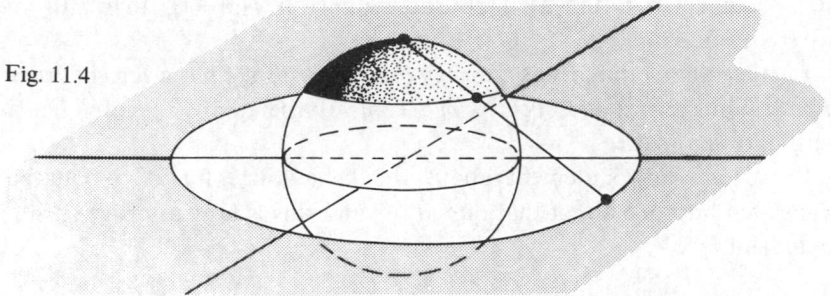

5. Behaviour of a differentiable function at ∞

Suppose that $f(z)$ is differentiable in $(z \in \mathbb{C}: |z| > R)$. Then we can define

$$g(z) = f(1/z) \quad (0 < |z| < 1/R).$$

Since $g'(z) = -z^{-2}f'(1/z)$ it follows that $g(z)$ is differentiable for $0 < |z| < 1/R$, and therefore has an isolated singularity at 0. We shall say that f *has a removable singularity, pole of order m,* or *isolated essential singularity at* ∞ if and only if $g(z)$ has the corresponding singularity at 0. (Note that this is a definition, not a theorem.)

EXAMPLE 1. $f(z) = 1/z$ $(|z| > 0)$. Then $g(z) = z$ $(|z| > 0)$ so that f has a removable singularity at ∞.

EXAMPLE 2. $f(z) = z$. Then $g(z) = 1/z$ $(|z| > 0)$ so that f has a simple pole at ∞.

EXAMPLE 3. $f(z) = e^z$. Then $g(z) = e^{1/z}$ $(|z| > 0)$ and f has an isolated essential singularity at ∞.

EXAMPLE 4. $f(z) = 1/\sin(z)$ $(z \neq n\pi, n \in \mathbb{Z})$. Then $g(z) = 1/\sin(1/z)$. We can't say that f has an isolated essential singularity at ∞, because f is not differentiable in $\{z: |z| > R\}$ for any $R > 0$. However, f certainly has some sort of singularity at ∞! In general if $f(z)$ has a sequence of isolated singularities $z_1, z_2, \ldots$ such that $z_n \to \infty$ then we say that f has an *essential* (but not isolated) singularity at ∞.

If f has a removable singularity at ∞ then g has a removable singularity at 0, so may be considered differentiable in $|z| < 1/R$. Thus we may say that f is *differentiable* (or *analytic*) at ∞, and define $f(\infty) = g(0) = \lim_{z \to 0}$

$g(z)$, so that $f(\infty)=\lim_{z\to\infty} f(z)$. For instance if $f(z)=1/z$ then $g(0)=0$ so $f(\infty)=0$.

Further we say that f has a *zero of order m at* ∞ if g has a zero of order m at 0. Thus if $f(z)=1/z^m(|z|>0,\ m\in\mathbb{Z},\ m>0)$ then $g(z)=z^m$ and f has a zero of order m at ∞.

In general any statement about the behaviour of f at ∞ can be translated into one about that of g at 0 – and this is how to prove such a statement.

6. Meromorphic functions

Poles are not, as singularities go, particularly nasty, and it is of interest to consider a class of functions more general than differentiable ones. If f is differentiable everywhere in a domain S except for points at which f has poles, then f is said to be *meromorphic*.

For instance, $f(z)=1/z$ is differentiable in $\mathbb{C}\setminus\{0\}$, and hence meromorphic in $\mathbb{C}$ because the singularity at 0 is a pole. Or $f(z)=1/\sin(z)$, differentiable in $\mathbb{C}\setminus\{n\pi:n\in\mathbb{Z}\}$, is meromorphic in $\mathbb{C}$.

We might go further, and consider functions meromorphic in the extended complex plane – by which we mean that the only singularities of f, including ∞ if necessary, are poles. It is not necessary to introduce a special term for such functions, because they turn out to have a simple description. Recall that a *rational* function is one of the form

$$f(z)=\phi(z)/\psi(z)$$

where $\phi(z)$ and $\psi(z)$ are polynomial functions. Then we have

PROPOSITION 11.8. A function is meromorphic in the extended complex plane if and only if it is rational.

Proof. Clearly a rational function is meromorphic. Suppose conversely that f is meromorphic in $\mathbb{C}\cup\{\infty\}$. Since f is differentiable at ∞ or has a pole at ∞ there exists $R>0$ such that f is differentiable in

$$\{z\in\mathbb{C}:|z|>R\}.$$

So the poles, other than ∞ of f occur inside the closed disc

$$\{z\in\mathbb{C}:|z|\leqslant R\}.$$

Since poles are isolated singularities each pole has a neighbourhood which contains no other poles, so by compactness this closed disc contains only finitely many poles, say $z_1, z_2, \ldots, z_k$. Let the orders of these poles be

$n_1, n_2, \ldots, n_k$. Define

$$g(z) = (z - z_1)^{n_1} \ldots (z - z_k)^{n_k} f(z)$$

which is differentiable throughout $\mathbb{C}$, hence analytic, having a Taylor series

$$g(z) = \sum_{n=0}^{\infty} a_n z^n \quad (z \in \mathbb{C}).$$

Now the polynomial

$$(z - z_1)^{n_1} \ldots (z - z_k)^{n_k} = \psi(z)$$

has a pole of order $n_1 + \cdots + n_k$ at ∞, since

$$\psi\left(\frac{1}{z}\right) = \left(\frac{1}{z} - z_1\right)^{n_1} \cdots \left(\frac{1}{z} - z_k\right)^{n_k}$$
$$= 1/z^{n_1 + \cdots + n_k} + \cdots$$

and since f has at worst a pole of order N at ∞ for some N it follows that $g(z)$ has a pole of order

$$M = n_1 + \cdots + n_k + N$$

at ∞. Now

$$g(1/z) = \sum_{n=0}^{\infty} a_n z^{-n} \quad (|z| > 0)$$

and since this has a pole of order M at 0 we must have $a_n = 0$ for $n > M$. Therefore

$$g(z) = a_0 + \cdots + a_M z^M$$

which is a polynomial; consequently $f(z) = g(z)/\psi(z)$ is a rational function. □

A function meromorphic only in $\mathbb{C}$ need not be rational, as the example $1/\sin(z)$ shows: it is not rational since it has infinitely many poles, but we have already noted that it is meromorphic in $\mathbb{C}$.

Exercises 11

1. Find Laurent expansions for the following around $z = 0$:
   (i) $(z-3)^{-1}$
   (ii) $(z-a)^{-k}$ $(a \in \mathbb{C}, k = 1, 2, 3, \ldots)$
   (iii) $1/z(1-z)$
   (iv) $1/(z-a)(z-b)$ $(a, b \in \mathbb{C})$
   (v) $z^3 e^{1/z}$
   (vi) $\exp(z + 1/z)$
   (vii) $\cos 1/z$
   (viii) $\exp(z^{-5})$.
   In each case, say in what annulus the expansion is valid.

2. Find Laurent expansions for the given functions on the stated annuli:

    (i) $(z-1)^{-2}(z-2)^{-1}$ on $0<|z|<1$

    (ii) $(z-1)^{-2}(z-2)^{-1}$ on $1<|z|<2$

    (iii) $(z-1)^{-2}(z-2)^{-1}$ on $2<|z|<3$

    (iv) $\exp(-z^{-2})$ on $|z|>0$

    (v) $(1-z+z^2)^{-1}$ in powers of $(z-1)$ on $0<|z-1|<1$

    (vi) $e^z/(1+z^2)$ on $|z|>1$.

3. Which of the following functions have (branches with) Laurent expansions around the given point z_0? (That is, in powers of $z-z_0$.)

    (i) $\sqrt{z},\ z_0=1$ (ii) $\sqrt{z},\ z_0=0$

    (iii) $\operatorname{Log} z,\ z_0=0$ (iv) $\operatorname{Log} z,\ z_0=3$

    (v) $\sqrt{1+\sqrt{z}},\ z_0=0$ (vi) $\tan^{-1}(1+z),\ z_0=0$

    (vii) $\sin^{-1}(z),\ z_0=0$ (viii) $\sqrt{(\pi/2)-\sin^{-1}(z)},\ z_0=1$

    (ix) $z^2\operatorname{cosec}(1/z),\ z_0=0$ (x) $\sqrt{(\pi/4)-\sin^{-1}(z)},\ z_0=1/\sqrt{2}$.

4. Prove the validity of the Laurent expansion

$$\frac{1}{(z-1)(2-z)}=\sum_{n=0}^{\infty}2^{-(n+1)}z^n+\sum_{n=1}^{\infty}z^{-n}.$$

5. Find Laurent series for $(z^2-1)^{-1}$ and $(z^2+1)^{-1}$ in powers of $z+i$ and $z-i$, and say in what annuli these are valid.

6. Let $a, b \in \mathbb{C}$. Show that

$$\exp(az+bz^{-1})=\sum_{-\infty}^{\infty}a_n z^n$$

where

$$a_n=\frac{1}{2\pi}\int_0^{2\pi}e^{(a+b)\cos\theta}\cos\left[(a-b)\sin\theta-n\theta\right]d\theta.$$

7. Let $a \in \mathbb{C}$. Show that

$$\sin(a(z+z^{-1}))=\sum_{-\infty}^{\infty}a_n z^n$$

where

$$a_n=\frac{1}{2\pi}\int_0^{2\pi}\sin(2a\cos\theta)\cos n\theta\, d\theta.$$

8. Find the poles of the functions

    (i) $1/(z^2+1)$ (ii) $1/(z^4+16)$

    (iii) $1/(z^4+2z^2+1)$ (iv) $1/(z^2+z-1)$.

9. Describe the type of singularity at 0 of each of the following functions:

    (i) $\sin(1/z)$ $(z\neq0)$ (ii) $z^{-3}\sin^2 z$

    (iii) $z\cot z$ $(z\neq n\pi)$ (iv) $\operatorname{cosec}^2 z-z^{-2}$ $(z\neq n\pi)$

    (v) $(\cos z-1)/z^2$ (vi) $(\sin z-z+z^3/6)/z^7$.

10. Let D be a disc centre z_0, let f be the differentiable on D except at z_0, and suppose that $|f(z)|$ is bounded on $D \smallsetminus \{z_0\}$. Then show that z_0 is a removable singularity of f. (Hint: negative Laurent terms are 0 – why?)

11. Find all singularities of the following functions, and say which are poles:

    (i) $(z + z^{-1})^{-1}$ (ii) $\dfrac{\cos \pi z}{1 - 4z^2}$ (iii) $\exp(z + z^{-1})$.

12. Construct a function defined on $\mathbb{C} \cup \{\infty\}$ having only the following singularities:

    (i) A pole of order 2 at ∞

    (ii) A simple pole at each of the points $e^{2\pi i k/p}$, $k = 0, 1, \ldots, p-1$, p an integer $\geqslant 2$

    (iii) A simple pole at $z = 2$, and a pole of order 5 at $z = \sqrt{2}$.

13. Let $p(z)$ and $q(z)$ be polynomials of degrees m, n respectively. Describe the behaviour at infinity of:

    (i) $p(z) + q(z)$ (ii) $p(z)q(z)$ (iii) $p(z)/q(z)$.

14. Find all singularities, and the behaviour at infinity, of:

    (i) $(z - z^3)^{-1}$ (ii) $z^5/(2 - z^2)^2$

    (iii) $(e^z - 1)^{-1} - 1/z$ (iv) $\cot 1/z$

    (v) $(\cos z)z^{-2}$ (vii) $[(\cos z) - 1]z^{-2}$

    (vii) $\cot(1/z) - 1/z$ (viii) $\sin(1/\cos(1/z))$.

15. Let Γ be a circle in the complex plane, or a straight line. Is its image in the Riemann sphere also a circle?

16. Find the poles and zeros of $\tan z$. Show that $\tan z$ is meromorphic in $\mathbb{C}$, but is not a rational function.

17. Show that $(z + 1 + z^{-1})^{-1}$ has a removable singularity at $z = 0$. Find its Taylor expansion, and the radius of convergence of this.

18. Verify Picard's Theorem directly for the functions

    (i) e^z (ii) $\tan^2 z$

    (iii) z^2 (iv) $\sin z$

    (v) $e^{(1/z)}$ (vi) $\cos z$

    (vii) $\tan z$ (viii) a function of your own choice.

19. Let $f(z)$ have a pole of order n at $z = a$. Define the *principal part* $\phi(z)$ of $f(z)$ to be the sum of the negative-power terms in the Laurent expansion of $f(z)$ in powers of $(z - a)$. Prove that $f(z) - \phi(z)$ is differentiable at $z = a$.

20. Show that, in the neighbourhood of a pole, a complex function is the sum of a rational function and a differentiable one.

21. Suppose f is analytic on $\mathbb{C}$ except at poles, and that ∞ is either a pole or a removable singularity of f. Show that

    (i) f has only finitely many poles.

    (ii) $f - \Sigma_j p_j$ is constant, where the poles of f are at points a_j and the p_j are the corresponding principal parts of f (question 19).

    (iii) f is a rational function, and $\Sigma_j p_j$ is its 'partial fraction' decomposition.

22. Let $f: \mathbb{C} \to \mathbb{C}$ be differentiable, with $f(z) \neq 0$ for all $z \in \mathbb{C}$. Suppose $\lim_{z \to \infty} f(z)$ exists and is non-zero. Prove that f is constant.

12
Residues

The tasks to which complex analysis may be set include the explicit computation of definite integrals and the summation of series. Although such problems are not as important a part of pure mathematics as they once were, they are still very useful in practical applications. Further, the power of the method and its wide applicability demonstrate the advantage of general principles and deep theorems over any amount of manipulative ingenuity. The techniques developed for these purposes also have theoretical applications, as we shall see in the later part of the chapter.

The basic idea is to use Cauchy's theorem to exploit the exceptional nature of the term $b_1/(z-z_0)$ in the Laurent expansion of an analytic function.

1. Cauchy's residue theorem

If f has an isolated singularity at z_0 and Laurent expansion

$$f(z_0+h)= \sum_{n=0}^{\infty} a_n h^n + \sum_{n=1}^{\infty} b_n h^{-n} \quad (0<|h|<R)$$

we define the *residue of f at* z_0 to be

$$\text{res}\,(f, z_0)=b_1.$$

From Theorem 11.1 we immediately deduce that

$$\text{res}\,(f, z_0)=\frac{1}{2\pi i} \int_{C_r} f(z)\,dz \tag{1}$$

where $C_r(t)=z_0+re^{it}\,(0\leqslant t\leqslant 2\pi)$ and $0<r<R$. This shows the relevance of residues to integration.

A closed path γ will be called a *simple loop* if every point z not on γ has $w(\gamma, z)=0$ or $w(\gamma, z)=1$. As usual, the set of points satisfying $w(\gamma, z)\neq 0$, which now means $w(\gamma, z)=1$, are said to be *inside* γ. In the applications the simple loops used will all be made up of straight line segments and parts of circles. (Fig. 12.1) All the simple loops encountered will be Jordan contours (that is, they don't self-intersect), but that is not essential for the theory. (Figure 12.2 shows a simple loop according to our definition which is not a Jordan contour.) What matters is that the points inside γ all have $w(\gamma, z) = +1$.

In such cases, we have

THEOREM 12.1 (Cauchy's Residue Theorem).
Let S be a domain containing a simple loop γ and the points inside γ. If f is differentiable in S except for finitely many isolated singularities at $z_1, \ldots, z_n$ inside γ, then

$$\int_\gamma f(z)\,dz = 2\pi i \sum_{r=1}^{n} \operatorname{res}(f, z_r).$$

Proof. Since S is open we can find circles $S_r(t) = z_r + \varepsilon_r e^{it}$ ($0 \leqslant t \leqslant 2\pi$) round z_r ($r = 1, \ldots, n$) such that S_r and points inside it lie within S, and such that S_r contains no singularity other than z_r. (Fig. 12.3)

Let

$$S' = S \setminus \{z_1, \ldots, z_n\}.$$

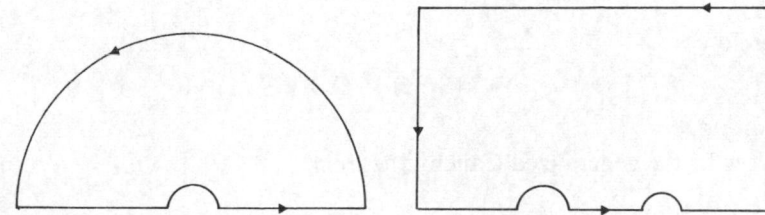

Fig. 12.1

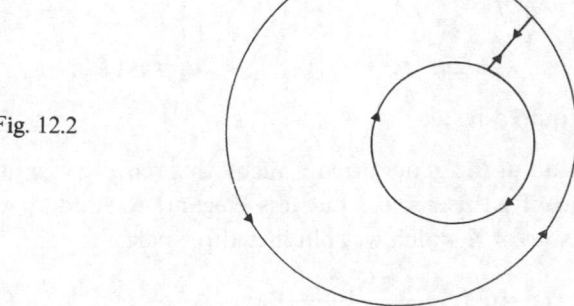

Fig. 12.2

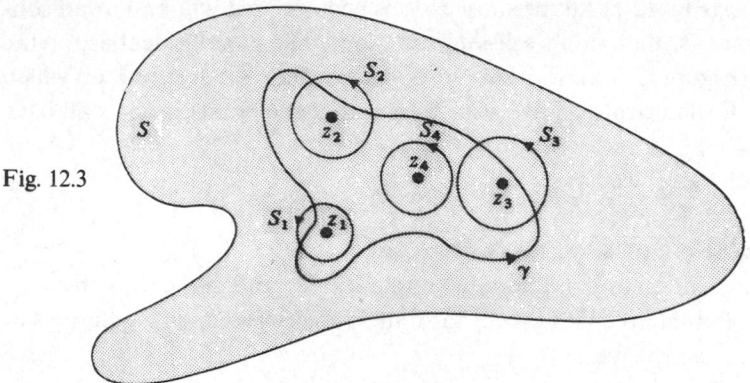

Fig. 12.3

Then the collection of paths

$$-\gamma, S_1, \ldots, S_n$$

satisfies the hypothesis of Theorem 8.9. For

$$w(-\gamma, z) = w(S_r, z) = 0 \quad (z \notin S)$$
$$w(-\gamma, z_r) = -w(\gamma, z_r) = 1 \quad \text{since } z_r \text{ is inside } \gamma$$
$$w(S_m, z_r) = 0 \quad \text{if } m \neq r, \quad 1 \text{ if } m = r.$$

Therefore

$$w(-\gamma, z) + w(S_1, z) + \cdots + w(S_n, z) = 0$$

for all $z \notin S'$.

Now by the generalized Cauchy theorem,

$$\int_{-\gamma} f + \int_{S_1} f + \cdots + \int_{S_n} f = 0$$

so that

$$\int_{\gamma} f = \int_{S_1} f + \cdots + \int_{S_n} f$$
$$= 2\pi i \cdot \text{res}\,(f, z_1) + \cdots + 2\pi i \cdot \text{res}\,(f, z_n)$$

from (1). This is the required result. □

Alternative proof. Instead of the generalized Cauchy theorem we can use Laurent series: the proof is instructive but less elegant. Around z_i we have a Laurent expansion of f, which we split into three parts:

$$f(z) = Q_j(z) + \frac{\text{res}\,(f, z_j)}{z - z_j} + P_j(z) \tag{2}$$

where

$$Q_j(z) = \sum_{n=2}^{\infty} b_n(z-z_j)^{-n}$$

$$P_j(z) = \sum_{n=0}^{\infty} a_n(z-z_j)^n.$$

Now $P_j(z)$ is differentiable in a neighbourhood of z_j, and $Q_j(z) + \text{res}\,(f, z_j)/(z-z_j)$ is differentiable for $z \neq z_j$ since all but a finite number of b_n are zero. It follows that

$$h(z) = f(z) - \sum_{j=1}^{n} Q_j(z) - \sum_{j=1}^{n} \frac{\text{res}\,(f, z_j)}{z - z_j} \tag{3}$$

is differentiable on S, except perhaps at $z_1, \ldots, z_n$. But from (2) it is also differentiable in a neighbourhood of each z_j. Hence $h(z)$ is differentiable in S, so by Cauchy's theorem

$$\int_\gamma h(z)\,\mathrm{d}z = 0. \tag{4}$$

But $Q_j(z) = T_j'(z)$ where

$$T_j(z) = \sum_{n=2}^{\infty} \frac{b_n}{-n+1} (z-z_j)^{-n+1}$$

so that

$$\int_\gamma Q_j(z)\,\mathrm{d}z = 0 \tag{5}$$

as well. Hence integrating (3) round γ and using (4) and (5) we get

$$0 = \int_\gamma f(z)\,\mathrm{d}z - 0 - \sum_{j=1}^{n} \text{res}\,(f, z_j) \int_\gamma (z-z_j)^{-1}\,\mathrm{d}z$$

$$= \int_\gamma f(z)\,\mathrm{d}z - 2\pi\mathrm{i} \sum_{j=1}^{n} \text{res}\,(f, z_j)$$

which finishes the alternative proof. ☐

2. Calculating residues

The residue theorem can be used to calculate integrals (and not just integrals round simple loops). For it to be of much use we must find ways of calculating residues. The following two lemmas are very useful in this respect.

LEMMA 12.2. If z_0 is a simple pole of f then

$$\text{res}\,(f,\,z_0)=\lim_{z\to z_0}\,(z-z_0)f(z).$$

If $f(z)=p(z)/q(z)$ where $p(z_0)\neq0$, $q(z_0)=0$, $q'(z_0)\neq0$, then

$$\text{res}\,(f,\,z_0)=p(z_0)/q'(z_0).$$

Proof. We've already done the first part in the previous chapter (Proposition 11.4), but to recap: we have

$$f(z)=\frac{b_1}{z-z_0}+\sum_{n=0}^{\infty}a_n(z-z_0)^n$$

and so

$$(z-z_0)f(z)=b_1+\sum_{n=0}^{\infty}a_n(z-z_0)^{n+1}$$

which tends to b_1 as $z\to z_0$.

For the second part, note that

$$\lim_{z\to z_0}\frac{(z-z_0)p(z)}{q(z)}=\lim_{z\to z_0}p(z)\bigg/\left(\frac{q(z)-q(z_0)}{z-z_0}\right)$$

since $q(z_0)=0$, and this is equal to $p(z_0)/q'(z_0)$.

For example, if

$$f(z)=\frac{\cos(\pi z)}{1-z^{976}}$$

then

$$\text{res}\,(f,\,1)=\frac{\cos(\pi)}{-976\cdot1^{975}}=\frac{1}{976}.$$

LEMMA 12.3. If z_0 is a pole of f of order m then

$$\text{res}\,(f,\,z_0)=\lim_{z\to z_0}\left\{\frac{1}{(m-1)!}\frac{d^{m-1}}{dz^{m-1}}\left((z-z_0)^mf\;(z)\right)\right\}.$$

Proof. We have

$$f(z)=\frac{b_m}{(z-z_0)^m}+\cdots+\frac{b_1}{z-z_0}+\sum_{n=0}^{\infty}a_n(z-z_0)^n$$

so that

$$(z-z_0)^mf(z)=b_m+\cdots+b_1(z-z_0)^{m-1}+\sum_{n=0}^{\infty}a_n(z-z_0)^{m+n}$$

and therefore

$$\frac{d^{m-1}}{dz^{m-1}}((z-z_0)^m f(z)) = (m-1)!b_1 + \sum_{n=0}^{\infty} \frac{(m+n)!}{(n+1)!} a_n(z-z_0)^{n+1}$$

and the result follows on taking limits.

For example, consider

$$f(z) = \left(\frac{z+1}{z-1}\right)^3$$

which has a triple pole at $z_0 = 1$. Then

$$(z-1)^3 f(z) = (z+1)^3$$

and so

$$\frac{1}{2!}\frac{d^2}{dz^2}((z-1)^3 f(z)) = \frac{6}{2!}(z+1)$$

which $\to 3 \cdot 2 = 6$ as $z \to 1$. So res $(f, 1) = 6$.

On occasion another technique may be brought into play: working out the appropriate part of the Laurent series. (It is a waste of time to work out the whole thing, because the whole point about residues is that we don't need the whole thing, but only b_1.) For instance,

$$f(z) = 1/(z^2 \sin(z))$$

$$= 1 \bigg/ \left(z^2 \left(z - \frac{z^3}{6} + \cdots\right)\right)$$

$$= \frac{1}{z^3}\left(1 - \frac{z^2}{6} + \cdots\right)^{-1}$$

$$= \frac{1}{z^3}\left(1 + \frac{z^2}{6} + \cdots\right)$$

$$= \frac{1}{z^3} + \frac{1}{6z} + \cdots$$

so that res $(f, 0) = 1/6$.

3. Evaluation of definite integrals

We now consider a number of techniques for the calculation of various kinds of definite integral.

(I) $$\int_0^{2\pi} Q(\cos(t), \sin(t)) \, dt$$

Let $C(t) = e^{it}$ ($0 \leqslant t \leqslant 2\pi$), the unit circle. If

$$z = C(t) = e^{it}$$

then

$$\cos(t) = \frac{1}{2}\left(z + \frac{1}{z}\right)$$

$$\sin(t) = \frac{1}{2i}\left(z - \frac{1}{z}\right)$$

from which we get

$$\int_0^{2\pi} Q(\cos(t), \sin(t))\, dt = \int_C Q\left(\frac{1}{2}\left(z + \frac{1}{z}\right), \frac{1}{2i}\left(z - \frac{1}{z}\right)\right) \frac{dz}{iz}$$

$$= 2\pi i \Sigma$$

where Σ is the sum of the residues of

$$\frac{1}{iz} Q\left(\frac{1}{2}\left(z + \frac{1}{z}\right), \frac{1}{2i}\left(z - \frac{1}{z}\right)\right) \tag{6}$$

inside C.

For example, consider

$$\int_0^{2\pi} (\cos^3(t) + \sin^2(t))\, dt.$$

Then (6) becomes

$$\frac{1}{iz}\left(\frac{1}{8}\left(z + \frac{1}{z}\right)^3 - \frac{1}{4}\left(z - \frac{1}{z}\right)^2\right)$$

$$= \frac{1}{8i}z^2 - \frac{1}{4i}z + \frac{3}{i} + \frac{1}{2iz} + \frac{3}{iz^2} - \frac{1}{4iz^3} + \frac{1}{iz^4}$$

which has just a pole inside C, of residue $1/2i$. So the integral is equal to $2\pi i/2i = \pi$.

If Q is at all complicated the computations can become very tedious. Sometimes integrals of this kind can be found from the real and imaginary parts of an integral

$$\int_C g(z)\, dz$$

with a suitable choice of g. For instance,

$$\int_C \frac{e^z}{z}\, dz = 2\pi i$$

since e^z/z has residue 1 at $z=0$. Therefore

$$\int_0^{2\pi} \frac{e^{\cos(t)+i\sin(t)}}{e^{it}} \, ie^{it} = 2\pi i$$

so

$$\int_0^{2\pi} e^{\cos(t)+i\sin(t)} \, dt = 2\pi,$$

so

$$\int_0^{2\pi} e^{\cos(t)} (\cos(\sin(t)) + i\sin(\sin(t))) \, dt = 2\pi$$

and equating real and imaginary parts yields

$$\int_0^{2\pi} e^{\cos(t)} \cos(\sin(t)) \, dt = 2\pi$$

$$\int_0^{2\pi} e^{\cos(t)} \sin(\sin(t)) \, dt = 0.$$

(II) $$\int_{-\infty}^{\infty} f(x) \, dx$$

The real integral

$$\int_{-\infty}^{\infty} f(x) \, dx \tag{7}$$

is defined to be equal to

$$\lim_{\substack{x_2 \to \infty \\ x_1 \to -\infty}} \int_{x_1}^{x_2} f(x) \, dx \tag{8}$$

provided the limit exists. The techniques we are about to discuss allow the calculation of

$$\lim_{R \to \infty} \int_{-R}^{R} f(x) \, dx \tag{9}$$

which is known as the *Cauchy principal value* of the integral, and will be written

$$P \int_{-\infty}^{\infty} f(x) \, dx.$$

If (8) exists then so does (9) and the two are equal. But the Cauchy principal value may exist when (8) does not. For example

$$\int_{-R}^{R} x \, dx = [x^2/2]_{-R}^{R} = 0$$

so that

$$P \int_{-\infty}^{\infty} x \, dx = 0.$$

But clearly (8) does not exist when $f(x) = x$.

It follows that when we use the technique below, we must take into account the convergence of (8). This, in part, leads to condition (ii) of the following.

Suppose (i) f is differentiable in the upper half-plane $\operatorname{im}(z) \geqslant 0$ except for a finite number of poles, none of which lies on the real axis.

(ii) If $S_R(t) = Re^{it}$ $(0 \leqslant t \leqslant \pi)$ then for large R there is a constant A such that

$$|f(z)| \leqslant A/R^2$$

when z lies on S_R.

Then

$$\int_{-\infty}^{\infty} f(x) \, dx = 2\pi i \Sigma$$

where Σ is the sum of the residues at the poles of f in the upper half-plane.

Proof. Choose R large enough for (ii) to be satisfied, and so that all the poles lie inside $S_R + [-R, R]$. (Fig. 12.4)

Then by Cauchy's theorem

$$\int_{-R}^{R} f(x) \, dx + \int_{S_R} f(z) \, dz = 2\pi i \Sigma$$

with Σ as stated. Now let $R \to \infty$. Then

$$\left| \int_{S_R} f(z) \, dz \right| \leqslant \frac{A}{R^2} \pi R = \pi A/R$$

which $\to 0$ as $R \to \infty$. Hence

$$\lim_{R \to \infty} \int_{-R}^{R} f(x) \, dx = 2\pi i \Sigma,$$

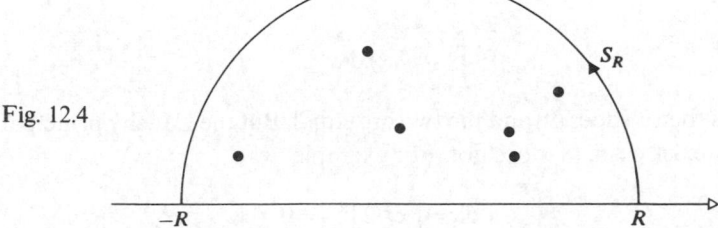

Fig. 12.4

that is,

$$\mathrm{P}\int_{-\infty}^{\infty} f(x)\,\mathrm{d}x = 2\pi\mathrm{i}\Sigma.$$

However, (ii) tells us that $|f(x)| \leqslant A/x^2$ for $|x|$ large, and so it follows from real analysis that

$$\int_{-\infty}^{\infty} f(x)\,\mathrm{d}x$$

exists. It is therefore equal to the Cauchy principal value, and so to $2\pi\mathrm{i}\Sigma$ as claimed.

Note A. Condition (ii) is certainly satisfied if $f(z) = p(z)/q(z)$ where p and q are polynomials such that q has no real zeros and the degree of q is greater than or equal to $2 +$ the degree of p.

   For example, consider

$$\int_{-\infty}^{\infty} \frac{\mathrm{d}x}{(x^2+a)(x^2+b)}$$

for $a>0$, $b>0$, $a \neq b$. By Note A this satisfies (ii), and (i) is obviously true. Now the only poles of $1/(z^2+a)(z^2+b)$ in the upper half-plane are simple poles at $\mathrm{i}a$, $\mathrm{i}b$. The residue at $\mathrm{i}a$ is

$$\lim_{z \to \mathrm{i}a} \frac{z-\mathrm{i}a}{(z^2+a^2)(z^2+b^2)} = \frac{1}{2\mathrm{i}a(b^2-a^2)}$$

and similarly that at $\mathrm{i}b$ is

$$\frac{1}{2\mathrm{i}b(a^2-b^2)}.$$

Hence the value of the integral is

$$2\pi\mathrm{i}\left(\frac{1}{2\mathrm{i}a(b^2-a^2)} + \frac{1}{2\mathrm{i}b(a^2-b^2)}\right)$$

$$= \frac{\pi}{ab(a+b)}.$$

Note B. In the proof we did not require $f(z)$ to be real for z on the real axis.
   For instance, let

$$f(z) = \mathrm{e}^{\mathrm{i}z}/(z^2+a^2)(z^2+b^2).$$

We see that on S_R we have $|\mathrm{e}^{\mathrm{i}z}| = |\mathrm{e}^{-y+\mathrm{i}x}| = \mathrm{e}^{-y} \leqslant 1$ for $y \geqslant 0$, so that (ii) holds. As before we have simple poles at $\mathrm{i}a$, $\mathrm{i}b$; but now the residues are

$$\mathrm{e}^{-a}/2\mathrm{i}a(b^2-a^2) \quad \text{and} \quad \mathrm{e}^{-b}/2\mathrm{i}b(a^2-b^2).$$

Hence

$$\int_{-\infty}^{\infty} \frac{e^{ix}}{(x^2+a^2)(x^2+b^2)}\,dx = \pi\left(\frac{e^{-a}}{a(b^2-a^2)} + \frac{e^{-b}}{b(a^2-b^2)}\right)$$

and equating real and imaginary parts we obtain

$$\int_{-\infty}^{\infty} \frac{\cos(x)}{(x^2+a^2)(x^2+b^2)}\,dx = \frac{\pi}{b^2-a^2}\left(\frac{e^{-a}}{a} - \frac{e^{-b}}{b}\right)$$

$$\int_{-\infty}^{\infty} \frac{\sin(x)}{(x^2+a^2)(x^2+b^2)}\,dx = 0.$$

Of these, the second is actually obvious (why?), but the first is not.

We can try to generalize this method in (at least) two ways: by getting a better estimate for $\int_{S_R} f(z)\,dz$; or by allowing f to have poles on the real axis. The first we deal with in (III), the second in (IV).

(III) $$\int_{-\infty}^{\infty} f(x)e^{ix}\,dx$$

Suppose (i) f is differentiable in a domain containing the upper half-plane, except for a finite number of poles, none on the real axis.

(ii) For large R there exists a constant A such that

$$|f(z)| \leqslant A/R \quad \text{for } |z| = R.$$

Then

$$\int_{-\infty}^{\infty} f(x)e^{ix}\,dx = 2\pi i\Sigma'$$

where Σ' is the sum of the residues of $f(z)e^{iz}$ at poles in the upper half-plane.

Proof. It is possible to use the same contour as in (II) and prove that

$$\lim_{R\to\infty} \int_{S_R} f(z)e^{iz}\,dz = 0,$$

but this only calculates the Cauchy principal value; and the convergence problem raised thereby is much harder because all we know is that f behaves like $1/x$ for large x, which on its own does not imply the existence of

$$\int_{-\infty}^{\infty} f(x)e^{ix}\,dx.$$

By more delicate arguments this obstacle may be overcome, but it is easier to sidestep the whole question by using a different contour, as shown in Figure 12.5.

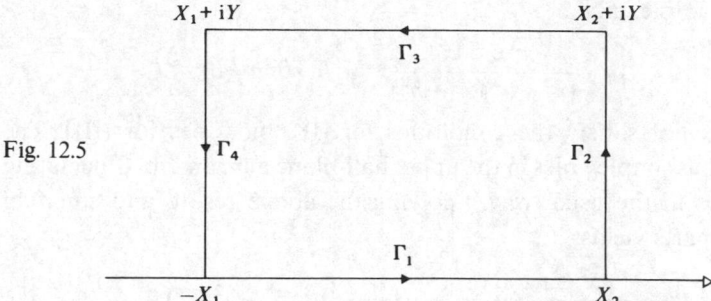

Fig. 12.5

We prove that as X_1, X_2, $Y \to \infty$, each of

$$\int_{\Gamma_r} f(z)e^{iz}\,dz$$

for $r = 2, 3, 4$, tends to zero. It will then follow as before that

$$\lim_{X_1, X_2 \to \infty} \int_{-X_1}^{X_2} f(x)e^{ix}\,dx = 2\pi i\Sigma'$$

which is what we want.

Now

$$\left| \int_{\Gamma_2} f(z)\,e^{iz}\,dz \right| = \left| \int_0^Y f(X_2 + it)\,e^{iX_2 - t}i\,dt \right|$$

$$\leqslant \int_0^Y \frac{A}{X_2}e^{-t}\,dt$$

$$\leqslant A/X_2$$

for X_2 large; and similarly

$$\left| \int_{\Gamma_4} f(z)e^{iz}\,dz \right| \leqslant A/X_1 \quad \text{for } X_1 \text{ large.}$$

Also

$$\left| \int_{\Gamma_3} f(z)e^{iz}\,dz \right| = \left| -\int_{-X_1}^{X_2} f(t + iY)e^{it - Y}\,dt \right|$$

$$\leqslant \left| \int_{-X_1}^{X_2} \frac{A}{Y}e^{-Y}\,dt \right|$$

$$= AY^{-1}e^{-Y}(X_1 + X_2).$$

Now for fixed X_1 and X_2 this $\to 0$ as $Y \to \infty$. Then letting X_1, $X_2 \to \infty$ we obtain the desired result.

As an example, take

$$\int_{-\infty}^{\infty} \frac{x^3 \, e^{ix} \, dx}{(x^2 + a^2)(x^2 + b^2)} \quad (a, b > 0, \, a \neq b)$$

which does not satisfy the conditions for (II), but does for (III). The integrand has simple poles in the upper half-plane at ia and ib. Calculating the residues in the usual way, applying the above result, and equating imaginary parts yields

$$\int_{-\infty}^{\infty} \frac{x^3 \sin x \, dx}{(x^2 + a^2)(x^2 + b^2)} = \frac{\pi}{b^2 - a^2} (b^2 e^{-b} - a^2 e^{-a}).$$

(Equating real parts shows that the corresponding cosine integral is zero, but once more this is obvious on other grounds.)

(IV)   *Poles on the real axis*

If $f(z)$ has poles on the real axis we make 'indentations' in the contour; i.e. draw small semicircles as in Figure 12.6. Suppose these semicircles have radii $\varepsilon_1, \varepsilon_2, \ldots$. Then we proceed as above, but letting $\varepsilon_1, \varepsilon_2, \ldots \to 0$ at the same time as $R, X_1, X_2, Y \to \infty$. There is a problem here similar to that in case (II): all we calculate is a Cauchy principal value

$$P \int_a^b f(x) \, dx = \lim_{\varepsilon \to 0} \left(\int_a^{x_0 - \varepsilon} f(x) \, dx + \int_{x_0 + \varepsilon}^b f(x) \, dx \right)$$

for a pole at x_0. This again may exist even though

$$\int_a^b f(x) \, dx = \lim_{\varepsilon_1 \to 0} \int_a^{x_0 - \varepsilon_1} f(x) \, dx + \lim_{\varepsilon_2 \to 0} \int_{x_0 + \varepsilon_2}^b f(x) \, dx$$

does not. Thus

$$\int_{-1}^1 \frac{dx}{x}$$

does not exist, but

$$P \int_{-1}^1 \frac{dx}{x} = 0.$$

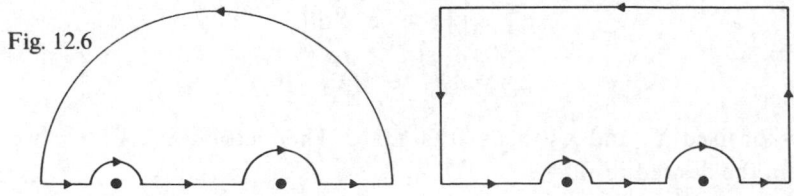

Fig. 12.6

Consequently we have a convergence problem to discuss, once the Cauchy principal value has been obtained. Apart from this, the hypotheses and conclusions of (II) and (III) remain valid even when poles occur on the real axis, provided that Σ and Σ' are summed only over the non-real poles in the upper half-plane. Rather than state a cumbersome general theorem we content ourselves with a typical example:

$$\int_{-\infty}^{\infty} \frac{e^{ix}}{x}\, dx.$$

There is a real pole at 0, so we take a contour as in Figure 12.7. As before, the integrals along Γ_2, Γ_3, and Γ_4 tend to zero. Since there are no poles inside the contour,

$$\lim_{X_1, X_2 \to 0} \left\{ \int_{-X_1}^{-\varepsilon} \frac{e^{ix}}{x}\, dx + \int_{\varepsilon}^{X_2} \frac{e^{ix}}{x}\, dx + \int_{C_\varepsilon} \frac{e^{iz}}{z}\, dz \right\} = 0.$$

But

$$\frac{e^{iz}}{z} = \frac{1}{z} + \sum_{n=1}^{\infty} \frac{z^{n-1}}{n!}$$

$$= \frac{1}{z} + \phi(z)$$

where ϕ is differentiable. Hence $|\phi(z)| \leq M$ in a neighbourhood of 0, and so

$$\lim_{\varepsilon \to 0} \int_{C_\varepsilon} \frac{e^{iz}}{z}\, dz = \lim_{\varepsilon \to 0} \int_{C_\varepsilon} \left(\frac{1}{z} + \phi(z) \right) dz$$

$$= \lim_{\varepsilon \to 0} \left\{ -\int_0^\pi \frac{1}{\varepsilon\, e^{it}}\, i\varepsilon\, e^{it}\, dt \right\} + \lim_{\varepsilon \to 0} M\pi\varepsilon$$

$$= -i\pi.$$

Therefore

$$P\int_{-\infty}^{\infty} \frac{e^{ix}}{x}\, dx = i\pi.$$

Fig. 12.7

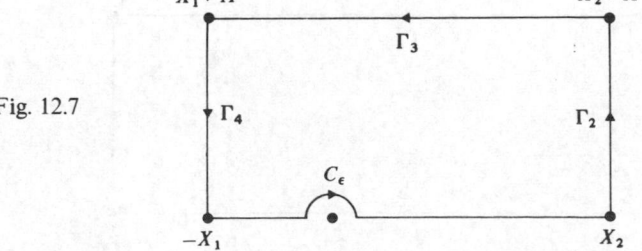

Equating real and imaginary parts,

$$P \int_{-\infty}^{\infty} \frac{\cos (x)}{x} \, dx = 0,$$

$$P \int_{-\infty}^{\infty} \frac{\sin (x)}{x} \, dx = \pi.$$

The first integral exists only as a principal value, since $\cos (x)/x$ behaves like $1/x$ for small x. But for the second we have

$$P \int_{-\infty}^{\infty} \frac{\sin (x)}{x} \, dx = \lim_{\varepsilon \to 0} \left(\int_{-\infty}^{-\varepsilon} \frac{\sin (x)}{x} \, dx + \int_{\varepsilon}^{\infty} \frac{\sin (x)}{x} \, dx \right)$$

$$= 2 \lim_{\varepsilon \to 0} \int_{\varepsilon}^{\infty} \frac{\sin (x)}{x} \, dx$$

and this limit does exist, and hence by definition equals

$$2 \int_{0}^{\infty} \frac{\sin (x)}{x} \, dx$$

because $\sin (x)/x \to 1$ as $x \to 0$. Hence we can remove the P from the second expression above. Further, we have proved that

$$\int_{0}^{\infty} \frac{\sin (x)}{x} \, dx = \frac{\pi}{2}.$$

(V) $$\int_{-\infty}^{\infty} \frac{e^{ax}}{\phi(e^x)} \, dx.$$

For integrals of this type we integrate around a contour of the form of Figure 12.8.

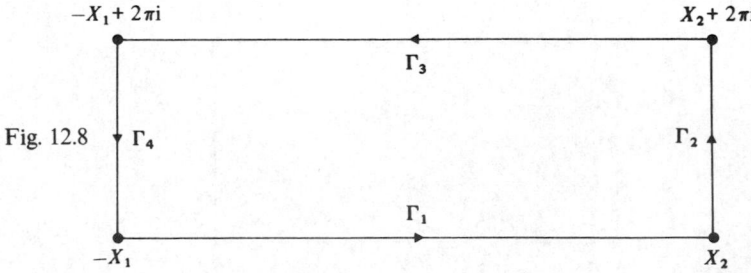

$-X_1 + 2\pi i$ $X_2 + 2\pi i$

Γ_3

Fig. 12.8 Γ_4 Γ_2

Γ_1

$-X_1$ X_2

Now on Γ_3, if we substitute $z = -t + 2\pi i$, then $-X_2 \leqslant t \leqslant X_1$, and so

$$\int_{\Gamma_3} \frac{e^{az}}{\phi(e^z)} dz = \int_{-X_2}^{X_1} \frac{e^{-at + 2\pi i a}}{\phi(e^{-t})} (-1) dt$$

$$= e^{2\pi i a} \int_{X_1}^{-X_2} \frac{e^{-at}}{\phi(e^{-t})} dt$$

which, putting $t = -x$, is equal to

$$-e^{2\pi i a} \int_{X_1}^{X_2} \frac{e^{ax}}{\phi(e^x)} dx.$$

If ϕ is such that

$$\int_{\Gamma_j} \frac{e^{az}}{\phi(e^z)} dz \to 0 \quad \text{as } X_1, X_2 \to \infty \quad (\text{for } j = 2, 4)$$

then we obtain

$$(1 - e^{2\pi a i}) \int_{-\infty}^{\infty} \frac{e^{ax}}{\phi(e^x)} dx = 2\pi i \Sigma''$$

where Σ'' is the sum of the residues of $e^{az}/\phi(z)$ at poles between the lines $\operatorname{im}(z) = 0$, $\operatorname{im}(z) = 2\pi$.

For example, consider

$$\int_{-\infty}^{\infty} \frac{e^{ax}}{e^{2x} + 1} dx \quad (0 < a < 1).$$

The relevant singularities are at $i\pi/2$, $3i\pi/2$, with corresponding residues

$$-\tfrac{1}{2} e^{i\pi a/2}, \quad -\tfrac{1}{2} e^{3i\pi a/2}.$$

Hence the value of the integral is

$$\frac{2\pi i}{1 - e^{2\pi i a}} (-\tfrac{1}{2} e^{i\pi a/2} - \tfrac{1}{2} e^{3\pi i a/2}).$$

Putting $k = e^{i\pi a/2}$ this is easily seen to be equal to

$$\frac{\pi}{2 \sin (\pi a/2)}.$$

(VI)   *Short cuts*

The student should always be on the lookout for quick methods, other than those given above. For example, an integral of great importance in applied mathematics is

$$\int_{-\infty}^{\infty} e^{i\lambda x} e^{-x^2} dx, \quad \lambda \in \mathbb{R}.$$

The function e^{-z^2} is differentiable throughout $\mathbb{C}$. If γ is the rectangle with vertices at $-R, R, R-(i\lambda)/2, -R-(i\lambda)/2$ then

$$\int_\gamma e^{-z^2}\, dz = 0.$$

As $R \to \infty$, the integrals over the vertical edges of the rectangle tend to zero; the other two converge. Hence

$$\int_{-\infty}^{\infty} e^{-x^2}\, dx = \int_{-\infty}^{\infty} e^{-(x-(i\lambda)/2)^2}\, dx$$

$$= \int_{-\infty}^{\infty} e^{-x^2} e^{i\lambda x} e^{\lambda^2/4}\, dx.$$

Therefore

$$\int_{-\infty}^{\infty} e^{i\lambda x} e^{-x^2}\, dx = e^{-\lambda^2/4} \int_{-\infty}^{\infty} e^{-x^2}\, dx.$$

The latter integral is just a constant: it may in fact be evaluated by other methods, and is equal to $1/\sqrt{\pi}$. So

$$\int_{-\infty}^{\infty} e^{i\lambda x} e^{-x^2}\, dx = \frac{e^{-\lambda^2/4}}{\sqrt{\pi}}.$$

4. Summation of series

Let f be a function which is differentiable at $z = n$ for an integer n. The functions $\cot(\pi z)$ and $\operatorname{cosec}(\pi z)$ have simple poles at $z = n$, $n \in \mathbb{Z}$; and one may check that

$$\operatorname{res}(f(z)\cot(\pi z), n) = \frac{f(n)}{\pi},$$

$$\operatorname{res}(f(z)\operatorname{cosec}(\pi z), n) = \frac{(-1)^n f(n)}{\pi}.$$

This suggests a method for summing certain series, as follows. We let C_N be the square whose vertices are

$$(N + \tfrac{1}{2})(\pm 1 \pm i)$$

parametrized, as usual, in the clockwise direction as shown. (Fig. 12.9) We claim that $\operatorname{cosec}(\pi z)$ and $\cot(\pi z)$ are bounded on C_N, where the bound is *independent of* N. First note that on the two sides parallel to the real axis

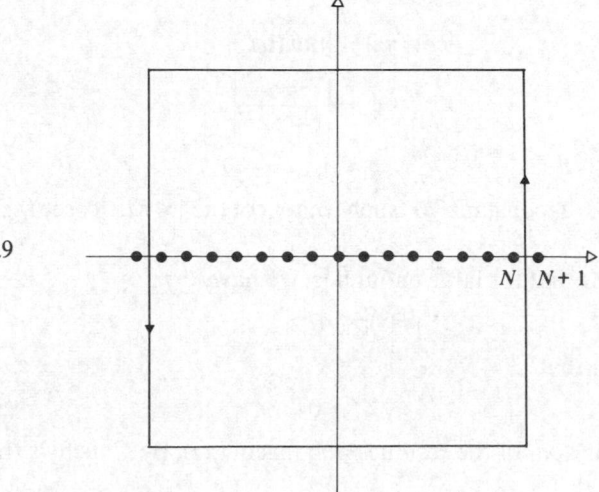

Fig. 12.9

we have $z = x + iy$ where $|y| \geqslant 1/2$. In this case

$$|\mathrm{cosec}\,(\pi z)| = (\tfrac{1}{2}|e^{i\pi z} - e^{-i\pi z}|)^{-1}$$
$$\leqslant (\tfrac{1}{2}|e^{-\pi y} - e^{\pi y}|)^{-1}$$
$$= (\sinh |\pi y|)^{-1}$$
$$\leqslant (\sinh (\pi/2))^{-1}.$$

And

$$|\cot\,(\pi z)| = \left|\frac{e^{i\pi z} + e^{-i\pi z}}{e^{i\pi z} - e^{-i\pi z}}\right|$$
$$\leqslant \frac{\left||e^{i\pi z}| + |e^{-i\pi z}|\right|}{\left||e^{i\pi z}| - |e^{-i\pi z}|\right|}$$
$$= \left|\frac{e^{-\pi y} + e^{\pi y}}{e^{-\pi y} - e^{\pi y}}\right|$$
$$= \coth |\pi y|$$
$$\leqslant \coth (\pi/2).$$

On the other two sides, we have $z = \pm N + \tfrac{1}{2} + it$, and so

$$|\mathrm{cosec}\,(\pi z)| = |\sin\,(\pi z)|^{-1}$$
$$= |\cos\,(i\pi t)|^{-1}$$
$$= (\cosh |\pi t|)^{-1}$$
$$\leqslant 1$$

and

$$|\cot(\pi z)| = |\tan(it)|$$

$$= \left| \frac{1 - e^{-2t}}{1 + e^{-2t}} \right|$$

$$\leqslant 1.$$

Hence there is a constant M such that $|\cot(\pi z)| \leqslant M$, $|\operatorname{cosec}(\pi z)| \leqslant M$, for z on any C_N.

Suppose now that for large enough $|z|$ we have

$$|f(z)| \leqslant A/|z|^2.$$

Then we claim that

$$\Sigma^* = 0$$

where Σ^* is the sum of the residues of $f(z)\cot(\pi z)$. By Cauchy's theorem we have

$$\int_{C_N} f(z)\cot(\pi z)\,\mathrm{d}z = 2\pi i \Sigma_N^*$$

where Σ_N^* is the sum of the residues of $f(z)\cot(\pi z)$ inside C_N. As $N \to \infty$, $\Sigma_N^* \to \Sigma^*$, so it is sufficient to show that the integral $\to 0$. But

$$\left| \int_{C_N} f(z)\cot(\pi z)\,\mathrm{d}z \right| \leqslant \frac{A}{N^2} M(8N + 4)$$

for large enough N, by the Estimation Lemma. As $N \to \infty$, this tends to 0 as claimed.

Now Σ^* usually forms an infinite series: the fact that it is zero allows us to sum certain related series. This is best illustrated by an example; and the obvious one to try is

$$f(z) = 1/z^2.$$

At an integer $n \neq 0$ the function $z^{-2}\cot(\pi z)$ has a simple pole with residue $1/(n^2\pi)$, whereas at the origin it has a triple pole with residue $-\pi/3$. Hence

$$0 = \Sigma^* = -\pi/3 + \sum_{n=-\infty}^{\infty} 1/(n^2\pi)$$

(where $n \neq 0$ in the infinite sum)

$$= -\pi/3 + \frac{2}{\pi} \sum_{n=1}^{\infty} 1/n^2.$$

Hence

$$\sum_{n=1}^{\infty} 1/n^2 = \pi^2/6,$$

a theorem originally proved by Euler by a different method.

If we use $\operatorname{cosec}(\pi z)$ instead of $\cot(\pi z)$ a similar theorem applies, and allows us to sum series of the form $\Sigma\,(-1)^n f(n)$. For instance, using $f(z) = z^{-2}$ and arguing much as above, we can prove that

$$\sum_{n=1}^{\infty}(-1)^{n+1}/n^2 = \pi^2/12.$$

5. Counting zeros

A rather different use for residues is the calculation of the number of zeros of a function analytic inside a simple loop. We begin with

THEOREM 12.4. Suppose that f is differentiable in a domain S containing a simple loop γ and all points inside γ, apart from a finite set of poles. If f has no zeros or poles on γ then

$$\frac{1}{2\pi i}\int_{\gamma}\frac{f'(z)}{f(z)}\,dz = N - P$$

where N is the number of zeros of f inside γ and P the number of poles of f inside γ, each counted according to multiplicity.

Proof. The integral is equal to the sum of the residues of $f'(z)/f(z)$ at poles inside γ. Now if z_0 is neither a zero nor a pole of f, then f'/f is differentiable at z_0. We show that

(i) If f has a zero of order k at z_1 then f'/f has a pole with residue k,

(ii) If f has a pole of order m at z_2 then f'/f has a pole with residue $-m$.

To prove (i) note that in that case

$$f(z) = (z - z_1)^k \phi(z)$$

where $\phi(z_1) \neq 0$ and ϕ is differentiable in a neighbourhood of z_1. Therefore

$$f'(z) = k(z - z_1)^{k-1}\phi(z) + (z - z_1)^k \phi'(z)$$

and hence

$$\frac{f'(z)}{f(z)} = \frac{k}{z - z_1} + \frac{\phi'(z)}{\phi(z)}$$

which has a simple pole with residue k at z_1 because ϕ'/ϕ is differentiable at z_1. This proves (i).

Similarly in case (ii)

$$f(z) = \psi(z)/(z - z_2)^m$$

where $\psi(z_2) \neq 0$ and ψ is differentiable in a neighbourhood of z_2. Hence

$$f'(z) = \frac{-m\psi(z)}{(z - z_2)^{m+1}} + \frac{\psi'(z)}{(z - z_2)^m}$$

and so

$$\frac{f'(z)}{f(z)} = \frac{-m}{z-z_2} + \frac{\psi'(z)}{\psi(z)}$$

which has a simple pole with residue $-m$ at z_2.

The theorem follows at once from (i) and (ii) by summing over all poles of f'/f. ☐

COROLLARY 12.5. Let γ be a simple loop in a domain S such that all points inside γ are within S. If f is differentiable in S and has no zeros on γ then the number of zeros of f inside γ is

$$\frac{1}{2\pi i} \int_\gamma f'(z)/f(z)\,dz.$$ ☐

From this we deduce another important theorem:

THEOREM 12.6 (Rouché's Theorem).

Suppose that f and g are differentiable in a domain S which contains a simple loop γ and all points inside γ. If

$$|f(z)-g(z)| < |f(z)| \tag{10}$$

for all $z=\gamma(t)$ $(a \leqslant t \leqslant b)$ then f and g have the same number of zeros inside γ.

Proof. Let $F(z)=g(z)/f(z)$, so that from (10) we have

$$|1-F(\gamma(t))| < 1 \tag{11}$$

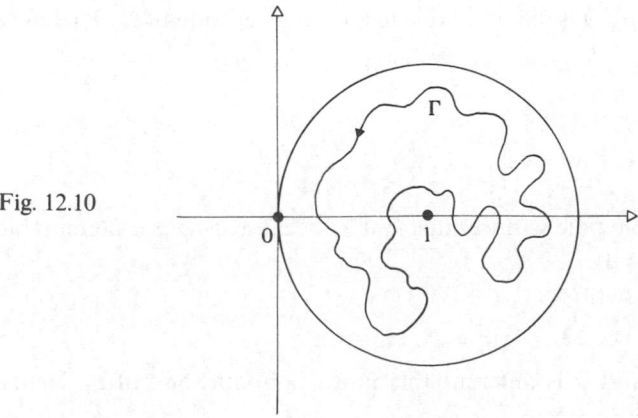

Fig. 12.10

for $a \leqslant t \leqslant b$. Inside S,

$$F \text{ has a zero} \Leftrightarrow g \text{ has a zero,}$$

$$F \text{ has a pole} \Leftrightarrow f \text{ has a zero.}$$

So to show that f and g have the same number of zeros inside γ, by Theorem 11.4 it is enough to prove that

$$\int_\gamma F'(z)/F(z) \, dz = 0.$$

Let $\Gamma(t) = F(\gamma(t))$. By (11) $|1 - \Gamma(t)| < 1$ for $a \leqslant t \leqslant b$, so that Γ lies inside the circle centre 1 radius 1. (Fig. 12.10) Now

$$\int_\gamma F'(z)/F(z) \, dz = \int_a^b \frac{F'(\gamma(t))}{F(\gamma(t))} \gamma'(t) \, dt$$

$$= \int_a^b \Gamma'(t)/\Gamma(t) \, dt$$

$$= \int_\Gamma \frac{dz}{z}$$

$$= w(\Gamma, 0)$$

$$= 0$$

from the diagram. $\qquad\qquad\qquad\qquad\qquad\qquad\qquad\qquad\qquad\qquad\quad\square$

As an example of Rouché's theorem in action, we shall give a second proof of the 'Fundamental Theorem of Algebra' (Theorem 10.7). Let

$$g(z) = z^m + a_1 z^{m-1} + \cdots + a_m$$

$$f(z) = z^m.$$

Then

$$|f(z) - g(z)| = |a_1 z^{m-1} + \cdots + a_m|$$

and for $z \neq 0$

$$\left| \frac{1}{z^m} \right| |f(z) - g(z)| = \left| \frac{a_1}{z} + \cdots + \frac{a_m}{z^m} \right|.$$

Since the right-hand side $\to 0$ as $|z| \to \infty$ it follows that there exists $R > 0$ such that if $|z| > R$ then

$$\left| \frac{a_1}{z} + \cdots + \frac{a_m}{z^m} \right| < 1,$$

and then

$$|f(z) - g(z)| < |z^m| = |f(z)|$$

so by Rouché's theorem f and g have the same number of zeros inside $\{z \in \mathbb{C} : |z| < R\}$. Since f has m zeros (counting multiplicities) so does g. Hence every polynomial of degree m over $\mathbb{C}$ has m zeros, and this is the 'Fundamental Theorem of Algebra'.

Exercises 12

1. Find the residue of f at z_0 in the following cases:
   (i) $f(z) = z^{-3} \sin z \ (z \neq 0), \ z_0 = 0$
   (ii) $f(z) = e^z z^{-n-1} \ (z \neq 0), \ z_0 = 0$
   (iii) $f(z) = \exp(1/z) \ (z \neq 0), \ z_0 = 0$
   (iv) $f(z) = z^2(z^2 + a^2)^{-3} \ (z \neq \pm ia), \ z_0 = ia, \ -ia$
   (v) $f(z) = (1 + z^2 + z^4)^{-1} \ (z \neq \exp(r\pi i/3), \ r = 1, 2, 4, 5), \ z_0 = \exp(\pi i/3)$.

2. Find the residue of the given function at each of its isolated singular points, including infinity (provided this is not the limit of a sequence of finite singularities, i.e., is isolated):
   (i) $1/(z^3 - z^5)$ (ii) $\sin(z) \sin(1/z)$
   (iii) $e^z/z^2(z^2 + 5)$ (iv) $\cot^3 z$
   (v) $(\sin z^{-1})^{-1}$ (vi) $(z \cos z^{-2})^{-1}$.

3. If f has a pole of order 2 at z_0, show that the residue of f at z_0 is $h'(z_0)$, where $h(z) = (z - z_0)^2 f(z)$.

4. Let $\gamma(t) = e^{it} \ (0 \leq t \leq 2\pi)$. Find, by residues, the value of

$$\int_\gamma \frac{dz}{z^2 - 2az + 1} \qquad (a > 1).$$

Hence calculate

$$\int_0^{2\pi} \frac{dt}{a - \cos t}.$$

What happens if $a < -1$? If $-1 \leq a \leq 1$?

5. Verify the following:

(i) $\displaystyle\int_0^\pi \frac{dt}{1 + b \cos^2 t} = \frac{\pi}{\sqrt{b+1}} \qquad (b > -1).$

(ii) $\displaystyle\int_0^\infty \frac{dx}{1 + x^4} = \pi/2\sqrt{2}$

(iii) $\displaystyle\int_0^\infty \frac{\log x}{1 + x^2}\, dx = 0$

(iv) $\displaystyle\int_{-\infty}^\infty \frac{x \sin(\pi x)}{1 - x^2}\, dx = \pi$

(v) $\displaystyle\int_0^\infty \frac{(\log x)^2}{1 + x^2}\, dx = \pi^3/8.$

6. Show that

$$\int_{-\infty}^{\infty} \frac{(10x)^2}{(x^2+4)^2(x^2+9)^2} \, dx = \pi.$$

7. Show that

$$\int_0^{\infty} \frac{\cos 5x}{x^4+a^4} \, dx = \frac{\pi}{2a^3} e^{-5a/\sqrt{2}} \sin\left(\frac{5a}{\sqrt{2}}+\frac{\pi}{4}\right) \quad (a>0).$$

8. Evaluate:

   (i) $\displaystyle\int_0^{2\pi} \cos^4 t + \sin^4 t \, dt$

   (ii) $\displaystyle\int_0^{2\pi} \sin^3 t \cdot \cos t + \cos^3 t \sin t \, dt$

   (iii) $\displaystyle\int_0^{2\pi} 2 \cos^3 t + 3 \cos^2 t \, dt.$

9. Prove:

   (i) $\displaystyle\int_0^{2\pi} \exp(\cos t) \cos(nt - \sin t) \, dt = 2\pi/n! \quad (n \in \mathbb{Z}, n > 0)$

   (ii) $\displaystyle\int_0^{2\pi} \exp(\cos t) \sin(nt - \sin t) \, dt = 0.$

10. Evaluate, by integrating suitable functions round a semicircle,

    (i) $\displaystyle\int_0^{\infty} \frac{dx}{1+x^2+x^4}$ (ii) $\displaystyle\int_0^{\infty} \frac{\cos mx}{x^2+a^2} \, dx \quad (a, m > 0).$

11. Prove that

$$\int_0^{\infty} \frac{x^2}{(x^2+a^2)^3} \, dx = \pi/16a^3 \quad (a>0).$$

12. By integrating round a rectangle whose vertices lie at $R, R+i, -R+i, -R$, and letting $R \to \infty$, show that

$$\int_{-\infty}^{\infty} \frac{\cosh(cx)}{\cosh(\pi x)} \, dx = \sec(c/2) \quad (-\pi < c < \pi).$$

13. Prove that

$$\int_0^{\infty} t^{a-1}(t+1)^{-1} \, dt = \frac{\pi}{\sin \pi a} \quad (0 < a < 1)$$

by making the substitution $t = e^x$ and integrating $ez^z(e^z + 1)^{-1}$ around the rectangle with vertices $\pm R, \pm R + 2\pi i$.

14. (Inversion formula for the Laplace Transform.) Suppose that F is differentiable in $\mathbb{C}$ except for a finite number of poles, of which $z_1, \ldots, z_n$ satisfy re $z_r < a$, and none lie on the line re $z = a$.

If there exist $M>0$, $b>0$, $c>0$ such that $|F(z)|<M/|z|^c$ for $|z|<b$, show that

$$f(t)=\lim_{R\to\infty}\int_{a-iR}^{a+iR}e^{zt}F(z)\,dz=2\pi i\sum_{r=1}^{n}\text{res}\,(e^{zt}F(z),z_r).$$

If $F(z)=\alpha(z^2+\alpha^2)^{-1}$ $(\alpha>0)$, show that $f(t)=\sin\alpha t$.

15. Use real analysis to prove Jordan's Inequality: $\sin t/t\geqslant 2/\pi$ for $0<t\leqslant\pi/2$. Hence show that if $S_R(t)=R\,e^{it}\,(0\leqslant t\leqslant\pi)$ then

$$\lim_{R\to\infty}\int_{S_R}e^{imz}/z\,dz=0\quad(m>0).$$

By integrating e^{imz}/z along the contours Γ_1, C_ε, Γ_2, S_R defined by

$$\Gamma_1(t)=t\quad(-R\leqslant t\leqslant-\varepsilon)$$
$$C_\varepsilon(t)=e^{i(\pi-t)}\quad(0\leqslant t\leqslant\pi)$$
$$\Gamma_2(t)=t\quad(\varepsilon\leqslant t\leqslant R),$$

prove Dirichlet's Discontinuous Factor:

$$\int_0^\infty\frac{\sin mx}{x}\,dx=\begin{cases}\pi/2 & \text{if } m>0\\0 & \text{if } m=0\\-\pi/2 & \text{if } m<0.\end{cases}$$

16. Show that

(i) $z/(e^z-1)=1-\dfrac{z}{2}+\displaystyle\sum_{n=1}^{\infty}\frac{2z^2}{z^2+4n^2\pi^2}$

(ii) $\operatorname{cosec} z=\dfrac{1}{z}+\displaystyle\sum_{n=1}^{\infty}\frac{(-1)^n}{z^2-n^2\pi^2}2z$

(iii) $\operatorname{cosec}^2 z=\displaystyle\sum_{n=-\infty}^{\infty}\frac{1}{(z-n\pi)^2}.$

17. Sum the following series, where $a\notin\mathbb{Z}$,

(i) $\displaystyle\sum_{-\infty}^{\infty}(n+a)^{-2}$ (ii) $\displaystyle\sum_{0}^{\infty}(n^2+a^2)^{-1}$

(iii) $\displaystyle\sum_{0}^{\infty}(2n+1)^{-2}$ (iv) $\displaystyle\sum_{0}^{\infty}(-1)^n(2n+1)^{-3}.$

18. By integrating

$$\frac{ze^{ibz}}{(a^2-z^2)\sin(\pi z)}$$

round a suitable contour, show that

$$\sum_{n=1}^{\infty}(-1)^n\frac{n\sin bn}{a^2-n^2}=\frac{\pi}{2}\cdot\frac{\sin ba}{\sin\pi a}\quad(|b|<\pi).$$

19. By considering $f(z)=1/(z-\xi)+1/z$, show that when $\xi\notin\mathbb{Z}$,

$$\pi\cot\pi\xi=\frac{1}{\xi}+\sum_{n=1}^{\infty}2\xi/(\xi^2-n^2).$$

20. Using the result of Exercise 19, integrate $\pi \cot \pi z$ along a suitable contour to show that

$$\log \sin \pi z = \log \pi z + \sum_{n=1}^{\infty} \log (1 - z^2/n^2)$$

where the log is chosen to make $\log (1) = 0$ in each term. Taking exponentials, obtain the infinite product expansion of the sine function (defined as the limit of suitable partial products, by analogy with infinite sums)

$$\sin \pi z = \pi z \prod_{n=1}^{\infty} (1 - z^2/n^2).$$

21. If $|a| > e$, use Rouché's Theorem to prove that the equation

$$e^z = az^n$$

has n roots in $|z| < 1$.

22. Find the number of zeros of the following polynomials lying inside the unit circle:
    (i) $z^9 - 2z^6 + z^2 - 8z - 2$
    (ii) $2z^5 - z^3 + 3z^2 - z + 8$
    (iii) $z^4 - 5z + 1$.

23. How many zeros of $z^4 + 4z^3 + 6z^2 - 4z + 3$ lie inside the circle $|z - 1| < 1$?

24. Prove that however small $\varepsilon > 0$ is chosen, for all large enough n the function

$$1 + z^{-1} + (2!z^2)^{-1} + (3!z^3)^{-1} + \cdots + (n!z^n)^{-1}$$

has all of its zeros inside the circle $|z| < \varepsilon$.

25. Let $p(z)$ be a polynomial of degree n, and suppose that $p(z_1) = p(z_2) = 0$. Show that there exists a zero of $p'(z)$ within the circle centre $\frac{1}{2}(z_1 + z_2)$ and radius $\frac{1}{2}|z_1 - z_2| \cot (\pi)/n$.

26. (Residues in Reverse.) Show that the residue of $\tan^{p-1} \pi z$ at $z = \frac{1}{2}$ is $(-1)^{p/2} \pi^{-1}$, for integer $p > 0$. [Hint: integrate it round the rectangle with vertices at $-iR$, $1 - iR$, $1 + iR$, iR; let $R \to \infty$ and estimate sizes.]

27. Show that

(i) $\displaystyle\int_0^{\infty} \frac{\log x}{1 + x^2}\, dx = 0$ 　　 (ii) $\displaystyle\int_0^{\infty} \frac{\log x}{(1 + x^2)^2}\, dx = -\frac{\pi}{4}.$

13

Conformal transformations

Often in mathematics one encounters functions which preserve some structure in which one is interested. For example, in Euclidean geometry all rigid motions preserve lengths and angles; changes of scale preserve the shape (but not the size) of geometrical figures; homomorphisms of groups preserve the group multiplication. Conversely, given an interesting class of functions, one may ask what structure such functions preserve. This chapter deals with a property preserved by all analytic (i.e. differentiable) functions, namely: angles between curves. Functions with this property are called 'conformal'. The conformal property can be used in two directions. By studying differentiable functions one may prove theorems about curves; by studying curves one may prove theorems about differentiable functions. The latter technique is of great importance in the advanced 'geometrical' theory of differentiable functions, but only the former falls within our present scope. The method has interesting applications to potential theory and fluid dynamics, and we shall outline the beginnings of these. We also consider in moderate detail several interesting special conformal functions, in particular the 'Möbius mappings' which have the remarkable property of mapping circles into circles.

1. Real numbers modulo 2π

The use of a real number to measure an angle is not entirely satisfactory, in that the same angle corresponds to many different real numbers. However, if θ and ϕ are reals representing the same angle, then $\theta - \phi$ is an integer multiple of 2π, and conversely: hence the ambiguity is not too great. For many purposes it can be avoided by making some artificial convention, such as our requirement that $-\pi < \arg(z) \leqslant \pi$. For other purposes it is more convenient to measure angles in a natural and unambiguous way, though less familiar than real numbers.

238

If $x, y \in \mathbb{R}$ we say that x and y are *congruent modulo* 2π, and write

$$x \equiv y \quad (\text{mod } 2\pi),$$

if there is an integer n such that $x - y = 2n\pi$. Congruence modulo 2π is an equivalence relation, and hence partitions $\mathbb{R}$ into mutually disjoint equivalence classes. We denote the set of all such equivalence classes by

$$\mathbb{R}/2\pi.$$

For each $x \in \mathbb{R}$ we let $p(x)$ be the equivalence class to which x belongs, thereby defining a function

$$p: \mathbb{R} \to \mathbb{R}/2\pi.$$

Then we have

$$p(x) = \{x + 2n\pi : n \in \mathbb{Z}\}.$$

Given an angle, measured by a real number θ, then the same angle is measured by all $\theta + 2n\pi$ $(n \in \mathbb{Z})$, and by these real numbers only. Instead of picking one of them, we can represent the angle by the whole collection, namely $p(\theta)$. In other words, the natural measure of an angle is not an element of $\mathbb{R}$, but of $\mathbb{R}/2\pi$.

For instance, the real numbers

$$\ldots, -11\pi/3, -5\pi/3, \pi/3, 7\pi/3, 13\pi/3, \ldots$$

all represent the same angle, whereas the set

$$\{\ldots, -11\pi/3, -5\pi/3, \pi/3, 7\pi/3, 13\pi/3, \ldots\}$$

is the unique element of $\mathbb{R}/2\pi$ corresponding to this angle.

The best way to think of $\mathbb{R}/2\pi$ geometrically is as a circle. To see this, let us define

$$q: \mathbb{R} \to \mathbb{C}$$

by

$$q(x) = e^{ix} \quad (x \in \mathbb{R}).$$

Then the image of q is the unit circle $S \subseteq \mathbb{C}$. Since $e^{2\pi i} = 1$, it is easy to see that $q(x) = q(y)$ if and only if $x \equiv y \pmod{2\pi}$, and so

$$q(x) = q(y) \Leftrightarrow p(x) = p(y). \tag{1}$$

Hence we may define

$$j: \mathbb{R}/2\pi \to S$$

as follows: if $r \in \mathbb{R}/2\pi$ then $r = p(x)$ for some $x \in \mathbb{R}$, and we set

$$j(r) = q(x).$$

By (1) this is independent of the choice of $x \in \mathbb{R}$, and j is a bijection. Hence the elements of $\mathbb{R}/2\pi$ are in natural one-to-one correspondence with the points of a circle.

We say that a function $f: X \to \mathbb{R}/2\pi$, where $X \subseteq \mathbb{C}$, is *continuous* if and only if the composite function $jf: X \to S$ is continuous in the usual sense. This accords with the intuitive idea of geometrical continuity if we think of $\mathbb{R}/2\pi$ as a circle. (In topological terms: we can use q to transfer the usual topology from S to $\mathbb{R}/2\pi$, saying that $G \subseteq \mathbb{R}/2\pi$ is open if and only if $q(G)$ is open in S. Continuity of maps $f: X \to \mathbb{R}/2\pi$ in this topology is equivalent to what we have just defined.)

We can also define addition and subtraction of elements of $\mathbb{R}/2\pi$, corresponding to geometrical addition and subtraction of angles. Let $r, s \in \mathbb{R}/2\pi$. Pick $x, y \in \mathbb{R}$ such that $p(x) = r$, $p(y) = s$, and define

$$r + s = p(x + y),$$
$$r - s = p(x - y).$$

One verifies as usual that these definitions do not depend on the choice of x and y. (In group-theoretic terms: the set $G = \{2n\pi : n \in \mathbb{Z}\}$ is a subgroup of the additive group of real numbers, hence a normal subgroup since the latter is abelian. The quotient group $\mathbb{R}/G$ is what we have called $\mathbb{R}/2\pi$. As an analogy, consider the definition of the integers modulo n, $\mathbb{Z}_n$, as a quotient group $\mathbb{Z}/H$ where $H = \{kn : k \in \mathbb{Z}\}$. In $\mathbb{Z}_n$ we can also define multiplication: the reader should verify that this does *not* work for $\mathbb{R}/2\pi$ since 2π is not an integer.)

For $z \in \mathbb{C} \setminus \{0\}$ we can now define a 'mod 2π' version of arg(z), namely

$$\text{arc}(z) = p(\arg(z)) \in \mathbb{R}/2\pi.$$

The notation is deliberately chosen to resemble 'arg' and further reminds one that angles are involved. The advantage of 'arc' over 'arg' is twofold. First, there is no ambiguity. Second, and more important, is the fact that

$$\text{arc} : \mathbb{C} \setminus \{0\} \to \mathbb{R}/2\pi$$

is a continuous function. This is false for arg, for as z moves across the negative real axis arg(z) jumps from near π to near $-\pi$, because of our convention that $-\pi < \arg(z) \le \pi$. However, since $p(-\pi) = p(\pi)$, the defect is not shared by arc. In fact, if $z = re^{i\theta}$, $r > 0$, then $j(\text{arc}(z)) = e^{i\theta}$. Using this, or directly, it is not hard to verify that

$$\text{arc}(z_1 z_2) = \text{arc}(z_1) + \text{arc}(z_2) \tag{2}$$

for all $z_1, z_2 \in \mathbb{C} \setminus \{0\}$. Again if we use arg this is true only up to integer multiples of 2π.

Thus we have two distinct ways of representing an angle: as a uniquely defined element of $\mathbb{R}/2\pi$, namely an equivalence class of reals modulo 2π; or as a real number, selected from the equivalence class, unique only up to integer multiples of 2π. It is advantageous to pass from one to the other at will, as we do in the proof of the next result.

LEMMA 13.1. If $\gamma:[a, b]\to\mathbb{C}$ is a path and $\gamma'(t_0)$ exists and is non-zero for some $t_0\in[a, b]$, then γ has a tangent at $z_0 = \gamma(t_0)$ making an angle arc $\gamma'(t_0)$ with the real axis.

Proof. Let $\gamma(t)=x(t)+iy(t)$. Then the required angle θ belongs to the congruence class modulo 2π of

$$\tan^{-1}(y'(t_0)/x'(t_0))$$
$$=\arg(x'(t_0)+iy'(t_0))$$
$$=\arg\gamma'(t_0)$$

and hence, on taking congruence classes, we get

$$\theta=\text{arc }\gamma'(t_0).$$

(In future we shall be less explicit in our passage from $\mathbb{R}$ to $\mathbb{R}/2\pi$ or back again.)

If γ_1 and γ_2 are two paths meeting at $z_0=\gamma_1(t_1)=\gamma_2(t_2)$ having derivatives $\gamma'_1(t_1)\neq0$, $\gamma'(t_2)\neq0$, then we define the *angle between γ_1 and γ_2 (at z_0)* to be (Fig. 13.1)

$$\theta=\text{arc }\gamma'_2(t_2)-\text{arc }\gamma'_1(t_1)\in\mathbb{R}/2\pi.$$

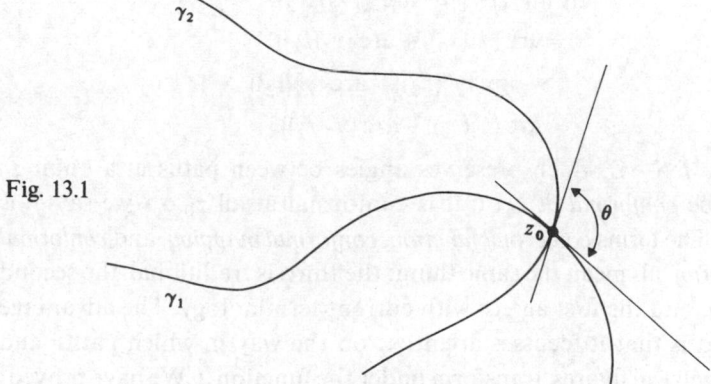

Fig. 13.1

2. Conformal transformations

In this section we shall consider functions $f:S\to\mathbb{C}$ where $S\subseteq\mathbb{C}$ is a domain. It is convenient to distinguish the two copies of $\mathbb{C}$ by using (x, y) as co-ordinates in S, and (u, v) as coordinates in the image $\mathbb{C}$. As usual we let $z=x+iy$, and we put $w=u+iv$. Then if f is differentiable on S we have

$$f(x+iy)=u(x, y)+iv(x, y)$$

where u and v are real-valued functions of two real variables x, y. Hence f defines a function from the subset S of the (x, y)-plane to the (u, v)-plane. A path γ in S, with

$$\gamma(t) = x(t) + iy(t) \quad (a \leqslant t \leqslant b)$$

is mapped by f to a path

$$f\gamma(t) = u(x(t), y(t)) + iv(x(t), y(t)) \quad (a \leqslant t \leqslant b)$$

in the (u, v)-plane.

Suppose that $z_0 = \gamma(t_0)$ and $\gamma'(t_0) \neq 0$, for some choice of $t_0 \in [a, b]$. Then

$$(f\gamma)'(t_0) = f'(\gamma(t_0))\gamma'(t_0)$$
$$= f'(z_0)\gamma'(t_0)$$

and hence

$$\operatorname{arc}((f\gamma)'(t_0)) = \operatorname{arc}(f'(z_0)\gamma'(t_0))$$
$$= \operatorname{arc}(f'(z_0)) + \operatorname{arc}(\gamma'(t_0)) \tag{3}$$

by (2). Hence if γ_1 and γ_2 are two paths through z_0, say $\gamma_1(t_1) = z_0 = \gamma_2(t_2)$, it follows that $f\gamma_1$ and $f\gamma_2$ meet at the same angle as γ_1 and γ_2. Geometrically, this is clear because (3) says that the tangents are both turned through an angle $\operatorname{arc}(f'(z_0))$. Alternatively, we compute

$$\operatorname{arc}((f\gamma_1)'(t_1)) - \operatorname{arc}((f\gamma_2)'(t_2))$$
$$= \operatorname{arc}(f'(z_0)) + \operatorname{arc}(\gamma_1'(t_1))$$
$$- \operatorname{arc}(f'(z_0)) - \operatorname{arc}(\gamma_2'(t_2))$$
$$= \operatorname{arc}(\gamma_1'(t_1)) - \operatorname{arc}(\gamma_2'(t_2)).$$

A function $f:S \to \mathbb{C}$ which preserves angles between paths at a point z_0 is said to be *conformal at* z_0; if it is conformal at all $z_0 \in S$ we say f is *conformal*. The terms *conformal function*, *conformal mapping*, and *conformal transformation* all mean the same thing: the third is traditional, the second convenient, and the first agrees with current terminology. The advantage of the third is that it focusses attention on the way in which paths, and other geometrical figures, transform under the function f. We have proved:

THEOREM 13.1. Let $f:S \to \mathbb{C}$ be differentiable. Then f is conformal at all $z_0 \in S$ such that $f'(z_0) \neq 0$. ☐

If $f'(z_0) = 0$ it is not true that f is conformal at z_0. For example if $f(z) = z^2$ then the positive half of the real axis and the 'positive' half (from 0 through i and outwards) of the imaginary axis transform, respectively, into the positive and negative halves of the real axis. Originally they met at an angle of $\pi/2$, but after transformation they meet at π. In fact, if z_0 is a zero of f' of order m, then the angles between paths meeting at z_0 is multiplied by $m+1$ on transforming by f.

We can find out a little about how f affects lengths. If z_0, $z \in \mathbb{C}$ and f is differentiable at z_0 then the ratio of the distances between $f(z)$ and $f(z_0)$, and z and z_0, is

$$\frac{|f(z)-f(z_0)|}{|z-z_0|} = \left|\frac{f(z)-f(z_0)}{z-z_0}\right|$$

which $\rightarrow |f'(z_0)|$ as $z \rightarrow z_0$. Hence near to z_0 the distances are multiplied by $|f'(z_0)|$.

Some special cases may help to make the preceding analysis clearer.

EXAMPLE 1. $f(z) = z^3$
We have

$$\begin{aligned} u(x,\ y) + iv(x,\ y) &= (x+iy)^3 \\ &= (x^3 - 3xy^2) + i(3x^2 y - y^3) \end{aligned}$$

and so

$$u(x,\ y) = x^3 - 3xy^2$$
$$v(x,\ y) = 3x^2 y - y^3.$$

Consider the paths

$$\gamma_1(t) = 1 + it$$
$$\gamma_2(t) = it + 1$$

which are, respectively, the lines $x = 1$, $y = 1$, and hence meet at right angles. The paths $\Gamma_1 = f\gamma_1$ and $\Gamma_2 = f\gamma_2$ are given by

$$\Gamma_1(t) = (1 - 3t^2) + i(3t - t^3)$$
$$\Gamma_2(t) = (t^3 - 3t) + i(3t^2 - 1).$$

If we sketch these curves we get Figure 13.2.

Now Γ_1 and Γ_2 meet at right angles at $(-2, 2)$.

EXAMPLE 2. $f(z) = 1/z$
This time we have

$$u(x,\ y) = x/(x^2 + y^2)$$
$$v(x,\ y) = -y/(x^2 + y^2).$$

If c is real and positive the circle

$$\gamma_c(t) = ce^{it} \quad (0 \leqslant t \leqslant 2\pi) \tag{4}$$

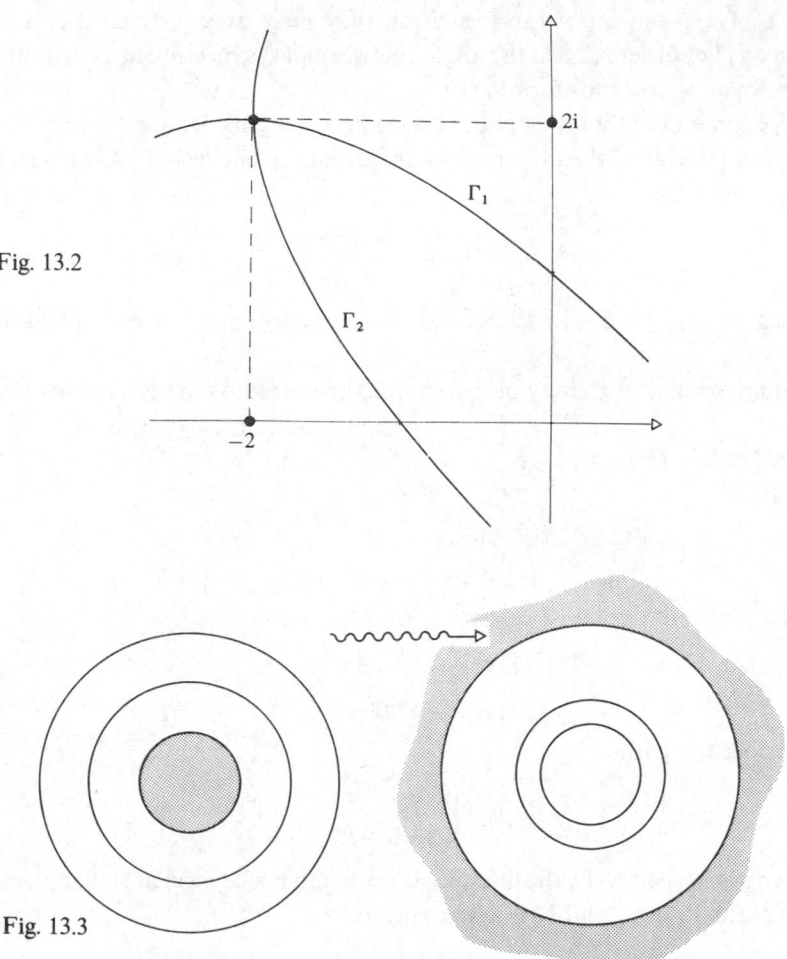

Fig. 13.2

Fig. 13.3

transforms into

$$\Gamma_c(t) = c^{-1}e^{-it} \quad (0 \leqslant t \leqslant 2\pi). \tag{5}$$

Hence the system of concentric circles (4), as c varies, maps into the system of concentric circles (5). However, points inside the circles of (4) map to points outside circles of (5). (Fig. 13.3) Further, the lines $x = ky \, (k \in \mathbb{R})$ through the origin are given by

$$\delta_k(t) = t + kit$$

and transform to

$$\Delta_k(t) = (t + kit)^{-1} = \frac{1}{1+k^2}\left(\frac{1}{t}\right) + \frac{ik}{1+k^2}\left(\frac{1}{t}\right)$$

which also represents lines through the origin. Now Γ_c and Δ_k meet at right angles, as do γ_c and δ_k. (Fig. 13.4)

Fig. 13.4

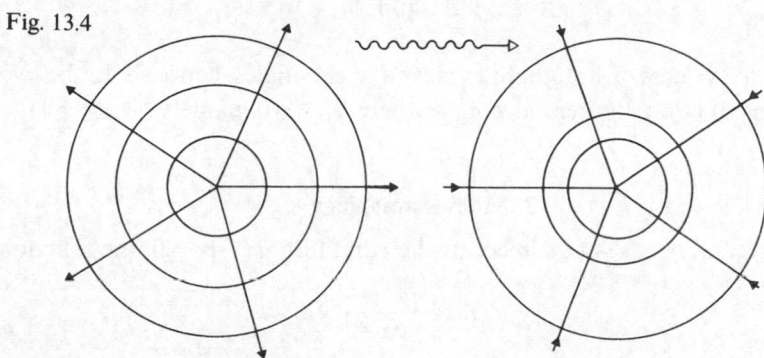

EXAMPLE 3. $f(z) = \sin(z)$
We have

$$u(x, y) = \sin(x) \cosh(y)$$
$$v(x, y) = \cos(x) \sinh(y).$$

Corresponding to the lines $x = c$ ($c \in \mathbb{R}$) we obtain the confocal hyperbolae

$$\frac{u^2}{\sin^2(c)} - \frac{v^2}{\cos^2(c)} = 1$$

Fig. 13.5

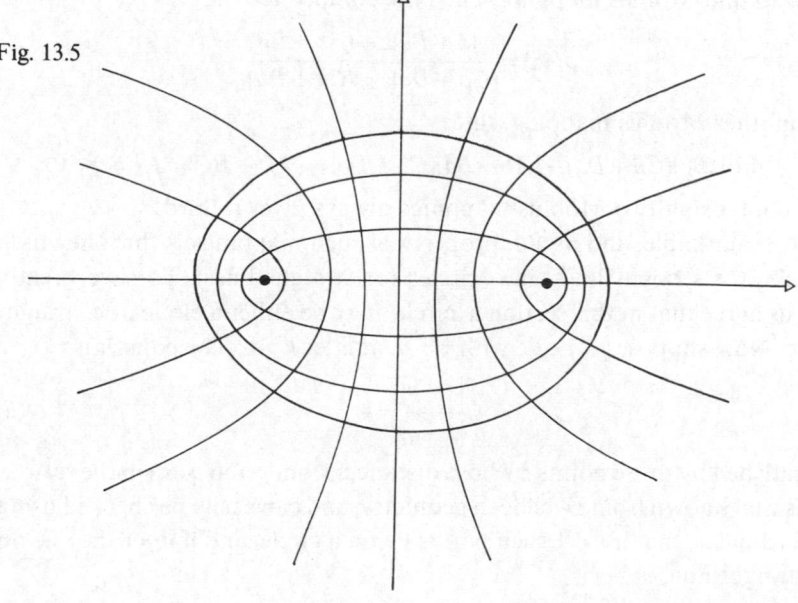

and corresponding to the lines $y = d$ ($d \in \mathbb{R}$) the confocal ellipses (Fig. 13.5)

$$\frac{u^2}{\cosh^2(d)} + \frac{v^2}{\sinh^2(d)} = 1.$$

The two systems of straight lines meet at right angles, hence so do the two systems of conics (except at points where $f'(z) = 0$, namely $f(z) = \pm 1$).

3. Möbius mappings

For fixed a, b, c, $d \in \mathbb{C}$, subject to the condition $ad - bc \neq 0$, the function

$$f(z) = \frac{az + b}{cz + d}$$

is known as a *Möbius mapping* (or *bilinear* mapping). These mappings have a number of important properties, some of which we shall indicate.

First note that $f(z)$ is differentiable for $z \neq -d/c$, and $f'(z) = (ad - bc)/(cz + d)^2 \neq 0$ because of the condition on $ad - bc$, also provided $z \neq -d/c$. Hence f is conformal throughout the domain $\mathbb{C} \setminus \{-d/c\}$ on which it is defined.

Now suppose that

$$g(z) = \frac{Az + B}{Cz + D} \quad (AD - BC \neq 0)$$

is a second Möbius mapping. Then the composite

$$gf(z) = \frac{(Aa + Bc)z + (Ab + Bd)}{(Ca + Dc)z + (Cb + Dd)}$$

is another Möbius mapping, since

$$(Aa + Bc)(Cb + Dd) - (Ab + Bd)(Ca + Dc) = (AD - BC)(ad - bc) \neq 0.$$

So composing two Möbius mappings always gives a third.

A remarkable, and useful, property of such mappings is that they map circles (or straight lines) into circles (or straight lines). To save breath, let us agree that in this section a 'circle' may be either a circle or a straight line. Now suppose p, $q \in \mathbb{C}$, with $p \neq q$, and let $k > 0$. The equation

$$\frac{|z - p|}{|z - q|} = k \tag{6}$$

is satisfied by those points z whose distances from p and q are in the ratio k. It is well known from Euclidean geometry, and can easily be checked using coordinates, that if $k \neq 1$ such points lie on a circle, and if $k = 1$ they lie on a straight line.

Here is a quick proof. We may choose coordinates in the plane so that p is $(0, 0)$ and q is $(1, 0)$. If (x, y) is distance kd from p, d from q, then

$$\sqrt{x^2 + y^2} = k\sqrt{(x-1)^2 + y^2}.$$

Hence

$$x^2 + y^2 = k^2(x^2 + y^2 - 2x + 1)$$

which implies

$$\left(x - \frac{k^2}{k^2 - 1}\right)^2 + y^2 + \frac{k^2}{k^2 - 1} - \left(\frac{k^2}{k^2 - 1}\right)^2 = 0$$

which is the equation of a circle.

If we put

$$w = f(z) = \frac{az + b}{cz + d} \quad (ad - bc \neq 0)$$

then it is easy to verify that

$$z = \frac{-dw + b}{cw + a}.$$

Now (6) takes the form

$$\left| \frac{\dfrac{-dw + b}{cw - a} - p}{\dfrac{-dw + b}{cw - a} - q} \right| = k$$

which simplifies to give

$$\left| \frac{w - P}{w - Q} \right| = K \tag{7}$$

where

$$P = (b + pa)/(d + pc)$$
$$Q = (b + qa)/(d + qc)$$
$$K = k|d + qc|/|d + pc|.$$

Hence (7) also represents a circle. This proves that the Möbius mapping f transforms circles into circles.

The above calculation, while it verifies this assertion, is not as instructive as one might hope, and the following approach has its advantages. We consider several special kinds of Möbius mappings, taking especially simple forms; and show that a general Möbius mapping can be obtained by composing mappings of these special types. This is perhaps not so surprising in view of the composition property of Möbius mappings mentioned above, and is analogous to the well-known fact that every rigid motion of Euclidean space (of two dimensions) can be obtained by composing a translation, a rotation, and a reflection. We begin with the special types.

(a) *Translation:* $w = z + k$ $(k \in \mathbb{C})$. This corresponds geometrically to moving points re(k) to the right and im(k) upwards, and clearly preserves the shape of geometrical figures.

(b) *Rotation:* $w = e^{i\theta}z$ $(\theta \in \mathbb{R})$. All points rotate around the origin through an angle θ.

(c) *Magnification:* $w = hz$ $(h > 0)$. This produces a change of scale (if $h < 1$ it shrinks rather than magnifies, but such pedantry is irrelevant), and so maps geometric figures to similar figures.

(d) *Inversion:* $w = 1/z$. Devotees of 'inversive geometry', now largely out of fashion, will recognize this as corresponding to 'geometrical inversion', which it is well-known preserves circles. However circles can map to straight lines, or straight lines to circles. Non-devotees must check this by a calculation similar to the more general one above.

We can now state:

THEOREM 13.2. Every Möbius mapping can be obtained by composing a translation, an inversion, a magnification, a rotation, and another translation (missing some out if necessary).

Proof. If $a, b, c, d \in \mathbb{C}$, $ad - bc \neq 0$, $c \neq 0$, define

$$t_1(z) = z + d/c \qquad \text{(translation)}$$

$$j(z) = 1/z \qquad \text{(inversion)}$$

$$m(z) = \left| \frac{ad - bc}{c^2} \right| z \qquad \text{(magnification)}$$

$$r(z) = \frac{ad - bc}{|ad - bc|} \frac{|c^2|}{c^2} z \qquad \text{(rotation)}$$

$$t_2(z) = z + a/c \qquad \text{(translation).}$$

It is routine to verify that

$$t_2 rmj t_1(z) = \frac{az + b}{cz + d}.$$

If $c = 0$ define

$$t_1(z) = z + b/a \quad \text{(translation)}$$

$$m(z) = \left|\frac{a}{d}\right| z \quad \text{(magnification)}$$

$$r(z) = \frac{a}{d}\left|\frac{d}{a}\right| z \quad \text{(reflection)}$$

(noting that $ad - bc \neq 0$ implies $ad \neq 0$ since $c = 0$, and hence $a \neq 0$, $d \neq 0$). Now we have

$$rmt_1(z) = \frac{az + b}{cz + d}. \qquad \square$$

COROLLARY 13.3. Every Möbius mapping transforms circles into circles.

Proof. Obviously each of the four special types transforms circles into circles. Now apply Theorem 13.2. $\qquad \square$

There are other functions $f: \mathbb{C} \to \mathbb{C}$ preserving circularity, for instance complex conjugation (of course not analytic). A theorem of Carathéodory asserts that every such function is either a Möbius mapping or a Möbius mapping composed with complex conjugation. No differentiability assumptions are needed for this theorem.

4. Potential theory

The two-dimensional *Laplace equation*

$$\frac{\partial^2 \phi}{\partial x^2} + \frac{\partial^2 \phi}{\partial y^2} = 0$$

for a function $\phi(x, y)$ is important in potential theory, with applications in particular to fluid dynamics. It is closely connected with complex function theory, as the following demonstrates. Let $f: S \to \mathbb{C}$ be differentiable, with $z = x + iy$,

$$f(z) = u(x, y) + iv(x, y).$$

Then we have

$$f'(z) = \frac{\partial u}{\partial x} + i\frac{\partial v}{\partial x} = \frac{\partial v}{\partial y} - i\frac{\partial u}{\partial y}$$

as in §4.2. By Theorem 10.3 f'' exists throughout S. If we let $f'(z) = U + iV$ then the Cauchy–Riemann equations (§4.2) show that

$$\frac{\partial U}{\partial x} = \frac{\partial V}{\partial v}, \quad \frac{\partial V}{\partial x} = -\frac{\partial U}{\partial y}.$$

Now

$$U = \frac{\partial u}{\partial x} = \frac{\partial v}{\partial y}, \quad V = \frac{\partial v}{\partial x} = -\frac{\partial u}{\partial y},$$

and so we get

$$\frac{\partial^2 u}{\partial x^2} = \frac{\partial}{\partial x}\left(\frac{\partial u}{\partial x}\right) = \frac{\partial U}{\partial x} = \frac{\partial V}{\partial y} = -\frac{\partial}{\partial y}\left(\frac{\partial u}{\partial y}\right) = -\frac{\partial^2 u}{\partial y^2}$$

and therefore $u(x, y)$ satisfies the Laplace equation. Similarly $v(x, y)$ does.

For instance, consider $f(z) = z\,e^z$. Then

$$u(x, y) = xe^x \cos(y) - ye^x \sin(y)$$
$$v(x, y) = ye^x \cos(y) + xe^x \sin(y)$$

and it may be verified directly that these functions satisfy Laplace's equation.

Solutions of Laplace's equation are called *harmonic* or *potential functions*. Pairs of functions u, v obtained from a differentiable function f as above are called *harmonic conjugates*.

Now the lines $u = $ constant, $v = $ constant are orthogonal (mutually perpendicular) in the (u, v)-plane, and so by conformality the lines

$$u(x, y) = \text{constant}$$
$$v(x, y) = \text{constant}$$

in the (x, y)-plane are orthogonal. In potential theory, if u is harmonic, the lines $u(x, y) = $ constant are called *equipotential lines*, and the set of orthogonal curves $v(x, y) = $ constant are called *stream-lines*. In the case of fluid flow described by the Laplace equation the stream-lines represent the paths along which the fluid flows. If we are given u in a domain S and wish to find the stream-lines given by v, we can often use complex integration. We have, for a fixed point $z_0 \in S$, and for any $z_1 \in S$,

$$f(z_1) = \int_{z_0}^{z_1} f'(z)\,dz$$

$$= \int_{z_0}^{z_1} \left(\frac{\partial u}{\partial x} - i\frac{\partial u}{\partial y}\right) dz.$$

For example, if we take $u(x, y) = x^2 - y^2$, which is harmonic, then picking $z_0 = 0$ for convenience we have

$$f(z_1) = \int_0^{z_1} (2x + 2iy)\, dz$$

$$= \int_0^{z_1} 2z\, dz$$

$$= z_1^2.$$

Hence $f(x+iy) = (x+iy)^2$, so that $v(x, y) = \text{im}\, (f(x+iy)) = 2xy$. So the streamlines are given by the equations $2xy = \text{constant}$, or equivalently

$$xy = \text{constant}.$$

Often, as in this case, one can *guess* what $v(x, y)$ ought to be – a process dignified by the name 'inspection'.

Conformal mapping and complex function theory played a part in the design of aircraft in the early days of aviation (and, in a more sophisticated way, still do). In particular the transformation from the z-plane to the w-plane given by

$$\frac{w-2}{w+2} = \left(\frac{z-1}{z+1}\right)^2$$

maps a circle in the z-plane, passing through -1 and containing $+1$ in its interior, into a 'bent teardrop' shape (Fig. 13.6), resembling the cross-section of an aeroplane's wing, and known as a *Joukowski aerofoil* after the discoverer of the transformation. This is used in the following way. It is quite easy to solve the Laplace equation and find stream-lines for the flow of a fluid around a circular disc. Now apply the Joukowski transformation: the disc maps to an aerofoil, the stream-lines round the disc

Fig. 13.6

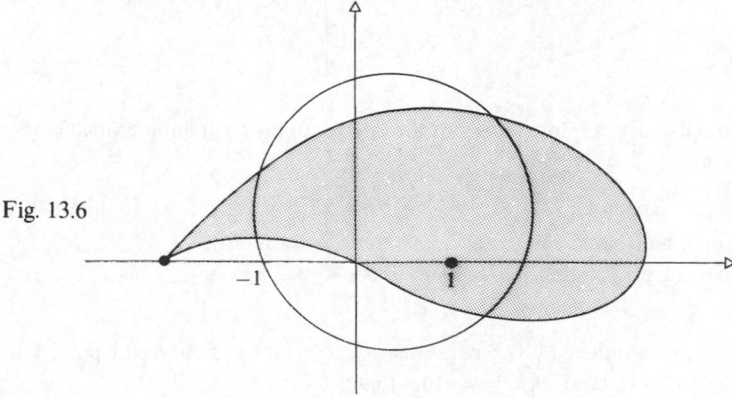

map to stream-lines around the aerofoil. From this one may calculate properties of the flow, in particular the amount of 'lift' imparted to the aircraft. More subtle transformations give more accurate information.

Exercises 13

1. Sketch the image of the set $D = \{z \in \mathbb{C} | \operatorname{re} z > 0, \operatorname{im} z > 0, |z| < 1\}$ under the conformal map $z \to 1/z$. Draw the images of the lines in D parallel to the coordinate axes.

2. Show that the mapping

$$f(z) = \frac{\sqrt{1+k^2}}{b} e^{-i((\pi)/2 + \tan^{-1}k)} z$$

transforms the strip between the lines $y = kx$, $y = kx + b$, into that between $x = 0$ and $x = 1$.

3. Find a conformal mapping f of the annulus $2 < |z| < 5$ on to the annulus $4 < |z| < 10$, such that $f(-5) = 10$.

    Find another one with $f(5) = -4$.

4. With a suitable choice of the square root, show that

$$f(z) = \sqrt{z - p} - i\sqrt{p}$$

maps the exterior of the parabola $y^2 = 4px$ $(p > 0)$ onto the right-hand half-plane $x > 0$. (Fig. 13.7)

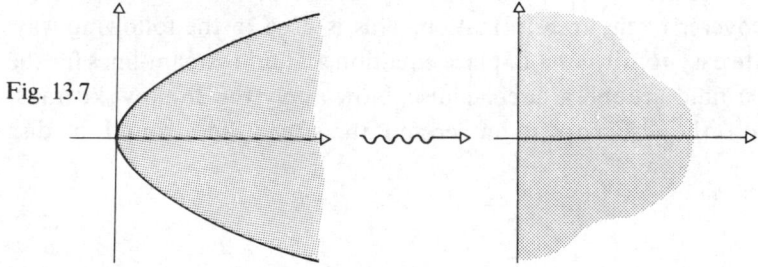

Fig. 13.7

5. Find a conformal mapping that sends the interior of the right hand branch of the hyperbola

$$x^2 - y^2 = \lambda^2$$

to the upper half-plane $y > 0$. (Fig. 13.8) (Hint: What is re (z^2)?)

6. Show that the semicircle $|z| < 1$, re $(z) > 0$, is mapped by

$$f(z) = z^2 + z$$

to the region bounded by the parabola $x = -y^2$ and the curve (in polar coordinates) $r = 2 \cos (\theta/3)$, $|\theta| \leqslant 3\pi/4$. (Fig. 13.9)

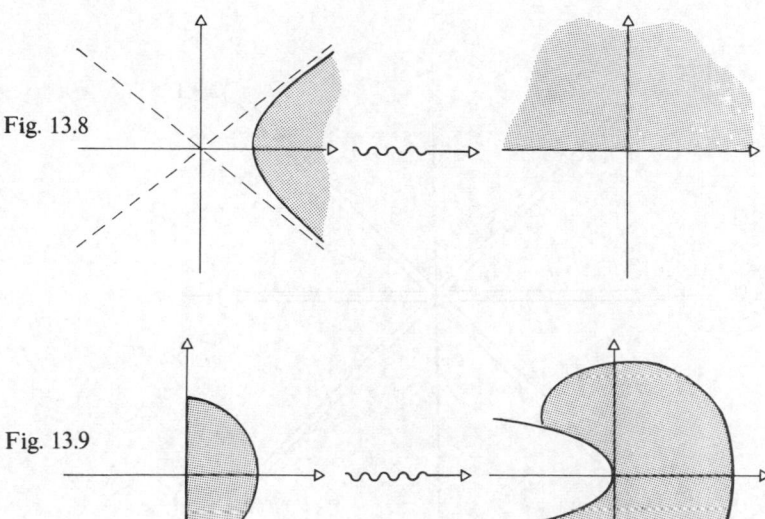

Fig. 13.8

Fig. 13.9

7. Show that (with suitably defined square roots)

$$f(z) = \sqrt{\dfrac{\sqrt{z^2 + c^2} + \sqrt{a^2 + c^2}}{\sqrt{b^2 + c^2} - \sqrt{z^2 + c^2}}}$$

maps the plane with the segments between $-a$ and b, $-ic$ and ic, removed to the upper half-plane, when $a, b, c > 0$. (Fig. 13.10)

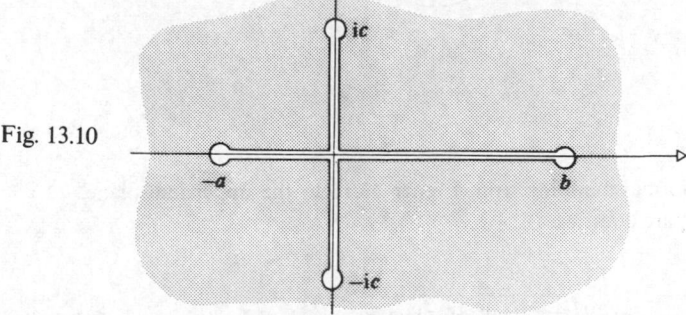

Fig. 13.10

8. Show that

$$f(z) = \dfrac{1}{\sqrt{2 + \sqrt{5}}} (\sqrt{(\sqrt{z^4 + 4} + 2)} + \sqrt{(\sqrt{z^4 + 4} - \sqrt{5})})$$

maps Figure 13.11 to the exterior of the unit circle Figure 13.12.

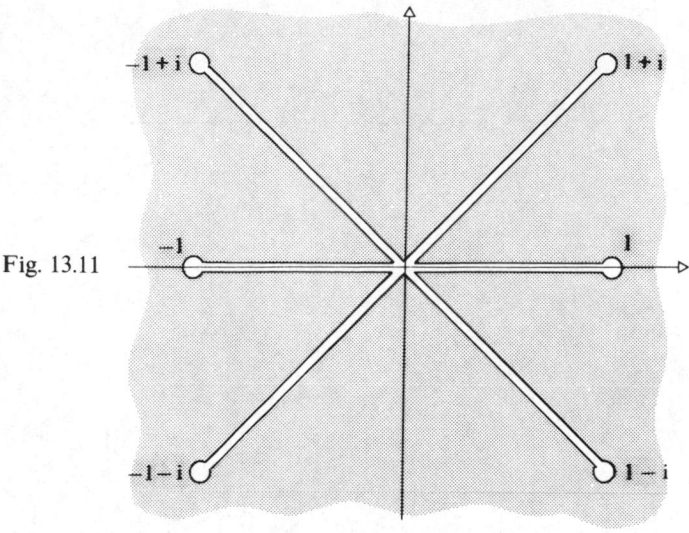

Fig. 13.11

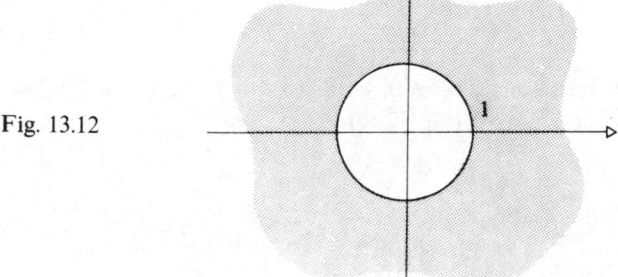

Fig. 13.12

9. Find a conformal mapping from Figure 13.13 to the upper half-plane.
   (Hint: Compare question 7.)
10. Show that

$$f(z) = \sqrt{\frac{\cos \pi z - \cos \pi h}{1 + \cos \pi z}}$$

   maps Figure 13.14 to the upper half-plane.
11. Find a conformal map sending Figure 13.15 to the upper half-plane.
12. Find the Möbius transformation which sends -1, ∞, i respectively to:
   (i) i, 1, 1+i   (ii) ∞, i, 1   (iii) 0, ∞, 1.

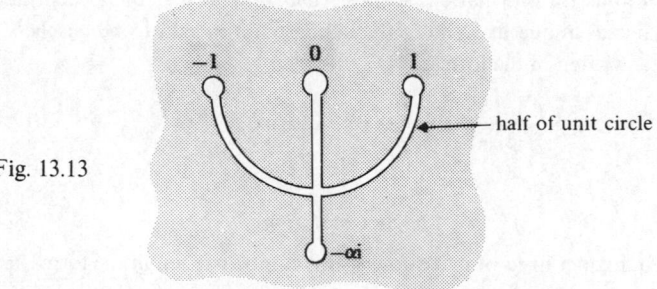

Fig. 13.13

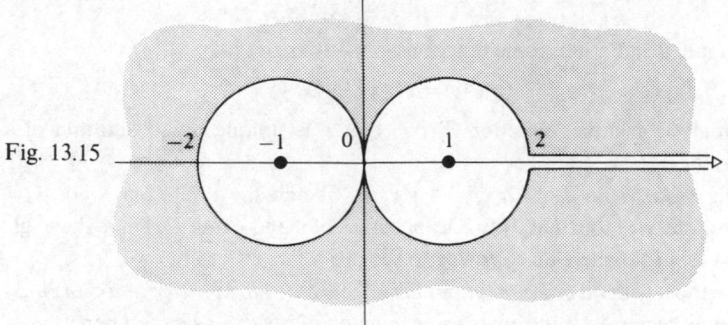

Fig. 13.14

Fig. 13.15

13. Find the general form of a Möbius transformation which:
  (i) Sends the upper half-plane to itself.
  (ii) Sends the upper half-plane to the lower half-plane.
  (iii) Sends the upper half-plane to the right half-plane.
  (iv) Preserves the unit circle.
  (v) Preserves the coordinate axes.
  (vi) Sends the upper half-plane to the interior of the unit circle.

14. Let f be a Möbius transformation. A *fixed point* of f is a point z such that $f(z)=z$. If f has a unique fixed point (including ∞) it is said to be *parabolic*. Prove it can be written in the form

$$\frac{1}{f(z)-z_0}=\frac{1}{z-z_0}+h \quad (z_0\neq\infty)$$

or

$$f(z)=z+h \quad (a\ translation).$$

If f has two distinct fixed points, show that it can be written in the form

$$\frac{f(z)-z_1}{f(z)-z_2}=k\frac{z-z_1}{z-z_2} \quad (z_1,z_2\neq\infty)$$

or

$$f(z)-z_1=k(z-z_1) \quad (z_2=\infty).$$

Such a transformation is said to be *hyperbolic* if $k>0$, *elliptic* if $k=e^{i\alpha}$ $(\alpha\neq0)$, *loxodromic* if $k=a\,e^{i\alpha}$ where $a\neq1$ is real, $\alpha\neq0$.

15. With the definitions of question 14, prove the following.
(i) Any Möbius transformation $(az+b)/(cz+d)$ is equal to one for which $ad-bc=$ $=1$.
(ii) Having ensured this, if $a+d$ is real then the transformation is elliptic if $|a+d|<2$, hyperbolic if $|a+d|>2$, and parabolic if $|a+d|=2$.
(iii) If $a+d$ is not real, the transformation is loxodromic.

16. Show that

$$u(x,\,y)=x^3-3xy^2$$

is harmonic. Find a harmonic function $v:\mathbb{R}^2\to\mathbb{R}$ such that

$$f(x+iy)=u(x,\,y)+iv(x,\,y)$$

is an analytic complex function. Prove that v is unique up to addition of a constant.

17. For $f(z)=1/z$, write $f(z)=u(x,\,y)+iv(x,\,y)$. Sketch the level curves $u(x,\,y)=$ constant, $v(x,\,y)=$ constant. If two level curves of u and v meet, what is the angle between them? Is there an easy way to see this?

18. Find the most general cubic polynomial $u(x,\,y)=ax^3+bx^2y+cxy^2+dy^3$ $(a,b,c,d$ real) that is harmonic. Find an analytic function of z with u as its real part.

19. Verify that the Joukowski transformation does, as claimed above, give rise to an aerofoil shape. Look up pp 131–4 of A. Kyrala, *Applied Functions of a Complex Variable*, Wiley-Interscience, New York 1972, and see how to compute flow-lines round it.

14

Analytic continuation

When Weierstrass started his programme to rigorize analysis he based it upon power series. Because these have well-behaved convergence properties, and can be differentiated or integrated term by term, they constitute a tool of great technical value. However, the tool has limitations, which we shall illustrate in the first section: many important functions cannot be represented by a *single* power series. This limitation can be overcome by the method of 'analytic continuation' which allows us, under the right conditions, to extend the domain of a given complex function. It turns out that such an extension is not always unique, and the problem of describing the different possibilities and the relations between them leads to a remarkable geometrical concept, known as a *Riemann surface* after its inventor. In this chapter we shall discuss these, and related topics. Here we prefer the term 'analytic' to 'differentiable', because we are emphasizing power series.

1. The limitations of power series

We illustrate the problem on the function

$$f(z) = 1/(1 - z^2).$$

(It would be possible to use a simpler example, such as $1/z$; but it is more appropriate to work with something a little less special where the general problem is more sharply defined.)

This function is analytic in $\mathbb{C} \setminus \{-1, 1\}$, and has simple poles at -1 and 1. Its power series expansion about $z_0 = 0$ is

$$1 + z^2 + z^4 + z^6 + \cdots \tag{1}$$

which diverges for $|z| > 1$. Thus if we wish our functions to be defined on domains, the series (1) at best represents $f(z)$ on the open unit disc

$$S_1 = \{z \in \mathbb{C} : |z| < 1\}.$$

Thus (1) tells us about only a small part of f.

257

We might make a different choice of z_0, say $z_0 = i$. To find the Taylor series for $f(z)$ around $z_0 = i$ it helps to note that

$$f(z) = \frac{1}{1-z^2} = \frac{1}{2}\left(\frac{1}{1+z} + \frac{1}{1-z}\right).$$

Let $w = z - i$, so that $z = w + i$. Then

$$
\begin{aligned}
f(z) &= \frac{1}{2}\left(\frac{1}{1+w+i} + \frac{1}{1-w-i}\right) \\
&= \frac{1}{2}\left\{\frac{1}{1+i}\left(1+\frac{w}{1+i}\right)^{-1} + \frac{1}{1-i}\left(1-\frac{w}{1-i}\right)^{-1}\right\} \\
&= \frac{1}{2(1+i)}\sum_{n=0}^{\infty}(-1)^n\left(\frac{w}{1+i}\right)^n + \frac{1}{2(1-i)}\sum_{n=0}^{\infty}\left(\frac{w}{1-i}\right)^n \\
&= \sum_{n=0}^{\infty}\frac{1}{2}\left\{\frac{1}{1+i}\left(\frac{-1}{1+i}\right)^n + \frac{1}{1-i}\left(\frac{1}{1-i}\right)^n\right\}(z-i)^n. \quad (2)
\end{aligned}
$$

The series (2) has radius of convergence $\sqrt{2}$ because that is the distance from $z_0 = i$ to the nearest pole of f, and hence it converges on

$$S_2 = \{z \in \mathbb{C} : |z - i| < \sqrt{2}\}.$$

From Figure 14.1 we see that (2) converges for some values of z for which (1) diverges, and conversely: thus not only do (1) and (2) represent small parts of f (namely the restrictions of f to S_1 and S_2) but they represent *different* parts.

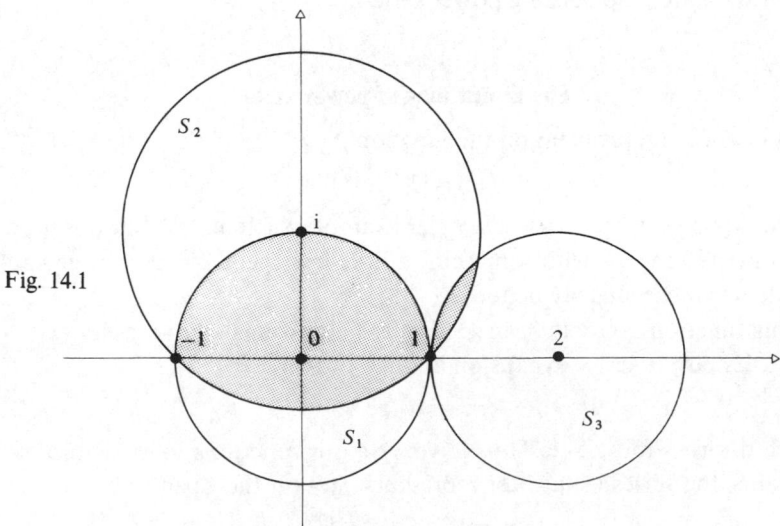

Fig. 14.1

Similarly if we take $z_0 = 2$ we obtain a third series

$$f(z) = \sum_{n=0}^{\infty} \{-\tfrac{1}{2}(-1)^n + \tfrac{1}{6}(-\tfrac{1}{3})^n\}(z-2)^n \qquad (3)$$

convergent on the disc

$$S_3 = \{z \in \mathbb{C} : |z-2| < 1\}$$

which represents still another part of f. And something of this kind will happen for any choice of z_0: the Taylor series

$$f(z) = \sum_{n=0}^{\infty} a_n(z-z_0)^n \qquad (4)$$

converges for

$$|z-z_0| < K = \min(|z_0 - 1|, |z_0 + 1|)$$

by Theorem 10.3. Thus no single choice of z_0 will give a power series expansion of $f(z)$ valid for all $z \in \mathbb{C} \setminus \{-1, 1\}$ even though f is analytic on this set. (The reader may verify that using Laurent series does not improve the situation.) This is more in the nature of a limitation of the tool – power series – than of analytic functions: it is no fault of $f(z)$ that our clumsy attempt to express it as a power series has apparently failed. The fact that a power series converges on a disc, so often a help, becomes a hindrance when we look at functions defined on domains other than a disc. We could give up at this stage and ignore power series, but this would hardly be enterprising: it is a cardinal principle in mathematics not to throw away a good idea just because it doesn't work. If a single power series is no good, why not try a whole collection of power series?

This solves part of the problem: we can certainly find, for each $z_0 \in \mathbb{C} \setminus \{-1, 1\}$, a power series expansion (4) around z_0, which converges to $f(z_0)$ for $z = z_0$. Thus using several power series gives us information about the whole of f.

But it raises another, more serious problem. In the discussion so far we started with f and worked out the power series. Weierstrass needed to work the other way: to use power series to *define* f. In the above example we know that the series (1), (2), (3) represent the same analytic function f, because of the way they were constructed. *If we are given two power series expansions around different points, how can we tell whether they represent the same analytic function, without knowing a priori what that function is?* This is the central problem in Weierstrass's approach.

2. Comparing power series

We wish to compare two power series

$$p(z) = \sum_{n=0}^{\infty} p_n(z - z_0)^n$$

$$q(z) = \sum_{n=0}^{\infty} q_n(z - z_1)^n$$

around z_0, z_1 respectively, and convergent on open discs P, Q. Everything is easy if P and Q *overlap* (i.e. $P \cap Q \neq \varnothing$), as in the case of (1) and (2), or (2) and (3) above. (Fig. 14.2) Suppose for all $z \in P \cap Q$ we have $p(z) = q(z)$, and that f and g are analytic functions defined on a domain S containing $P \cap Q$, such that $f(z) = p(z)$ for $z \in P$, and $g(z) = q(z)$ for $z \in Q$. Then for z in the non-empty open set $P \cap Q$ we have

$$f(z) = p(z) = q(z) = g(z)$$

and so by Theorem 10.11 we have $f(z) = g(z)$ for all $z \in S$. In this sense, then, p and q represent the same analytic function. Obviously the converse applies: if p and q represent the same analytic function on S then they must agree on the overlap.

The existence of a non-empty overlap allows *direct* comparison of the two series: all we have to do is look at their values. In general to compare power series defined on non-overlapping discs (such as (1) and (3) above) we have to construct a chain of overlapping discs from one to the other, on each of which a power series converges, and such that the power series agree on overlaps. Thus S_1 and S_2 overlap, and (1) agrees with (2) on $S_1 \cap S_2$; and S_2 and S_3 overlap, and (2) agrees with (3) on $S_2 \cap S_3$.

It is fairly clear that for $f(z)$ as above we can get from S_1 to any point $z_0 \neq \pm 1$ by a chain of at most three discs (two unless z_0 is real and $|z_0| > 1$). For a more complicated function with more poles (or other singularities)

Fig. 14.2

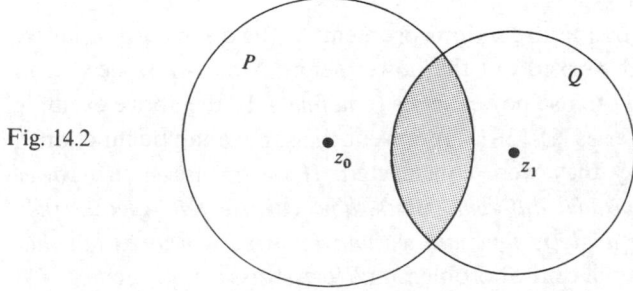

we may need more discs in a chain (Fig. 14.3), because the discs have to 'push between' the singularities. In a sense this is the limitation of power series: discs are too blunt an instrument to penetrate efficiently beyond the singularities into virgin territory.

If one formalizes these ideas (as in the next section) it becomes clear that the restriction to power series and discs is inessential. This is often the way in mathematics: the solution to a special problem turns out to apply in a much more general setting. The special problem fulfils the invaluable function of a psychological springboard, hurling our thoughts into higher realms.

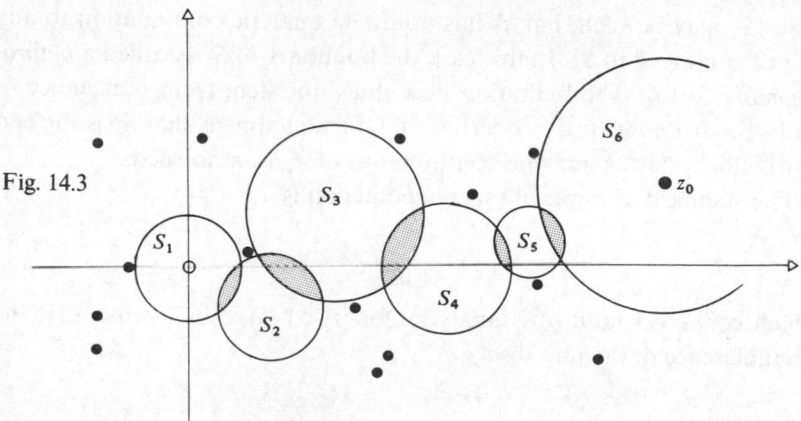

Fig. 14.3

3. Analytic continuation

If f_1 is analytic on a domain S_1 and f_2 is analytic on a domain S_2, where $S_1 \cap S_2 \neq \varnothing$ and $f_1(z) = f_2(z)$ for all $z \in S_1 \cap S_2$ (Fig. 14.4) then we say that f_2 is a *direct analytic continuation* of f_1 to the domain S_2. As we remarked

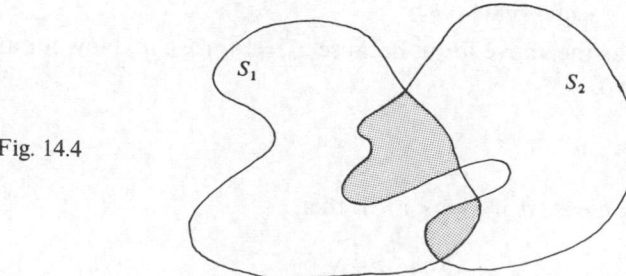

Fig. 14.4

before, such an f_2 must be unique, for if g is also analytic on S_2 and if $g(z) = f_1(z)$ for all $z \in S_1 \cap S_2$ then $f_2(z) = g(z)$ for all z in the non-empty open set $S_1 \cap S_2$, and so $f_2(z) = g(z)$ for all $z \in S_2$ by Theorem 10.1.

As a simpler example than that in the previous section, take

$$f_1(z) = \sum_{n=0}^{\infty} z^n \qquad (|z| < 1)$$

$$f_2(z) = 1/(1-z) \quad (z \in \mathbb{C} \setminus \{1\}).$$

Then f_2 is a direct analytic continuation of f_1. Whereas f_1 is defined only on the interior of the unit disc, f_2 is defined on the whole of $\mathbb{C} \setminus \{1\}$.

Before going on to the general case, corresponding to a series of overlapping domains, it is worth dealing with another phenomenon. Sometimes S_1 may be such that f_1 has no direct analytic continuation to any S_2 not contained in S_1. In this case the boundary of S_1 is called a *natural boundary* for f_1. The limitation now does not stem from our choice of tools, but from intrinsic properties of f_1: it so happens that S_1 is the end of the line as far as analytic continuation of f_1 is concerned.

The standard example of this phenomenon is

$$f(z) = \sum_{n=0}^{\infty} z^{n!}$$

which converges (and so is analytic) for $|z| < 1$. We shall prove that the circumference of the unit disc,

$$\{z \in \mathbb{C} : |z| = 1\}$$

is a natural boundary for f.

If $z_0 = e^{\pi i p/q}$ for $p, q \in \mathbb{Z}$, $q \geq 1$, then $z_0^{n!} = 1$ for all $n \geq q$. We show first that as $z \to z_0$ we have $f(z) \to \infty$. Let $z = r z_0$ where $0 < r < 1$. Then

$$f(z) = \sum_{n=0}^{\infty} (r z_0)^{n!}$$

$$= (1 + r z_0 + \cdots + r^{(q-1)!} z_0^{(q-1)!}) + \sum_{n=q}^{\infty} r^{n!}$$

$$= g(r) + h(r), \quad \text{say.}$$

(Note that $h(r)$ has the above form, because $z_0^{n!} = 1$ for $n \geq q$.) Now for any fixed integer $N \geq 0$,

$$\sum_{n=q}^{q+N} r^{n!} \to N + 1$$

as $r \to 1$. So for some $\varepsilon > 0$, if $1 - \varepsilon < r < 1$, then

$$\sum_{n=q}^{q+N} r^{n!} \geq \tfrac{1}{2} N.$$

Then

$$h(r) = \sum_{n=q}^{\infty} r^{n!} \geqslant \sum_{n=q}^{q+N} r^{n!} \geqslant \tfrac{1}{2}N$$

and hence $h(r) \to \infty$ as $r \to 1$, so that $h(r) \to \infty$ as $z \to z_0$. On the other hand, $g(z) \to 1 + z_0 + \cdots + z_0^{(n-1)!}$ as $z \to z_0$. Therefore

$$\lim_{z \to z_0} f(z) = \infty. \tag{5}$$

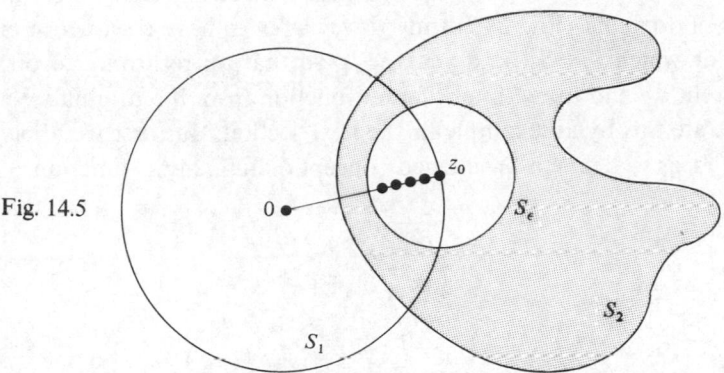

Fig. 14.5

Now suppose that F is a direct analytic continuation of f to a domain S_2 not contained in $S_1 = \{z \in \mathbb{C} : |z| < 1\}$. Then $\partial S_1 \cap S_2$ is open in ∂S_1, and so contains some point $z_0 = e^{\pi i p/q}$ for $p, q \in \mathbb{Z}$ and $q \geqslant 1$, since such points are dense on the unit circle. There is a small disc S_ε round z_0 such that $S_\varepsilon \subseteq S_2$. Now for $1 > r > 1 - \varepsilon$ we have $r z_0 \in S_\varepsilon$, so that

$$F(rz_0) = f(rz_0).$$

As $r \to 1$ we have

$$F(rz_0) \to \infty$$

by (5). But F is analytic in S_2, so that

$$F(rz_0) \to F(z_0)$$

as $r \to 1$, by continuity. This is a contradiction; and it shows that no direct analytic continuation outside S_1 is possible.

(Informally the point is this: the singularities $e^{\pi i p/q}$ are so close together that there is no room to squeeze a disc between them onto which f might be analytically continued.)

We now return to the question of performing a sequence of direct analytic continuations, one after the other. Let $S_1, \ldots, S_n$ be domains such that (Fig. 14.6)

$$S_r \cap S_{r+1} \neq \varnothing \quad (r = 1, \ldots, n-1).$$

If we have a sequence of analytic functions f_r defined on S_r such that f_{r+1} is a direct analytic continuation of f_r to S_{r+1} $(r = 1, \ldots, n-1)$, then we say that f_n is an *analytic continuation* of f_1 from S_1 to S_n. An analytic continuation which is not direct we shall call an *indirect* analytic continuation.

Unlike the direct case, we may obtain different results by using different sequences of domains. (Fig. 14.7) Indeed we can even have a sequence of domains for which $S_n = S_1$, and yet $f_n \neq f_1$, so that on returning to our starting point we end up with a different function from the original. We shall illustrate this by an example in the next section. But first we apply the above ideas to define a broadened concept of an analytic function.

Fig. 14.6

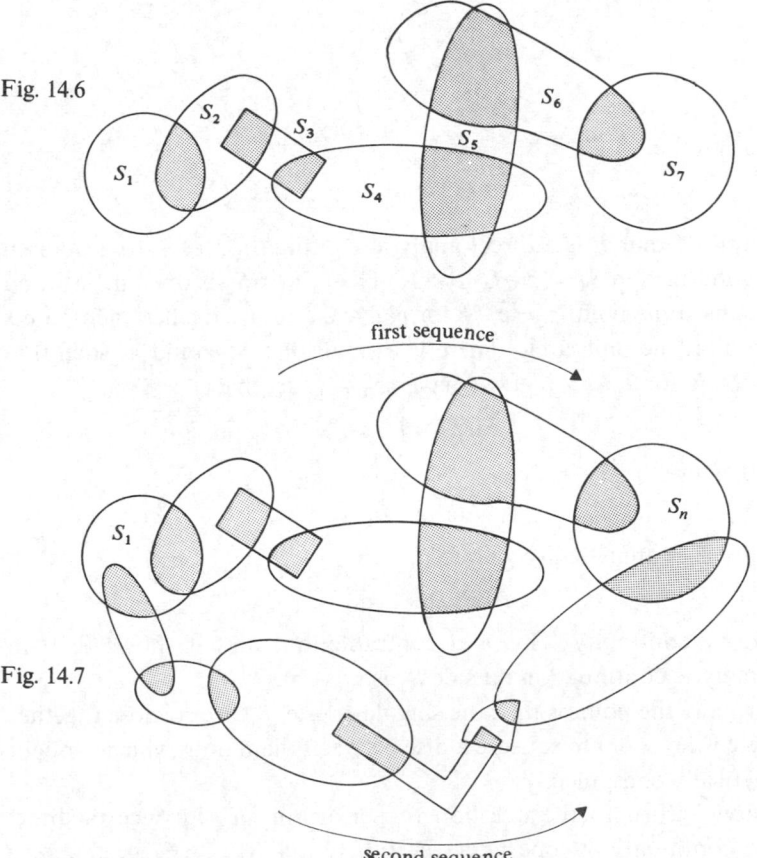

first sequence

Fig. 14.7

second sequence

If f is analytic in a domain S we call the pair (f, S) a *function element*. We define a relation $\sim$ on the set of all function elements, by

$$(f_1, S_1) \sim (f_2, S_2)$$

if f_2 is an analytic continuation of f_1 from S_1 to S_2. Since indirect continuations are allowed, it follows easily that $\sim$ is an equivalence relation. A *complete analytic function* is an equivalence class (under $\sim$) of function elements.

In other words, a complete analytic function in the new sense is an analytic function in the old sense, together with all of its analytic continuations. It is clearly more convenient to have the continuations 'built in' than to have to fit them together as the occasion warrants.

If there exist function elements (f_1, S_1) and (f_2, S_2) of the complete analytic function F such that for some $z \in S_1 \cap S_2$ we have $f_1(z) \neq f_2(z)$ then we say that F is *multiform*. If not, then F is *uniform*. All of the examples we have considered so far in this chapter are uniform, and so we shall give some multiform examples in the next section. A multiform F is a formalized version of the idea of a 'multi-valued function' with the advantage that it is broken down into pieces on each of which it is a genuine single-valued function. The multiformity comes from the manner in which the pieces fit together. A geometric approach to this leads to the concept of a Riemann surface, and an informal discussion of this may be found in §5.

4. Multiform functions

A simple instance of a multiform function is

$$f(z) = \sqrt{z}.$$

If $z = r \, e^{i\theta}$ then we can choose for $\sqrt{z}$ either

$$\sqrt{r} \, e^{i\theta/2} \quad \text{or} \quad \sqrt{r} \, e^{i(\theta/2 + \pi)}$$

(where $\sqrt{r}$ is real and positive.) With the old concept of an analytic function we must choose arbitrarily one of these, and then $f(z)$ is analytic only if we make a cut in the complex plane. From the present viewpoint, we can do better. We introduce four domains

$$H_1 = \{z \in \mathbb{C} : \mathrm{re}\,(z) > 0\}$$
$$H_2 = \{z \in \mathbb{C} : \mathrm{im}\,(z) > 0\}$$
$$H_3 = \{z \in \mathbb{C} : \mathrm{re}\,(z) < 0\}$$
$$H_4 = \{z \in \mathbb{C} : \mathrm{im}\,(z) < 0\}$$

which are the open half-planes to the right, top, left, and bottom of the complex plane. Let $z = r\,e^{i\theta}$ where $r > 0$, $-\pi < \theta \leqslant \pi$. Define

$$f_1(z) = \sqrt{r}\,e^{i\theta/2} \qquad \text{for } z \in S_1 = H_1$$

$$f_2(z) = \sqrt{r}\,e^{i\theta/2} \qquad \text{for } z \in S_2 = H_2$$

$$f_3(z) = \begin{cases} \sqrt{r}\,e^{i\theta/2} & \text{for } z \in S_3 = H_3,\ \text{im}\,(z) \geqslant 0 \\ \sqrt{r}\,e^{i(\theta/2 + \pi)} & \text{for } z \in S_3 = H_3,\ \text{im}\,(z) < 0 \end{cases}$$

$$f_4(z) = \sqrt{r}\,e^{i(\theta/2 + \pi)} \qquad \text{for } z \in S_4 = H_4$$

$$f_5(z) = \sqrt{r}\,e^{i(\theta/2 + \pi)} \qquad \text{for } z \in S_5 = H_1$$

$$f_6(z) = \sqrt{r}\,e^{i(\theta/2 + \pi)} \qquad \text{for } z \in S_6 = H_2$$

$$f_7(z) = \begin{cases} \sqrt{r}\,e^{i(\theta/2 + \pi)} & \text{for } z \in S_7 = H_3,\ \text{im}\,(z) \geqslant 0 \\ \sqrt{r}\,e^{i\theta/2} & \text{for } z \in S_7 = H_3,\ \text{im}\,(z) < 0 \end{cases}$$

$$f_8(z) = \sqrt{r}\,e^{i\theta/2} \qquad \text{for } z \in S_8 = H_4.$$

Each of these eight functions is analytic in its domain of definition: this is why for f_3 and f_7 we have to change our choice of $\sqrt{z}$ as we cross the imaginary axis, because our choice of θ as lying between $-\pi$ and π would otherwise introduce a discontinuity. Further, f_{r+1} is a direct analytic continuation of f_r ($r = 1, \ldots, 7$) and f_1 is a direct analytic continuation of f_8. For each $z \in \mathbb{C} \setminus \{0\}$ the values $f_r(z)$, where defined, will be one or other of the two possible square roots. Further, for each r, $f_r(z)$ will take one of the two values and $f_{r+4}(z)$ the other (where $r + 4$ is to be interpreted mod 8, of course).

Thus we have a multiform function, taking two values at each $z \neq 0$. Going around the origin once, from S_1 to S_2 to S_3 to S_4 to $S_5 = S_1$, we reach a different value of $f(z)$ from the original. However in this case going round again returns us to the original value on the second tour round the origin. This last effect is a special property of $\sqrt{z}$, as the next example shows.

An immensely important multiform function is

$$f(z) = \log(z).$$

The multiformity of this was Euler's great discovery, arising from the Bernouilli/Leibniz controversy mentioned in Chapter 0. In terms of the present discussion we introduce domains

$$S_{4k+r} = H_r \qquad (k \in \mathbb{Z},\ r = 1, 2, 3, 4).$$

To avoid the kind of two-piece definition which occurred for f_3 and f_7 above, we proceed as follows: for $z \in S_n$ and $z = r\,e^{i\theta}$ where

$$\frac{n-2}{2}\pi < \theta \leqslant \frac{n}{2}\pi$$

we define

$$f_n(z) = \log{(r)} + i\theta.$$

Then f_n is analytic on S_n. On S_1 we have

$$f_1(z) = \text{Log}\,(z),$$

the principal value of the logarithm; on S_5

$$f_5(z) = \text{Log}\,(z) + 2\pi i,$$

on S_{4k+1}

$$f_{4k+1}(z) = \text{Log}\,(z) + 2k\pi i.$$

It is not hard to check that f_{s+1} is a direct analytic continuation of f_s from S_s to S_{s+1} for all $s \in \mathbb{Z}$. For each $r = 1, 2, 3, 4$ and $k \in \mathbb{Z}$ the values of $f_{4k+r}(z)$ $(z \in H_r)$ give all the infinitely many possible values

$$\log{|z|} + (2k\pi + \arg{(z)})i$$

of the logarithm.

We can now give a general definition of a singularity. If no analytic continuation of f can be defined at a point z_0 then we say that z_0 is a *singularity* of the corresponding complete analytic function. We have met several kinds of singularity before: poles, isolated essential singularities. Removable singularities are not really singularities at all according to the new definition. For $\Sigma\, z^{n!}$ every point with $|z| \geqslant 1$ is a singularity. For $\sqrt{z}$ and $\log{(z)}$ we encounter a new kind of singularity, known as a *branch-point*: analytic continuation around such a point leads to changes in value. Notice that for $\sqrt{z}$ there is even a natural definition of $f(z)$ at the branch-point, namely 0, and yet f is not analytic there.

Multiform functions make an appearance in contour integration whenever a different choice of paths from z_0 to z_1 leads to a different value of the integral

$$F(z_1) = \int_{z_0}^{z_1} f(z)\,dz$$

(as may occur, for instance, if f has a pole in the region between the two paths). Hence quite nice singularities of f, such as poles, give rise to much nastier singularities of F, namely branch-points. This can occur even though f is uniform: thus $1/z$ is uniform but its integral $\log(z)$ is multiform, and the pole of f at 0 has become a branch-point of F.

5. Riemann surfaces

Riemann invented a geometrical way to envisage multiform functions, much more intuitively appealing than an equivalence class of function elements, which involves replacing $\mathbb{C}$ by a more complicated 'Riemann surface'. In the case of the logarithm we can describe it informally in the following terms, which should not be subjected to too deep scrutiny of a logic-chopping kind. We are not attempting a rigorous definition at this point: the informal description, though it may sound far-fetched, is in fact capable of a precise and rigorous rendering.

Consider a collection of copies $\mathbb{C}_k$ of $\mathbb{C}$, one for each $k \in \mathbb{Z}$. Slit each $\mathbb{C}_k$ along the negative real axis from 0 to $-\infty$. For each k join the top left-hand quadrant of $\mathbb{C}_k$ to the bottom left-hand quadrant of $\mathbb{C}_{k+1}$ along the slit. The resulting surface (Fig. 14.8) resembles a spiral staircase or a concertina, with the planes stacked on top of each other in order relative to k, and with a continuous spiral path from any $\mathbb{C}_k$ to any other $\mathbb{C}_{k'}$ going up the 'steps' where the slits were made. These planes $\mathbb{C}_k$ are the *sheets* of the Riemann surface, and it is convenient to imagine the whole collection to be stacked on top of $\mathbb{C}$ as shown in the picture.

We define the logarithm for points on the Riemann surface by

$$\log(\hat{z}) = \log|z| + i(\arg(z) + 2k\pi)$$

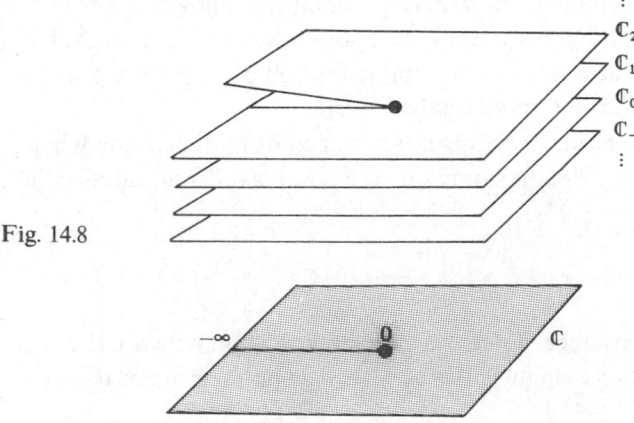

Fig. 14.8

where $\hat{z}$ is the point of $\mathbb{C}_k$ lying directly above the point $z \in \mathbb{C}$. This is a *single*-valued function on the Riemann surface, and it is continuous in the sense that the values join up correctly across the cuts. (In a similar sense we can even say that it is differentiable: the derivatives join up correctly as well.)

Now we shall see how to relate analytic continuation on $\mathbb{C}$ to the corresponding operation on the Riemann surface. Corresponding to domains $S_1, \ldots, S_n = S_1 \subseteq \mathbb{C}$, we have $S'_1, \ldots, S'_n$ on the surface. In $\mathbb{C}$ the values of the functions $f_1, \ldots, f_n$ agree on the relevant overlaps $S_1 \cap S_2, \ldots, S_{n-1} \cap S_n$, but because of the multiformity they do not agree on S_1 and S_n. On the Riemann surface we find that if we make S'_1 and S'_2 overlap, then S'_2 and S'_3, and so on, then by the time we get to S'_n it is one sheet further up the staircase than S'_1. (Fig. 14.9) Hence S'_1 and S'_n do not overlap, and the fact that f_1 and f_n are different is only to be expected. The multiform nature of analytic continuation is automatically taken care of by the multiplicity of sheets in the Riemann surface. Even the nature of the singularity at 0 is apparent from the geometry of the surface, because here all the levels meet.

Similarly for $\sqrt{z}$ we obtain the Riemann surface by taking two copies $\mathbb{C}_1$ and $\mathbb{C}_2$, slit from 0 to $-\infty$; joining the top left-hand quadrant of $\mathbb{C}_1$ to the bottom left-hand quadrant of $\mathbb{C}_2$; and the top left-hand quadrant of $\mathbb{C}_2$ to the bottom left-hand quadrant of $\mathbb{C}_1$. (To do this in three-dimensional space involves allowing the surface to intersect itself, but there is no conceptual problem here.) (Fig. 14.10) Again the phenomena noted above are clear from the geometry of the Riemann surface: the

Fig. 14.9

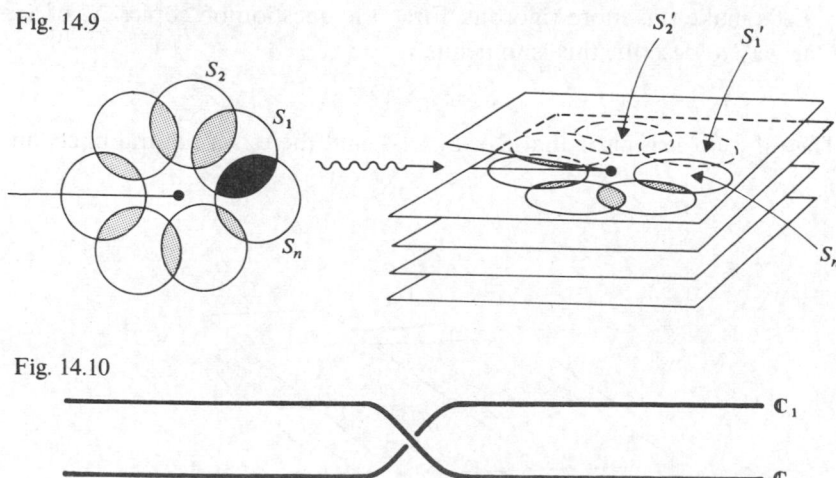

Fig. 14.10

two values of $f(z)$, the fact that going round once changes the value but going round twice does not, the branch-point at 0.

Obviously one cannot rely on such *ad hoc* methods for constructing the Riemann surface in general. We can approach the general method through the same example of the logarithm, but building up the surface in a different way. Consider the sequence of half-planes S_s and functions f_s used for analytic continuation above. Suppose we take half-planes S_s' corresponding to the S_s, but pairwise disjoint. (The S_s are not, in the sense that $S_1 = S_5$, and so on.) Suppose we 'glue' the S_s' together in the following way: S_s and S_{s+1} overlap in a quadrant on which $f_s = f_{s+1}$, and we glue together the corresponding quadrants of S_s' and S_{s+1}'. Then S_2' glues on top of S_1' at the top right-hand quadrant, S_3' glues on to the top left of S_2', S_4' glues on to the bottom left of S_3', S_5' glues to the bottom right of S_4'. At this point S_5' is lying directly over S_1', but we do not glue them together since f_1 and f_5 are different. (Fig. 14.11) Continuing with S_6', S_7', ... (and, in the downward direction, S_0', S_{-1}', S_{-2}', ...) we build up the Riemann surface again. (As a practical exercise, try this with real pieces of paper and real glue: the result will bring home the idea far more vividly than a written explanation.)

The general method for constructing a Riemann surface of a complete analytic function F follows the same lines. Recall that F is an equivalence class of pairs (f, S) where f is analytic on the domain S, and $(f_1, S_1) \sim (f_2, S_2)$ if f_1 and f_2 are equal on the non-empty set $S_1 \cap S_2$. We take disjoint copies S_λ' of all the domains S_λ occurring in pairs (f_λ, S_λ) belonging to F. If $S_\lambda \cap S_\mu \neq \varnothing$ and $f_\lambda = f_\mu$ on it, then we glue S_λ' to S_μ' at the points corresponding to the overlap $S_\lambda \cap S_\mu$.

Let's make this more rigorous. First, the question of 'copies' S_λ' of S_λ. One way to describe this is to define

$$S_\lambda' = S_\lambda \times \{\lambda\}.$$

Then if $\lambda \neq \mu$ it is clear that $S_\lambda' \cap S_\mu' = \varnothing$, and there is a natural bijection

$$j_\lambda : S_\lambda \to S_\lambda'$$

Fig. 14.11

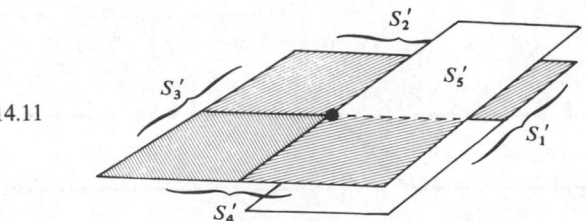

defined by

$$j_\lambda(s) = (s, \lambda) \quad (s \in S_\lambda).$$

Now that 'gluing': this is accomplished by a simple trick, using yet another equivalence relation; which however requires a certain sophistication on of the part of the reader to be appreciated. We let

$$(z_\lambda, \lambda) \approx (z_\mu, \mu)$$

for $z_\lambda \in S_\lambda$, $z_\mu \in S_\mu$, if

(i) $z_\lambda = z_\mu$,

(ii) $f_\lambda(z_\lambda) = f_\mu(z_\mu)$.

Then the set of equivalence classes under $\approx$ of points in the union of all the S_λ' is defined to be the Riemann surface of F. The equivalence relation $\approx$ acts as the glue. The definition is quite elegant, but not the sort of thing that appeals at first meeting.

The advantage of this apparently esoteric construction is that one can now think of an analytic function as a genuine function, defined on the Riemann surface and taking complex values, and not as an equivalence class of function elements; and yet still deal satisfactorily with multiform functions. We get a 'global' view of the function, instead of having to take it to pieces. The last few sections of this book explore the kinds of insight that may be gained in this way, in relation to material already discussed.

6. Complex powers

So far we have considered powers z^a of a complex number z only for rational a, where $z^{p/q}$ is a qth root of z^p. We shall now define it for arbitrary $a \in \mathbb{C}$. We would like to do this in such a way that the laws $z^{a+b} = z^a z^b$, $(z^a)^b = z^{ab}$, $(zw)^a = z^a w^a$, continue to hold. It turns out that we can 'almost' do this: however z^a is in general multiform, and the formulae hold only if we choose the values appropriately.

A natural approach is to write

$$z = r\, e^{i\theta} = e^{\log r + i\theta} \tag{6}$$

where $\log r \in \mathbb{R}$, and let

$$a = \alpha + i\beta.$$

We then, on the assumption that the above laws hold, have

$$z^a = (r\, e^{i\theta})^a$$

$$= e^{a(\log r + i\theta)}$$

$$= e^{(\alpha + i\beta)(\log r + i\theta)}$$

$$= e^{\alpha \log r - \beta\theta} e^{i(\beta \log r + \alpha\theta)}. \tag{7}$$

Accordingly, we *define* z^a by (7). This makes sense for all $z \neq 0$, and for all $a \in \mathbb{C}$. It amounts to requiring that

$$z^a = e^{a \log z} \tag{8}$$

which is an equally natural way to define z^a. It follows that in any domain for which we may define a unique branch of log (such as the cut plane $\mathbb{C}_\rho$ for any ρ) we may also define z^a as a single-valued differentiable function.

In general, however, since log is multiform, so is z^a. To see what the possibilities are, choose a particular value θ_0 for θ (the principal value of arg z is the obvious choice). Then the possible θ that make (1) hold are of the form

$$\theta = \theta_0 + 2n\pi \quad (n \in \mathbb{Z}). \tag{9}$$

Write

$$(z^a)_n = e^{a(\log r + i(\theta_0 + 2n\pi))}$$

which is the 'nth branch' of z^a, obtained by substituting θ as defined by (9) into (7). The question now is: how does $(z^a)_n$ depend on n?

From (7) we have

$$(z^a)_n = (e^{-2n\pi\beta} \, e^{2n\pi i \alpha})(z^a)_0 \tag{10}$$

where $(z^a)_0$ is the '0th branch'. From (10) we may answer the question above: there are three cases.

(a) If $\beta \neq 0$ then z^a has infinitely many values, one for each n. For

$$|(z^a)_n / (z^a)_0| = (e^{-2\pi\beta})^n$$

which takes distinct values for distinct n. The Riemann surface for z^a in this case is the same as that for the logarithm: an 'infinite staircase' on whose nth layer the function is given by the branch $(z^a)_n$.

(b) If $\beta = 0$, so $a = \alpha \in \mathbb{R}$, and α is irrational, then $z^a = z^\alpha$ is infinitely-valued, with distinct values for distinct n. For if

$$(z^a)_m = (z^a)_n$$

then

$$e^{2m\pi i \alpha} = e^{2n\pi i \alpha}$$

so that

$$e^{(m-n)\alpha \cdot 2\pi i} = 1.$$

This implies that

$$(m - n)\alpha \in \mathbb{Z} \quad (\S5.6)$$

which implies $m = n$ since α is irrational.

The Riemann surface is the same as for case 1. But, unlike case 1, the *moduli* of distinct branches are now equal: only the arguments vary.

(c) Let $\beta = 0$, and suppose that $a = \alpha$ is rational. Write $\alpha = p/q$ in lowest terms, $p, q \in \mathbb{Z}$. Then $z^a = \sqrt[q]{(z^p)}$.

Following the discussion in (b), two branches give the same values if and only if

$$(m-n)\frac{p}{q} \in \mathbb{Z}.$$

This happens if and only if q divides $m-n$. To see this, let $m-n = qd + k$ $(0 \leqslant k < q)$. Then $\mathbb{Z} \ni (m-n)(p)/q = pd + kp/q$, so $kp/q \in \mathbb{Z}$. If $k \neq 0$ then $kp/q = l \in \mathbb{Z}$, so $p/q = l/k$, contrary to p/q being in lowest terms.

It follows that the value of $(z^a)_n$ depends only on $n \bmod q$. The branches

$$(z^a)_0, \quad (z^a)_1, \ldots, \quad (z^a)_{q-1}$$

are distinct, but the qth branch repeats

$$(z^a)_q = (z^a)_0$$

and thereafter, repetitions continue.

The Riemann surface, of course, is a q-sheeted spiral, with its top identified with its bottom (as in our discussion of $z^{1/2}$). See Figure 14.12, for the case $q = 5$.

Fig. 14.12

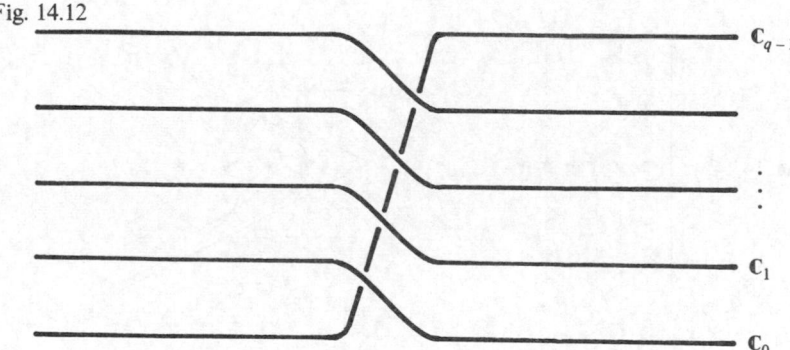

$$c_{q-1}$$

$$\vdots$$

$$c_1$$

$$c_0$$

7. Conformal mapping using multiform functions

Conformal mapping with multiform functions requires careful attention to the way values are specified; again the simplest approach is to put everything on to a Riemann surface. Here we shall avoid the issue by considering only one case, of practical importance: the map $z \to z^\alpha$. Further, we shall assume that the domain is the cut-plane $\mathbb{C}_\pi$, on which the map is single-valued.

The great virtue of this map is that it transforms a half-plane into a wedge, as in Figure 14.13. The vertex angle of the wedge is $\alpha\pi$, so we have to assume $\alpha < 2$ to prevent the image from overlapping itself. The result is conformal except at the origin.

For example, uniform fluid flow, given by re (z) = constant, transforms to the flow round a corner of angle α, as in Figure 14.14.

If we take $\alpha = 1/2$, the corner is right-angled. Then $z = x + iy$, and $z^{1/2} = w = u + iv$; the flow-lines are x = constant. Now $z = w^2$ so $x = u^2 - v^2$, $y = 2uv$; so the flow-lines in the transformed w-plane are given by $u^2 - v^2 =$ constant: they are branches of rectangular hyperbolae.

In combination with other 'standard' conformal maps, z^α provides a useful weapon for the applied mathematician.

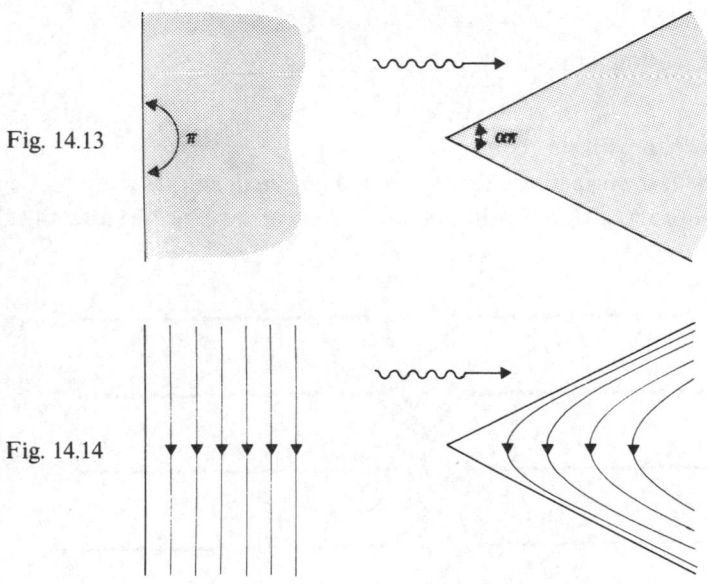

Fig. 14.13

Fig. 14.14

8. Contour integration of multiform functions

It is conventional to interpret a contour integral $\int_\gamma f$ of a *multiform* f by choosing the values of $f(z)$ so that they vary continuously as z traverses the contour. (This still leaves an arbitrary initial choice to be made, which must be specified.) A more civilized approach is to break the contour γ up as a sum $\gamma = \gamma_1 + \cdots + \gamma_n$ such that

(a) Each γ_j lies inside a domain D_j on which f may be defined as a single-valued function, (*b*) On each D_j we choose branches of f so that the values agree at the joining point of γ_j and γ_{j+1}.

In terms of Riemann surfaces we may interpret this process as the definition of a contour integral *when the contour lies on the Riemann surface*. The way the contour wanders up and down the staircases on the surface automatically takes care of the choice of values of the multiform function. (Fig. 14.15) In practice this is all much more straightforward than it may sound, and the computations are no harder than in the uniform case – provided you use your head!

EXAMPLE 1. Let γ be the contour

$$\gamma(t) = (1+t)\,e^{it} \quad (0 \leqslant t \leqslant 6\pi).$$

Find

$$\int_{\gamma} z^{1/5}\,dz.$$

Method 1 (Stupid). Defining $z^{1/5}$ by $(\gamma(t))^{1/5} = (1+t)^{1/5}\,e^{it/5}$ makes it vary continuously with t. By Theorem 6.4 we have

$$\int_{\gamma} f(z)\,dz = \int_{0}^{6\pi} f(\gamma(t))\gamma'(t)\,dt$$

$$= \int_{0}^{6\pi} (1+t)^{1/5}\,e^{it/5}(i\,e^{it} + it\,e^{it} + e^{it})\,dt$$

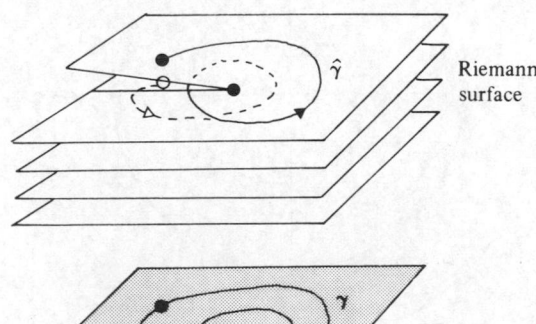

Fig. 14.15

Riemann surface

which leads the hopeful calculator along some fascinating byways inhabited by integrals such as

$$\int_0^{6\pi} t(1+t)^{1/5} \cos{(6t/5)}\, dt.$$

A wise man should know when to cut his losses.

Method 2 (*Sound but pedestrian*). Cover the contour γ by domains in which $z^{1/5}$ may be given a unique value for each z, and is then differentiable. For example, we can find six such domains $D_1, \ldots, D_6$; one for each of the intervals *AB, BC, CD, DE, EF, FG* shown on Figure 14.16. To make the values match on overlaps we choose the definition of $z^{1/5}$ as follows. Write z in the form $r\,e^{i\theta}$ where the choice of θ is given by the table below.

Domain	θ chosen in interval
D_1	$[-\pi/4, 5\pi/4]$
D_2	$[3\pi/4, 9\pi/4]$
D_3	$[7\pi/4, 13\pi/4]$
D_4	$[11\pi/4, 17\pi/4]$
D_5	$[15\pi/4, 21\pi/4]$
D_6	$[19\pi/4, 25\pi/4].$

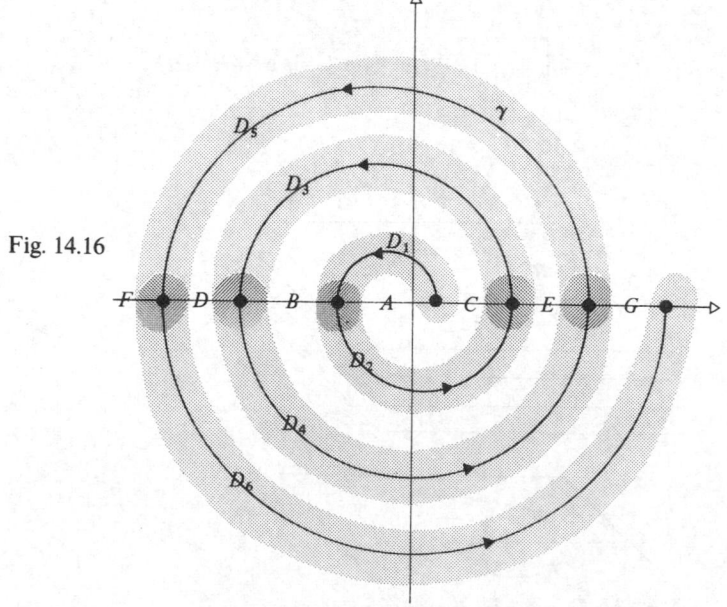

Fig. 14.16

On each domain choose the branch of $z^{1/5}$ given by $r^{1/5} e^{i\theta/5}$. The choices of θ make these agree on overlaps.

Break up the integral as

$$\int_\gamma = \int_{AB} + \int_{BC} + \int_{CD} + \int_{DE} + \int_{EF} + \int_{FG}.$$

In each domain D_i the function $z^{1/5}$ has an antiderivative $\frac{5}{6}z^{6/5}$ where the latter is given by the corresponding branch $z^{6/5} = r^{6/5}e^{6i\theta/5}$. By Theorem 6.7, if PQ is any of the arcs AB, BC, etc., then we have

$$\int_{PQ} z^{1/5}\, dz = \tfrac{5}{6}Q^{6/5} - \tfrac{5}{6}P^{6/5}.$$

When we add the six expressions of this type, all terms but two cancel; so the integral is

$$\tfrac{5}{6}G^{6/5} - \tfrac{5}{6}A^{6/5}.$$

Here $A = \gamma(0) = 1$, $G = \gamma(6\pi) = 1 + 6\pi$. We must evaluate the 6/5th power according to the above prescription: we find that

$$A^{6/5} = (1 \cdot e^{i0})^{6/5} = 1,$$
$$G^{6/5} = ((1+6\pi)e^{i6\pi})^{6/5} = (1+6\pi)^{6/5}e^{36\pi i/5} = -(1+6\pi)^{6/5}e^{i\pi/5}.$$

Therefore the integral is equal to

$$-\tfrac{5}{6}((1+6\pi)^{6/5}\, e^{i\pi/5} + 1).$$

Method 3 (Slicker). In terms of the parameter t we can define the required branch of $z^{1/5}$ to vary continuously *along the contour* γ if we put

$$z^{1/5} = (\gamma(t))^{1/5} = (1+t)^{1/5}\, e^{it/5}.$$

There is a local antiderivative

$$\tfrac{5}{6}z^{6/5}$$

whose value varies continuously along the path γ if we choose branches so that

$$(\gamma(t))^{6/5} = (1+t)^{6/5}\, e^{6it/5}.$$

Now pave γ by a finite number of domains, on each of which $z^{1/5}$ may be rendered single-valued, choosing the branches to agree with those already chosen on the path. The integral, as above, may be expressed as a sum of terms of the form $\tfrac{5}{6}((1+t_{j+1})^{6/5}e^{6it_{j+1}/5} - (1+t_j)^{6/5}e^{6it_j/5})$, where the t_j subdivide the interval over which t runs. All but two terms cancel, leaving the answer

$$\tfrac{5}{6}((1+6\pi)^{6/5}e^{36\pi i/5} - 1)$$

as before.

The advantage of this approach is that by choosing branches in an obvious way for $\gamma(t)$ we may leave the prescription of the domains, and the branches on those domains, implicit.

Method 4 (Smart). Let the Riemann surface do the work. The above analysis easily generalizes to give a version of Theorem 6.7 for multiform functions, integrated along a contour *in the Riemann surface*. Given a global antiderivative F for f on the Riemann surface, we have

$$\int_\gamma f = F(z_1) - F(z_0)$$

where z_0 is the initial point, z_1 the final point, and the branches are chosen with reference to the end points of γ *on the surface* (cf. Fig. 14.15). This gives the above result immediately on noting that γ winds three times anticlockwise, so the 6/5th power winds $3 \cdot 6/5 = 18/5$ times, giving for $F(z_1)$ the argument $2\pi \cdot 18/5 = 36\pi/5$, if we choose the argument 0 for z_0.

We leave it to the reader, as an exercise, to formulate and prove generalizations of Cauchy's theorem for integrals along contours in the Riemann surface of a multiform function. For those who prefer not to use the insight that the Riemann surface can provide, however, we recommend Method 3 above as a relatively direct and simple technique. The above example is of course a little artificial, and its main interest is pedagogical. More typical of the integrals likely to be encountered in practice is the following:

EXAMPLE 2. Suppose that a is real, with $1 < a < 2$. Find

$$\int_0^\infty \frac{x^a}{1+x^2}\, dx.$$

First, we should point out that the restrictions on a make this integral converge, so the question is a sensible one. We begin with the complex function

$$f(z) = z^a(1+z^2)^{-1}$$

which is differentiable for all $z \neq 0$, i, $-$i, and is multiform for all a in the given range.

We make a cut along the positive real axis and work in the cut plane $\mathbb{C}_0$ (in the notation of §7.2). On this domain z^a may be rendered single-valued: we set $(re^{i\theta})^a = r^a e^{ai\theta}$ where $0 \leqslant \theta < 2\pi$. Now we integrate $f(z)$ along a contour γ shown in Figure 14.17. This runs from the real point ρ to the real point R, then once anticlockwise round the circle of radius R, then back to ρ, then once clockwise round the circle radius ρ.

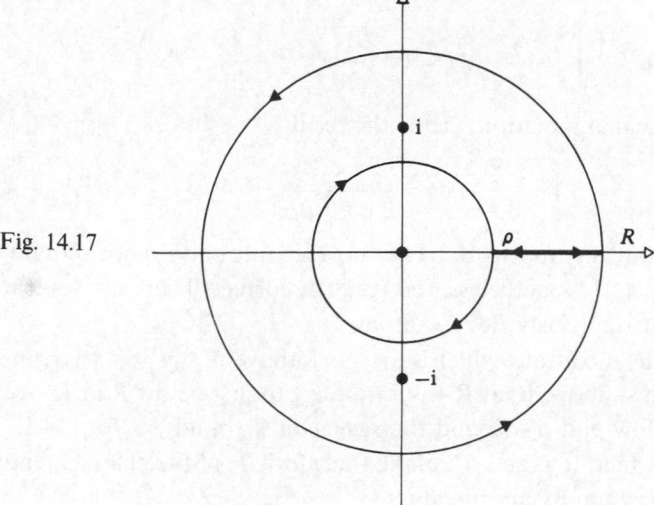

Fig. 14.17

By the 'residue' version of Cauchy's Theorem, we have

$$\int_\gamma f = 2\pi i (\Sigma \text{ residues inside } \gamma).$$

We intend to let $\rho \to 0$ and $R \to \infty$; so we may as well assume that $\rho < 1$, $R > 1$. Then the singularities inside γ are $z = i$, $z = -i$. The residues at these points are found as follows:

At $z = i$ the residue is $\lim_{z \to i} z^a/(z+i) = (1)/2i\, e^{ia\pi/2} = \alpha$, say.

At $z = -i$ the residue is $\lim_{z \to -i} z^a/(z-i) = -(1)/2i\, e^{ia3\pi/2} = \beta$, say. Let the circle radius R be γ_1, that radius ρ be γ_2. Then

$$\int_\rho^R \frac{x^a}{1+x^2}\,dx + \int_{\gamma_1} \frac{z^a}{1+z^2}\,dz + \int_R^\rho \frac{(x\,e^{2\pi i})^a}{1+x^2}\,dx + \int_{\gamma_2} \frac{z^a}{1+z^2} = 2\pi i(\alpha + \beta).$$

Note that in the third integral we have to choose the branch of z^a corresponding to argument 2π, because we must retain continuity. Now let ρ and R tend to 0, ∞ respectively. Easy estimates show that the first and third integrals tend to zero. The first tends to

$$\int_0^\infty \frac{x^a}{1+x^2}\,dx$$

and the third to

$$-e^{2\pi i a} \int_0^\infty \frac{x^a}{1+x^2}\,dx$$

So we have

$$(1 - e^{2\pi i a}) \int_0^\infty \frac{x^a}{1+x^2} \, dx = 2\pi i \left(\frac{1}{2i} e^{i a \pi/2} - \frac{1}{2i} e^{i a 3\pi/2} \right)$$

which, after some manipulation, yields the result:

$$\int_0^\infty \frac{x^a}{1+x^2} \, dx = \frac{\pi}{2 \cos (a)/z}.$$

Did you spot the nifty footwork? The chosen contour does not actually lie in the domain $\mathbb{C}_0$! Nonetheless, the result is correct. There are several ways to justify it rigorously; for example:

(1) Replace γ by a contour which starts just above ρ, say at $\rho + i\varepsilon$, runs horizontally to just above R, at $R + i\varepsilon$, circles γ_1 to just below R at $R - i\varepsilon$, runs back to below ρ at $\rho - i\varepsilon$, and then goes back round γ_2. (Fig. 14.18) Now, as ρ and R tend to 0 and ∞, make the width 2ε of the channel tend to zero and use continuity arguments.

(2) The choice of $\mathbb{C}_0$ causes problems, so reject it. Work in two over-lapping domains, say $\mathbb{C}_{-\pi/4}$ and $\mathbb{C}_{\pi/4}$, and switch from one to the other when you come to define the integral from R back to ρ (and bits of $\int_{\gamma_1}, \int_{\gamma_2}$).

(3) Put it all on a Riemann surface: the contour γ then winds round the staircase and up one step, and then winds back down one step to get back to the start. Everything generalizes to such contours, because they are

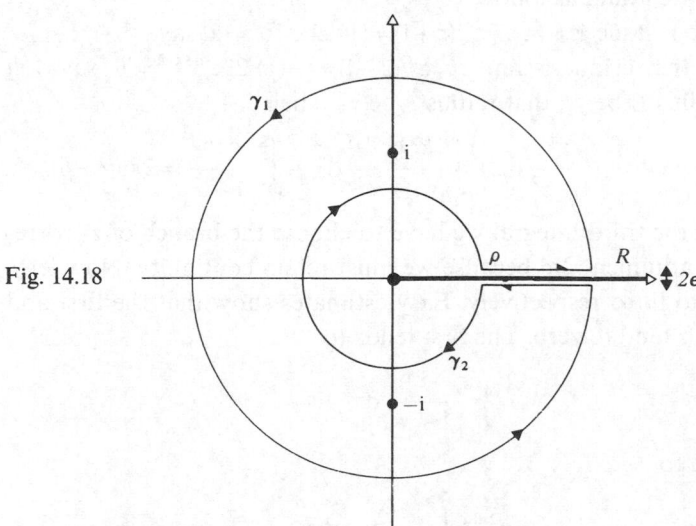

Fig. 14.18

obviously boundaries of rectangles in the Riemann surface, as in Figure 14.19.

These are rigorous justifications of the given method of computation. In actually using the method, of course, it is not necessary to trot out the justification every time – provided you understand how it would go. In fact, everything will work out fine provided you follow the golden rule: make the multiform function vary *continuously* as you walk round the contour.

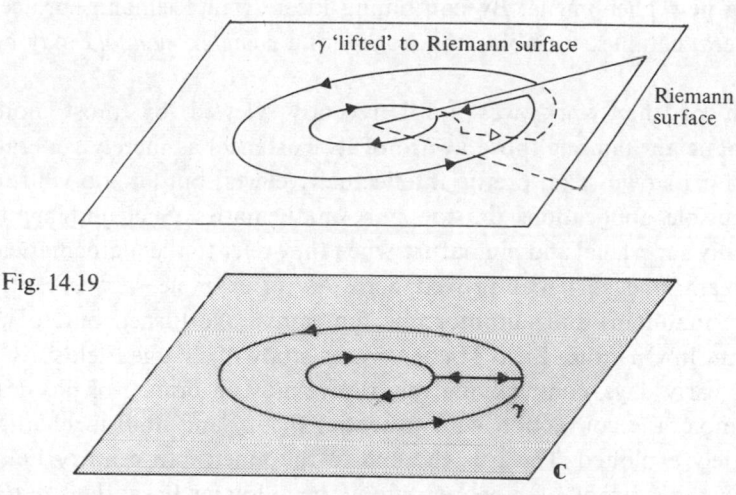

γ 'lifted' to Riemann surface

Riemann surface

Fig. 14.19

C

9. The road goes ever on . . .

The end of this book is fast upon us, but there is as yet no discernible end to complex analysis itself. It remains a vigorous and growing part of mainstream mathematics.

The Riemann surface alone has opened up broad vistas, by capturing the essential structure of a complex function in a single geometric object. All kinds of information, such as the presence of branch-points and singularities, can be read from it; and the 'rigidity' of analytic functions means that they are essentially determined by the position and nature of their singularities. Many questions which are quite baffling without Riemann surfaces become transparent when this extra geometry is invoked.

Extending the idea of periodic functions (such as exp) leads to *doubly periodic functions*, which satisfy $f(z)+f(z+p)=f(z+q)$ for two different (rationally independent) complex numbers p and q. Generalizing further, one obtains the notion of an automorphic function, which preoccupied a great many mathematicians around the turn of the century, uniting in one package group theory, differential equations, algebraic function theory, topology, and complex analysis. This is still an important field of research.

Functions of *several* complex variables may be studied; they turn out to be much trickier to handle than one might imagine, and there are startling new phenomena. By combining ideas from Riemann surfaces and several complex variables, the notion of a *complex manifold* may be derived.

Much of this work was until recently viewed by most non-mathematicians (among those aware of its existence) as merely generalization for its own sake: pretty, intellectually clever, but far too wild to have sensible applications outside pure mathematics. Such judgments are usually superficial and premature when they refer to the mathematical mainstream, and so it has proved here. As an example, very recently complex manifolds and automorphic functions have turned out to be important in Quantum Field Theory, in the study of 'Gauge Fields'.

In its early days, complex analysis was (almost) a branch of physics: for example, the connection with potential theory and fluid mechanics was widely exploited. Towards the end of the nineteenth century Felix Klein offered a 'proof' of a theorem along the following lines: *think of the Riemann surface as being made of thin metal, and an electric current flowing through it....* It would not, today, be considered a logically convincing argument; but the physical intuition certainly revealed some important mathematical ideas. Today we are seeing the converse process, with mathematical intuition providing important concepts for physics. It is a two-way trade. And, whatever the attractions of beauty for its own sake, it is vital to the health of both mathematics and science that this trade be maintained.

Exercises 14

1. Define three power series by

$$a(z)=1+z+z^2+z^3+\cdots=\sum_{n=0}^{\infty} z^n,$$

$$b(z)=i-(z-i-1)-i(z-i-1)^2+(z-i-1)^3-\cdots=\sum_{n=0}^{\infty}i^{n+1}(z-i-1)^n,$$

$$c(z)=-1+(z-2)-(z-2)^2+\cdots=\sum_{n=0}^{\infty}(-1)^{n+1}(z-2)^n.$$

Find their discs of convergence, and sketch them. Prove that $a(z)=b(z)$ on the overlap of their discs of convergence, and similarly $b(z)=c(z)$ on the corresponding overlap. Do the discs of convergence of $a(z)$ and $c(z)$ intersect?

2. Let $f(z)=\sum_{n=0}^{\infty}a_n(z-z_1)^n$, $g(z)=\sum_{n=0}^{\infty}b_n(z-z_2)^n$ be two power series, and assume that their discs of convergence have non-empty intersection. Prove that g is a direct analytic continuation of f if and only if there exists z_0 in both discs of convergence, such that for $m=0, 1, 2, 3, \ldots$ the equations

$$\sum_{n=0}^{\infty}n(n-1)\ldots(n-m)[a_n(z_0-z_1)^{n-m}-b_n(z_0-z_2)^{n-m}]=0$$

hold.

3. A certain function has singularities at precisely the points z such that $\mathrm{re}\,(z)$ and $\mathrm{im}\,(z)$ are integers. You are given its Taylor expansion around the point $\frac{1}{2}+\frac{1}{2}i$. What is the smallest number of stages needed to find an indirect analytic continuation, by power series, to a domain that contains $\frac{7}{2}+\frac{5}{2}i$?

4. Show that the functions defined by

(i) $f(z)=1+z^2+z^4+z^8+\cdots+z^{2^n}+\cdots$

(ii) $g(z)=1+z^3+z^9+z^{27}+\cdots+z^{3^n}+\cdots$

have natural boundaries at $|z|=1$.

5. Suppose that $f(z)=\sum a_n z^n$ has radius of convergence 1. Set

$$z=w/(1+w)=(w-w^2+w^3-w^4+\cdots)$$

and transform $f(z)$ to a power series in w, say $\sum b_n w^n=F(w)$.

Prove that this latter series has radius of convergence $\geqslant\frac{1}{2}$; and if -1 is a singular point of f, the radius is exactly $\frac{1}{2}$.

If this radius is strictly between $\frac{1}{2}$ and 1, show that the equation $f(z)=F(z/(1-z))$ defines an analytic continuation of f beyond the disc $|z|\leqslant 1$.

6. Show that the series

$$\sum_{n=1}^{\infty}((1-z^{n+1})^{-1}-(1-z^n)^{-1})$$

converges either for $|z|<1$, or for $|z|>1$; but that the two functions so represented are *not* analytic continuations of each other.

7. Suppose that f and g are defined, and have no singularities, on the whole of $\mathbb{C}$. Define

$$\Phi(z)=\sum_{n=1}^{\infty}\left(\frac{1-z^n}{1+z^n}-\frac{1-z^{n-1}}{1+z^{n-1}}\right).$$

Show that

$$\tfrac{1}{2}(f(z)+g(z))+\tfrac{1}{2}\Phi(z)(f(z)-g(z))$$

is equal to $f(z)$ when $|z|<1$, and to $g(z)$ when $|z|>1$.

8. Let $f(z)$ be the (multiform) function $z\sqrt{z}$. Show that at $z=0$ there exists a first derivative which is the same for all branches, but a finite second derivative does not exist. What about $z^2 \log z$?

9. Describe the Riemann surfaces for the multiform functions
   - (i) $\sqrt[7]{z+43}$
   - (ii) $\sqrt{(1-z^3)}$
   - (iii) $\cos^{-1}(z)$
   - (iv) $\tan^{-1}(z)$.

10. Describe the Riemann surface of

$$((z-1)(z-2)^{-2})^{1/3}+(z-3)^{1/2}.$$

11. Let $\omega=\mathbb{C}$. Show that the values of 1^ω form a subgroup U_ω of the multiplicative group of non-zero complex numbers: Show that U_ω is cyclic of order q if ω is a rational number p/q in lowest terms; otherwise U_ω is infinite cyclic.

Show that U_ω is a subset of the unit circle if and only if ω is real; lies on the positive real axis if and only if $\text{re}(\omega)$ is an integer; and otherwise lies on an Archimedean spiral parametrized by $t \in \mathbb{R}$ in the form $e^{\alpha t}$ ($\alpha \in \mathbb{C}$ some fixed number).

Also show that for any $z\neq 0$, the values of z^ω form a coset of U_ω in $\mathbb{C}\setminus\{0\}$.

12. Show that the function

$$f(z)=e^{-\pi i/8}\sqrt{(z-i)}$$

defines a conformal mapping from the domain in Figure 14.20 to the upper half-plane.

Fig. 14.20

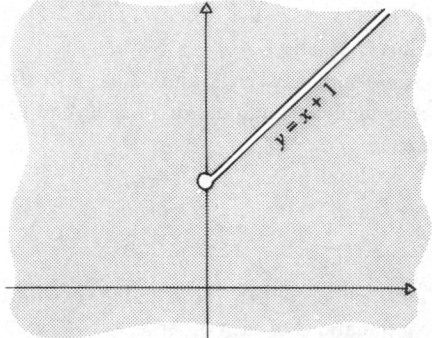

13. Show that the function

$$f(z)=e^{i\pi/3}\left(\frac{z+1}{z-1}\right)^{2/3}$$

defines a conformal mapping of the domain in Figure 14.21 onto the upper half-plane.

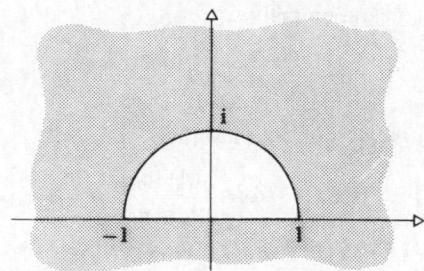

Fig. 14.21

14. Find the image of the sector $-\pi/n < \arg z < \pi/n$, $|z| < 1$ (Fig. 14.22) under the conformal transformation

$$f(z) = z(1 + z^n)^{2/n}$$

where n is a positive integer.

Fig. 14.22

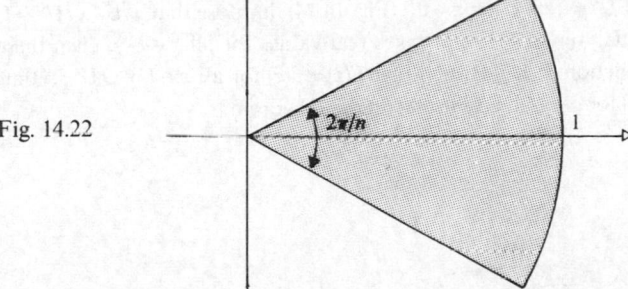

15. Let $\gamma(t) = e^{it}$, $0 \leqslant t \leqslant 4\pi$. Calculate $\int_{\gamma} \sqrt{z}\, dz$ where at $t = 0$ we take $\sqrt{1} = 1$.

16. Using the contour illustrated in Figure 14.23, show that

$$\int_0^\infty x^{-k}(1+x)^{-1}\, dx = \pi \operatorname{cosec}(k\pi) \quad (0 < k < 1).$$

Fig. 14.23

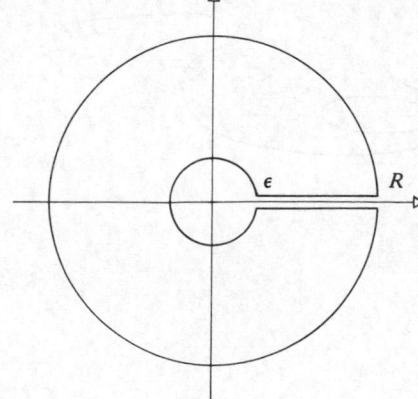

17. Show by contour integration that

$$\int_0^1 \frac{dx}{(1+ax^2)\sqrt{1-x^2}} = \frac{\pi}{2\sqrt{(1+a)}} \quad (a>0).$$

18. Using the contour of question 16, show that

$$\int_0^\infty x^a(1+x^2)^{-2}\, dx = \frac{\pi(1-a)}{4\cos\frac{1}{2}\pi a} \quad \text{if } -1<a<3.$$

19. Let $\gamma(t) = (1 + \frac{1}{2}\cos t + i \sin t)^5$, $0 \leqslant t \leqslant 2\pi$. Find

$$\int_\gamma z^{99} \log(z) - \sqrt{z} + (z - \tfrac{1}{4})^{5/17}\, dz.$$

20. Describe the Riemann surface of the function
    $$f(z) = \sqrt{(z + z^2 + z^4 + \cdots + z^{2^n} + \cdots)}.$$

21. (*Schwartz Reflection Principle.*) Let U be a domain in $\mathbb{C}$ which is symmetric about the real axis (that is, if $z \in U$ then $\bar{z} \in U$). Let $U^+ = \{z \in U \mid \text{im } z > 0\}$, $U^- = \{z \in U \mid \text{im } z < 0\}$, $U^0 = \{z \in U \mid \text{im } z = 0\}$. (Fig. 14.24) Suppose that $f: U^+ \cup U^0 \to \mathbb{C}$ is continuous, analytic on U^+, and takes real values for all $z \in U^0$. Then there is an analytic function $F: U \to \mathbb{C}$ such that $F(z) = f(z)$ for all $z \in U^+ \cup U^0$. (Hint: define $F(z) = \overline{f(\bar{z})}$ for $z \in U^-$, and use Morera's Theorem.)

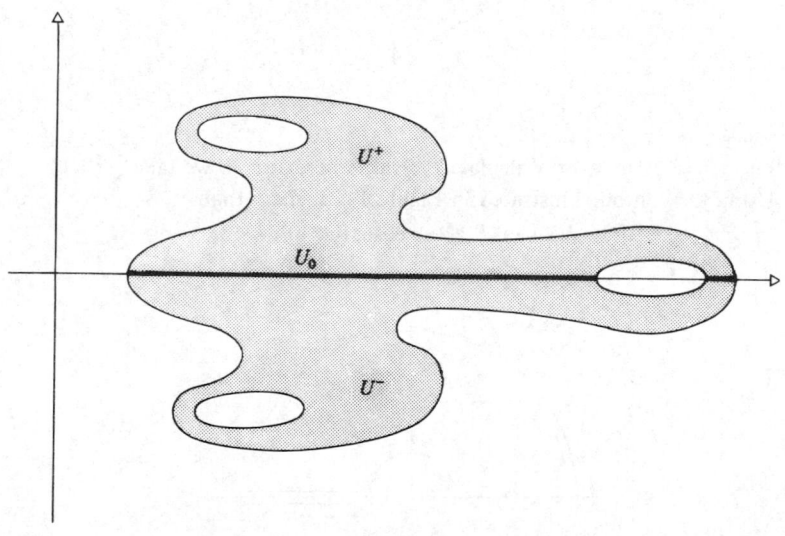

Index

absolute convergence 54
algebra 10
analytic function 182
angle 121
  modulo 2π 238
  between two contours 241
antiderivative 108
  existence in a star domain 146
  local 149
arc 240
analytic continuation 257, 261, 264, 265
  direct 261
  indirect 264
  multiform 265
  uniform 265
annulus 196
Argand 4, 7
Argand diagram 14
argument 17, 122
  principal value 17, 122
  continuous choice 127
Ars Magna 1
axis
  imaginary 14
  real 14

Bernoulli 3, 5, 266
Bernoulli Paradox 137
Bessel 4, 5
Bessel function 62
bilinear mapping 246
Bolyai 5
Bombelli 2, 7
boundary 47, 162, 191
boundary, natural 262
boundary point 47, 191
branch point 267
Brooke Taylor 182

Cardano 1
calculus, fundamental theorem of 96
Cartesian coordinates 17
Cauchy 5, 68, 141, 182

Cauchy principal value 219, 224
Cauchy-Riemann equations 5, 67, 68, 70
Cauchy's Estimate 184
Cauchy's Integral Formula 178
Cauchy's Residue Theorem 213
Cauchy's Theorem 5, 141, 153
  comparison of versions 172
  for a boundary 162, 164
  for a triangle 143
  generalized version 154
  homotopy versions 168, 169
chain rule 66
closed set 24
compact set 44
comparison test 55
complete analytic function 256
complex conjugate 16
complex manifold 282
complex numbers 10
  construction 10
  geometry 13
complex powers 271
complex plane 14
conjugate, complex 16
  harmonic 250
connected component 42, 46
connected set 22, 40
continuity 22, 28, 29, 30
contour 106
  Jordan 155
  inside 156
  outside 156
contour integral 5, 106
  of multiform function 274
convergence, general principle 50–1
  radius of 56
  disc of 56
convergent sequence 49
  series 52
cut plane 122

D'Alembert 4, 68
De Moivre 7

De Moivre's formula (theorem) 20, 21, 85
derivative 64
Descartes 2
differentiable function 64
differentiation 64
direct analytic continuation 261
Dirichlet's discontinuous factor 236
disc of convergence 56
divergent series 52
domain 22, 44
  simply connected 157
  star 141
doubly periodic function 282

equipotential lines 250
essential singularity 202, 207
estimation lemma 111
Euler 3, 5, 7, 12, 266
Euler's formula 85
exponential function 82
extended complex plane 206
extension function 188

Fibonacci numbers 192
field 11
fixed point 256
Frankenstein's monster 81
function,
  analytic 182
  complete analytic 265
  conformal 242
  continuous 29
  differentiable 64
  exponential 82
  extension 188
  harmonic 250
  hybrid 72
  hyperbolic 91
  meromorphic 208
  multiform 265, 273, 274
  periodic 88
  potential 250
  rational 208
  trigonometric 84
  uniform 264
fundamental group 176
Fundamental Theorem
  of Algebra 185, 233
  of Calculus 96
  of Contour Integration 109

Gauss 4, 5, 7, 12, 141
Gauss plane 14
general principle of convergence 50–1
Gregory 182
Gregory's series 94

Hamilton 4, 7
harmonic conjugate 250
harmonic function 250
homology 174
homotopic to zero 170
homotopy 159, 165
  fixed endpoint 167
  closed path 169

Identity theorem 187
infinite product 237
infinity 205
integral value 175
integral, complex 99
  contour 106
  Riemann 96
  Riemann-Stieltjes 97
integration 95
  along an arbitrary path 160
  contour 105
  by parts 119
inversion 248
isolated singularity 201

Joke 53
Jordan contour 155
Jordan contour theorem 157
Jordan's inequality 236
Joukowski aerofoil 251

Kipling 172
Klein 282
Kline 7
Kronecker 11

Lagrange 182
Laplace equation 249
Laplace transform 235
Laurent 6, 195
Laurent expansion 199
  series 6, 195, 199
Laurent's Theorem 196
Leibniz 3, 5, 10, 266
limit 22
  of function 25
  of sequence 49
  of series 52
limit point 24
Liouville 184
Liouville's theorem 184
local antiderivative 149
local maximum 190
local minimum 191
logarithm 3, 9, 120, 125, 148
  natural 84
loop, simple 212

magnification 248
maximum, local 190
maximum modulus theorem 191
meromorphic function 208
minimum, local 191
minimum modulus theorem 192
Möbius mapping 238, 246
modulus 14
Moivre, De *see* De Moivre
Moore 143
Morera 183
Morera's theorem

natural boundary 262
natural logarithm 84
neighbourhood 24
Newton 182

order 18
order
    of pole 202
    of zero 186
Osborne's Rule 92

Parseval's inequality 194
partial sum 52
path 33
    in S 37
    length 100
path,
    opposite 36
    smooth 98
    step 22
    sub 34
path connected set 40
paving lemma 22, 37, 39
Picard 204
plane,
    complex 14
    cut 122
    Gauss 14
point,
    boundary 47
    final 33
    fixed 256
    initial 33
    isolated 25
    limit 24
pole 202
    simple 202
    double 202
    triple 202
pole of order m 202
potential function 250
potential theory 249

radian measure of angle 120
radius of convergence 56
ratio test 55
rational function 208
relevant rectangle 151
removable singularity 201
residue 212
    calculations 215
Riemann 5, 6, 68, 205, 268
Riemann integral 96
Riemann sphere 205
Riemann surface 257, 268
Riemann-Stieltjes integral 97
root of unity 137
rotation 248
Rouché's theorem 232

Schwartz reflection principle 286
sequence 48
    convergent 49
series 52
    absolutely convergent 54
    convergent 52
    divergent 52
    Laurent 199
    power 48, 56, 73
    Taylor 76, 177, 179, 181
set,
    closed 24
    compact 44
    connected 22
    open 24
    relatively open 29
    step connected 42
signed area 118
simple loop 212
simply connected domain 157
singularity 189, 267
    essential 207
    isolated 201
    isolated essential 202
    removable 201
singularity at infinity 207
star-centre 146
star-domain 141, 146
stream lines 250
sub path 34
sum, partial 52
summation of series by integration 228

Tartaglia 2
Taylor 182
Taylor series 76, 177, 179, 181
test, comparison 55
    ratio 55

transformation, conformal 238
  elliptic 256
  hyperbolic 256
  loxodromic 256
translation 248
triangle inequality 15

unity, root of 137

Vandermonde 7

Wallis 2, 7
Weierstrass 259
Weierstrass-Casorati theorem 204
Wessell 4, 7
winding number 120, 126
  as an integral 130
  computed by eye 133
  round an arbitrary point 131

zero 185
  isolated 186
  of order m 186

Bargain Books
Princeton, NJ
Mon 8 May 2000
$5.98 + .36 tax
sticker price: $14.95